RENEWALS 458-4574

WITHDRAWN
UTSA LIBRARIES

Analytical Methods for Problems of Molecular Transport

FLUID MECHANICS AND ITS APPLICATIONS
Volume 83

Series Editor: R. MOREAU
MADYLAM
Ecole Nationale Supérieure d'Hydraulique de Grenoble
Boîte Postale 95
38402 Saint Martin d'Hères Cedex, France

Aims and Scope of the Series

The purpose of this series is to focus on subjects in which fluid mechanics plays a fundamental role.

As well as the more traditional applications of aeronautics, hydraulics, heat and mass transfer etc., books will be published dealing with topics which are currently in a state of rapid development, such as turbulence, suspensions and multiphase fluids, super and hypersonic flows and numerical modeling techniques.

It is a widely held view that it is the interdisciplinary subjects that will receive intense scientific attention, bringing them to the forefront of technological advancement. Fluids have the ability to transport matter and its properties as well as to transmit force, therefore fluid mechanics is a subject that is particularly open to cross fertilization with other sciences and disciplines of engineering. The subject of fluid mechanics will be highly relevant in domains such as chemical, metallurgical, biological and ecological engineering. This series is particularly open to such new multidisciplinary domains.

The median level of presentation is the first year graduate student. Some texts are monographs defining the current state of a field; others are accessible to final year undergraduates; but essentially the emphasis is on readability and clarity.

For a list of related mechanics titles, see final pages.

Analytical Methods for Problems of Molecular Transport

by

I.N. IVCHENKO
University of Missouri-Columbia,
Columbia, Missouri, U.S.A.

S.K. LOYALKA
University of Missouri-Columbia,
Columbia, Missouri, U.S.A.

and

R.V. TOMPSON, JR.
University of Missouri-Columbia,
Columbia, Missouri, U.S.A.

A C.I.P. Catalogue record for this book is available from the Library of Congress.

ISBN-10 1-4020-5864-0 (HB)
ISBN-13 978-1-4020-5864-6 (HB)
ISBN-10 1-4020-5865-9 (e-book)
ISBN-13 978-1-4020-5865-3 (e-book)

Published by Springer,
P.O. Box 17, 3300 AA Dordrecht, The Netherlands.

www.springer.com

Printed on acid-free paper

Library
University of Texas
at San Antonio

All Rights Reserved
© 2007 Springer
No part of this work may be reproduced, stored in a retrieval system, or transmitted
in any form or by any means, electronic, mechanical, photocopying, microfilming, recording
or otherwise, without written permission from the Publisher, with the exception
of any material supplied specifically for the purpose of being entered
and executed on a computer system, for exclusive use by the purchaser of the work.

Dedications

This book is dedicated to the memory of my son Yaroslav.

I. N. Ivchenko

This book is dedicated to the memory of Joel H. Ferziger, Heinz Lang, and Lloyd B. Thomas who inspired me to work in this area.

S. K. Loyalka

This book is dedicated to my family; my parents, my siblings, my nephews and nieces, and especially to my wife Elena and my son Arseniy.

R. V. Tompson, Jr.

Special Dedication

This book is especially dedicated to the memory of our very dear friend and colleague

Igor Nikolaevich Ivchenko

without whom this book would not have been possible.

S. K. Loyalka

R. V. Tompson, Jr.

Contents

Table of Tables

Table of Figures

Preface

The transport of a given species (atoms, molecules, neutrons, photons, etc.), either through its own kind or through some other host medium, is a problem of considerable interest. Practical applications may be found in many technologically and environmentally relevant areas such as the transport of neutrons in a nuclear power reactor or in a nuclear weapon, the transport of ions and electrons in plasma, the transport of photons which constitutes radiative heat transfer in various industrial, environmental and space applications, the transport of atoms or molecules of one species either through itself or as one component of a multi-component gas mixture, and the interactions of such gas mixtures with various solid and liquid surfaces such as one might find associated with capillary tubes, aerosol particles, interstellar dust grains, etc.. These application areas are obviously quite broad and it is readily apparent that there are, indeed, few scientific activities that do not require some level of understanding of transport processes.

One of the most important and influential texts in the area of transport theory has been *The Mathematical Theory of Non-Uniform Gases* by Sidney Chapman and T.G. Cowling that was first printed in 1939. This book, along with several other more recent texts (Hirschfelder, J.O., Curtiss, C.F. and Bird, R.B., *Molecular Theory of Gases and Liquids*, John Wiley and Sons, NY, 1954; Kogan, M.N., *Rarefied Gas Dynamics,* Plenum Press, NY, 1969; Ferziger, J.H. and Kaper, H.G., *Mathematical Theory of Transport Processes in Gases*, North-Holland, Amsterdam-London, 1972; Cercignani, C., *Theory and Application of the Boltzmann Equation*, Scottish Academic Press, Edinburg, UK, 1975.), have provided students and researchers in the area of transport theory with an excellent basis for pursuing in-depth research in the area and have made possible some limited, albeit very useful, applications of the theory. These texts have focused primarily on the development and simplification of the Boltzmann equations and on the derivation of general expressions for the coefficients associated with the transport phenomena of viscosity, thermal conduction, and diffusion in essentially unbounded gases.

In recent years, however, it has been recognized that not only do intermolecular collisions within the bulk of the gas play an important role in determining the various transport coefficients but that interactions at boundaries are also significant. Hence, a simultaneous understanding of both is required in order to obtain reasonably complete descriptions of, and ultimately solutions to, any given transport problem. The relative importance of a boundary in a given transport problem is generally specified by Knudsen number, $Kn = \lambda/d$, where λ represents the mean free path of the species of interest in the host medium while d is some characteristic dimension of the system. Thus, in transport problems in general, there exist several somewhat arbitrarily defined regimes of interest. These cover the range from $Kn \gg 1$ where a given boundary exerts a strong influence on the transport processes to $Kn \ll 1$ where a given boundary exerts an essentially negligible influence on the transport processes.

Accounts of the effects of boundaries on transport problems have been discussed in a variety of texts on subjects such as Neutron Transport Theory, Radiative Heat Transfer, and the Kinetic Theory of Gases. These, and the basic transport theory books mentioned above, are generally written at a level suitable only for specialists in the field or very advanced students. Thus, even though many have tried to use them as teaching texts, their application in this capacity has remained somewhat constrained with a clear need continuing to exist for a more student-friendly text having a clearer emphasis on basic problem solving. At the University of Missouri-Columbia, we became clearly aware of this need in the Winter of 1992 when one of us (INI), as a part of teaching obligations associated with his status as a visiting professor to Columbia from Russia under the Fulbright exchange program, taught a class on the subject of transport theory. Based on this experience, we felt that the time had arrived for a new text that would combine a concise, but basically complete, introduction to the field of transport theory with a fairly tight focus on a few recently successful analytical solution techniques. Further, we recognized that any successful text of this type would, of necessity, include a good selection of easily applicable, representative problems that were either fully solved or which included sufficient basic guidance to assist the problem solver in reaching the correct solution. In this context, we have been strongly influenced by the organization and style of the excellent series of texts by Landau and Lifshitz.

The subject material in the book divides itself quite naturally into two parts: the non-equilibrium properties of an infinite expanse of gas and those of a bounded gas. The first part (Chapters 1-5) contains the basis of the Kinetic Theory of Gases developed along the traditional lines of Chapman and Cowling. Some simplifications have been made during this development but only to facilitate the reader's understanding of the basic principles of the kinetic theory. For example, the derivation of the Boltzmann collision operator is substantially simpler than that described in the above mentioned texts. The analysis presented in these chapters is based on the framework of the scattering cross section being the basic quantity characteristic of molecular encounters. In description of the Chapman-Enskog method, we focus our attention on the first-order approximation for rigid-sphere molecules to facilitate one's understanding of this method. Nevertheless, the second-order Chapman-Enskog solutions along with tabulated values of all quantities necessary for their complete evaluation, are also provided for the Lennard-Jones (6-12) intermolecular potential as this development is potentially very important for practical applications. The presentation of this material here is in the most complete form currently available.

The second part of the book (Chapters 6-12) presents focused descriptions of a limited selection of analytical methods that may be used to solve boundary value problems of transport with reasonable accuracy. The typical features found in the statements of boundary value problems for various gas flow regimes are discussed in Chapter 6. Within the framework of the traditional Maxwellian boundary model, new expressions for the reflected distribution functions at surfaces are presented which are then adapted to numerous applied problems of a practical nature. The physical features associated with statements of the boundary conditions for condensable and non-condensable gases are discussed and appropriate definitions for the basic accommodation coefficients are described in detail.

Chapter 7 contains a kinetic theory treatment of gas transport problems for extremely high Knudsen numbers. The theory presented here starts with a Boltzmann equation that takes its simplest form due to the absence of the collision operator. Under stationary conditions, the constancy of the distribution function along molecular trajectories results in its discontinuity in velocity space; an effect that may be described in terms of the classic 'cone of influence.' For the spherical geometry, a new analytical representation of the distribution function in the full velocity space is given. This representation is a generalization of that previously proposed for the planar geometry. A number of basic problems are considered which show, in detail, the features of a mathematical solution technique that is presented here, in monographic form, for the first time and which otherwise could only be found by perusing the original journal articles in which it was developed.

The next two chapters, Chapters 8 and 9, present selected methods of solution of planar transport problems. Detailed analytical solutions of some applied problems are given for the special case when molecules can be approximated as rigid spheres. In our description of selected methods, we have concentrated on various moment approaches in which we have emphasized the importance of using conservation laws to construct exact moment solutions. In Chapter 8, our development of the various moment approaches starts with the classic Maxwellian method and proceeds on to a generalized form of the Maxwellian method, which we term the Loyalka method after the one of us (SKL) who developed it. A key feature of this latter method that we emphasize is the opportunity that it represents to construct an accurate theory of slip phenomena for arbitrary models of the intermolecular potential. We follow these with a description of the characteristic features of the half-range moment method which, in combination with the mathematical solution technique described in Appendix A, is presented here in monographic form for the first time. Also in Chapter 8 we discuss in detail various boundary models and introduce a new model that allows for the incorporation of two conservation laws.

We continue our discussion of planar transport problems in Chapter 9 with a discussion of the variational method which is followed by a discussion of the accuracy of all the various solution methods described for planar transport problems. Features of the integral representation of the distribution function and of the construction of basic functionals for slip problems are discussed in detail. We touch briefly on a new approximate method (Prob. 9.8) that may be used to describe gas behavior in the Knudsen layer. This comparative analysis of various approximate methods and estimates of their accuracy are compiled here in monographic form for the first time allowing the reader to make easy comparisons between the various methods.

Chapter 10 contains straightforward applications of the slip-flow theory to various classical problems of interest. Derivations of the basic equations and statements of the boundary conditions in the slip-flow regime are discussed as well. While the subject treated here may be found in various places in the literature, several gaps in the subject have been bridged in order to make possible a systematic description of the various linearized problems.

Chapter 11 contains original material that has been worked out only recently and which, in some cases, is still under development (such as the sphere drag problem). In many cases, methods and results for arbitrary Knudsen number and different geometries are described for the first time in such a complete manner. For instance, analytical expressions for the bracket integrals associated with each transport problem as well as numerical and various approximate methods of calculation of these bracket integrals for different curvilinear geometries are original and have not been reported previously. The authors were the first to suggest the use of the Chapman-Enskog solutions as the molecular properties used to construct moment systems. This approach yields a very simple procedure that may be used to calculate bracket integrals for arbitrary intermolecular potentials. The advantage of this method is shown in analyses of some classical spherical transport problems such as the torque and thermal conduction problems.

In Chapter 12 the analytical methods presented earlier for the Maxwell and Loyalka methods are again used to solve planar boundary value problems involving the various boundary slip phenomena. Here, however, the emphasis is on binary gas mixtures where the preceding chapters have focused only on simple one-component gases. The relevant boundary slip phenomena are discussed in detail and the accuracy of the various methods is evaluated further in the context of gas mixtures. This chapter also contains all of the material necessary to completely specify the transport coefficients for a binary gas mixture for both the first- and second-order Chapman-Enskog approximations.

In addition to the material in Chapters 1-12, a number of appendices have been included to facilitate the ease of the reader in finding and using such information as is necessary to effectively make use of the analytical methods described in this book. Appendices A-C contain descriptions of analytical methods needed to evaluate the necessary bracket integrals associated with the various transport problems in both the planar and curvilinear geometries. Of particular importance, the methods of calculation of bracket integrals containing the discontinuous distribution function have all been collected together in one monograph for the first time. Appendix D provides additional information about the variational principle for use in planar transport problems. Here, the reader can find greater detail regarding the general principles behind the construction of the necessary variational functionals needed for specific boundary value transport problems. Appendix E contains an extensive listing of definite integrals that are most frequently encountered in boundary value transport problems and Appendix F contains tabulated numerical values of the Ω-integrals that are encountered when using the Lennard-Jones (6-12) potential model. This set of tabulated Ω-integrals is the complete set that it is necessary to use when employing either the first-order or the second-order Chapman-Enskog approximations.

The problems following each chapter may be useful to both students and instructors. While all of these problems have been solved, sometimes only an outline of the solution is given and the reader must supply the details.

The authors recommend that students of the subject seek to obtain their own solutions to these problems in order to gain familiarity with the techniques. In addition to basic experience, this will also help students of the subject to understand some of the subtleties of the analytical methods described. While some of the problems presented at the ends of the chapters only require the reader to utilize basic information that may be readily obtained in the necessary form directly from the book, other problems will require the reader to engage in some significant extension and generalization of the basic material presented in the book.

This book is designed to serve a dual function. It is intended that it be capable of serving as a teaching instrument, either in a classroom environment or independently, for the study of basic analytical methods and mathematical techniques that may be used in the Kinetic Theory of Gases. It is primarily suitable for use in graduate level physics and engineering courses on the Kinetic Theory of Gases. This book should also prove to be useful as a reference for scientists and engineers working in the fields of Rarefied Gas Dynamics and Aerosol Mechanics. In addition, the material in this book may prove to be of interest to individuals working in the areas of Physical Chemistry, Chemical Engineering, or any other applied discipline in which gas-surface interactions can be expected to play a significant role.

INI
SKL
RVT

Acknowledgments

The authors gratefully acknowledge the invaluable assistance that they have received from many sources including:

Dr. Robert L. Buckley for extensive help in typing and proof reading.
Dr. James L. Griffin for *Mathematica*® assistance in some calculations.
Professor Mikhail N. Kogan for helpful discussions relating to arbitrary Knudsen number in Chapter 11.
Professor Yuri I. Yalamov for helpful discussions relating to boundary conditions for the slip-flow regime in Chapters 8 and 10.
Dr. David Gabis for discussions relating to the spinning sphere problem at the end of Chapter 10.
Dr. Perapong Tekasakul for help with figures and equations.
Mr. Earl L. Tipton for extensive help with *Adobe Illustrator*® and for proof reading the final version.
Mr. Ryan Meyer, Mr. Zeb Smith, and Mr. Earl L. Tipton for contributions made in preparing the omega-integral program of Appendix F.
The Council for the International Exchange of Scholars (CIES) for the Fulbright grant that supported INI during his first visit to Columbia Missouri, 1/15/92-7/4/92 and which sponsored two subsequent visits, 8/25/92-1/15/93 and 7/29/93-2/18/94.
The U.S. National Aeronautics and Space Administration (NASA) for funding from grant NAG-3-1420 that was used to support INI during his second and third Fulbright visits to Columbia as well as during his fourth visit, 7/01/94-3/15/95.
This manuscript was initially prepared on a Macintosh® IIci using System 6.0.5. The text was initially formulated using *Microsoft Word*® (ver.

4.0). Equations were initially prepared using *MathType*® (ver. 2.03), plots using *Cricket Graph*® (ver. 1.3.2), and figures using *MacDraft*® (ver. 1.2b). The manuscript was subsequently converted to *Microsoft Word*® (ver. 2002) employing *MathType*® (ver. 5.2a) on a Dell Dimension® 2400 configured with *Microsoft Windows*® *XP* (ver. 2002 including Service Pack 2). Plots were converted to *Microsoft Excel*® (ver. 2002) and figures were converted to *Adobe Illustrator*® *CS* (ver. 11.0.0). Omega-integral calculations in Appendix F were performed using *Compact Visual Fortran*® 77 (ver. 6.4) and *Mathematica*® (ver. 5.2) on a number of different machines.

Chapter 1

THE GENERAL DESCRIPTION OF A RAREFIED GAS

1. SOME INTRODUCTORY REMARKS.

Systems consisting of what are usually called rarefied gases will be considered in this book. Such systems contain large numbers of molecules; specifically, about 2.7×10^{19} cm^{-3} for a gas at standard temperature and pressure (STP) defined to be 0 °C (273.15 K) and 1.0 atm (760 torr). How may the behavior of such huge collections of particles be described? If one were to try to apply the methods of classical mechanics, one would have to construct and solve a system of equations of motion containing as many equations and initial conditions as there are numbers of interacting particles. Obviously, such a problem could not be solved today even with the help of the most advanced computers. On this basis, one might naively suspect that the greater the number of particles, the more difficult the problem. This is not strictly true, however, as it turns out that useful results can be obtained by applying statistical descriptions to such systems.

The state of a gas can be analyzed by employing statistical laws which allow one to determine average values for the different macroscopic quantities that characterize the behavior of the gas. It has been proven in statistical mechanics [1-3] that the relative fluctuations of additive quantities (i.e. quantities whose values for the body as a whole are equal to the sum of the values for its separate parts) are proportional to $N^{-1/2}$, where N is the number of molecules of the gas. In accordance with this theorem, the additive quantities are really equal to their average values to an extremely high degree of accuracy and, therefore, deviations of the actual quantities from the average values do not have a practical influence on the trustworthiness of the statistical description.

The mean values mentioned above may be found by using a probability framework in which the main interest is in the statistical distribution function for the molecules. It is very important to note that the extreme accuracy of this type of probabilistic analysis, due to the relatively small deviations of the macroscopic quantities from their mean values, is much greater than the accuracy of actual experimental measurements and, hence, small deviations of the macroscopic quantities from their mean values are typically neglected.

2. DENSITY AND MEAN MOTION.

First, some macroscopic quantities for a 'simple' gas composed entirely of identical molecules [4-6] will be determined. Let the mass of any molecule be m and let $d\tilde{\mathbf{r}}$ denote a small volume element surrounding the point, $\tilde{\mathbf{r}}$. This volume element is assumed to be large enough to contain a great number of molecules while still possessing dimensions small compared to the scale of variation of the macroscopic quantities of interest.

Let the mass contained in $d\tilde{\mathbf{r}}$ be averaged over a time interval, dt, which is long compared to the average time needed for a molecule to traverse $d\tilde{\mathbf{r}}$ yet short compared to the scale of the time variations in the macroscopic properties of the gas. Then, the average value of the mass contained by $d\tilde{\mathbf{r}}$ will be proportional only to its volume and will not depend upon its shape. This mass will be denoted by $\rho d\tilde{\mathbf{r}}$, where $\rho(\tilde{\mathbf{r}},t)$ may be termed the mass density (or typically just 'density') of the gas at $(\tilde{\mathbf{r}},t)$. The number density of the gas, $n(\tilde{\mathbf{r}},t)$, is identified by analogy. These densities are connected by the relationship:

$$\rho(\tilde{\mathbf{r}},t) = mn(\tilde{\mathbf{r}},t) ,$$

where m is the mass of a single gas molecule.

If the velocity of a molecule is denoted by $\mathbf{v}$, then the mean velocity of a gas at $(\tilde{\mathbf{r}},t)$ may be denoted by $\mathbf{u}(\tilde{\mathbf{r}},t)$ which is defined by the vector equation:

$$(nd\tilde{\mathbf{r}})\mathbf{u} = \sum \mathbf{v} ,$$

where the summation over $\mathbf{v}$ on the right-hand-side extends over all of the molecules in $d\tilde{\mathbf{r}}$, and both $nd\tilde{\mathbf{r}}$ and $\sum\mathbf{v}$ are averaged over the time interval, dt.

The translational motion of an individual molecule in $d\tilde{\mathbf{r}}$ may be specified either by its 'actual' velocity, $\mathbf{v}$, relative to some standard frame of reference, or by its velocity, $\mathbf{V}'$, relative to axes moving with a velocity, $\mathbf{u}'$, so that:

$$\mathbf{V}' = \mathbf{v} - \mathbf{u}' \ .$$

If $\mathbf{u}' = \mathbf{u}$, then the quantity, $\mathbf{V} = \mathbf{v} - \mathbf{u}$, is called the peculiar velocity of the molecule.

3. THE DISTRIBUTION FUNCTION OF MOLECULAR VELOCITIES.

Within a statistical framework, the state of a rarefied gas is described by the distribution function, $f\left(\mathbf{v},\tilde{\mathbf{r}},t\right)$, so that the quantity, $fd\mathbf{v}d\tilde{\mathbf{r}}$, gives the probable number of molecules which, at time, t, are situated around $\tilde{\mathbf{r}}$ in the volume element, $d\tilde{\mathbf{r}}$, and have velocities near $\mathbf{v}$ lying in the range, $d\mathbf{v}$. It is also important to note that $fd\mathbf{v}d\tilde{\mathbf{r}}$ is the number of molecules in $d\mathbf{v}d\tilde{\mathbf{r}}$ averaged over the time interval, dt. The distribution function may be specified as the probability density of points in the phase space (i.e. the six-dimensional space of $\tilde{\mathbf{r}}$ and $\mathbf{v}$) of a molecule.

The description of a gas by its distribution function is excessively detailed. For practical purposes this amount of detail is not always needed. For instance, the gas as a continuum may be characterized by various macroscopic quantities which may be experimentally measured. These quantities may be considered to be moments of the distribution function and these moment relationships will be discussed in detail later.

It is clear that the number density may be obtained by integrating the distribution function throughout the entire velocity space. This integration may be performed over either the actual or peculiar velocities of the molecules since the distribution of velocity points is unaffected if the origin in the velocity space is shifted to the point, $\mathbf{u}$. The same integration concept also applies with respect to all similar integrals over the distribution function (moments of the distribution function). Hence, the number density may be expressed in the form:

$$n\left(\tilde{\mathbf{r}},t\right) = \int f\left(\mathbf{v},\tilde{\mathbf{r}},t\right)d\mathbf{v} = \int f\left(\mathbf{V}+\mathbf{u},\tilde{\mathbf{r}},t\right)d\mathbf{V} \ . \tag{1-1}$$

4. MEAN VALUES OF FUNCTIONS OF MOLECULAR VELOCITIES.

Let $\phi(\mathbf{v},\tilde{\mathbf{r}},t)$ be any function of the molecular velocity, $\mathbf{v}$, position, $\tilde{\mathbf{r}}$, and time, t. This function will be called a molecular property. The mean value of $\phi(\mathbf{v},\tilde{\mathbf{r}},t)$ in the volume element, $d\tilde{\mathbf{r}}$, is defined by the relation:

$$\bar{\phi} = (nd\tilde{\mathbf{r}})^{-1} \sum \phi,$$

where $\sum \phi$ denotes the time average during dt of the sum of the values of $\phi(\mathbf{v},\tilde{\mathbf{r}},t)$ for all the molecules in $d\tilde{\mathbf{r}}$ and can be expressed in terms of the distribution function, $f(\mathbf{v},\tilde{\mathbf{r}},t)$. For this purpose, consider the set of molecules in the volume element of the phase space, $d\mathbf{v}d\tilde{\mathbf{r}}$. Molecules of this set all have the same velocity and thus their contribution to the sum, $\sum \phi$, is $\phi f d\mathbf{v}d\tilde{\mathbf{r}}$. The net contribution of all molecules in the volume element, $d\tilde{\mathbf{r}}$, may then be expressed in the form:

$$\sum \phi = d\tilde{\mathbf{r}} \int \phi f d\mathbf{v} \ .$$

Hence $\bar{\phi}$, the mean value of any molecular property, $\phi(\mathbf{v},\tilde{\mathbf{r}},t)$, may be represented by a moment of the distribution function, i.e.:

$$\bar{\phi} = n^{-1} \int \phi f d\mathbf{v} \ . \tag{1-2}$$

In particular, the mean velocity of a gas is given by:

$$\mathbf{u} = n^{-1} \int \mathbf{v} f d\mathbf{v} \ . \tag{1-3}$$

From Eq. (1-3), it then follows directly that the mean value of the peculiar molecular velocity is equal to zero, i.e.:

$$\bar{\mathbf{V}} = \overline{\mathbf{v} - \mathbf{u}} = \bar{\mathbf{v}} - \mathbf{u} = 0 \ .$$

5. TRANSPORT OF MOLECULAR PROPERTIES.

Consider the passage of molecules, as shown in Fig. 1-1, across a small element of surface, dS, moving in the gas with arbitrary velocity, $\mathbf{u}'$. Let $\mathbf{n}$ be a unit vector positioned normal to the surface element and pointing in

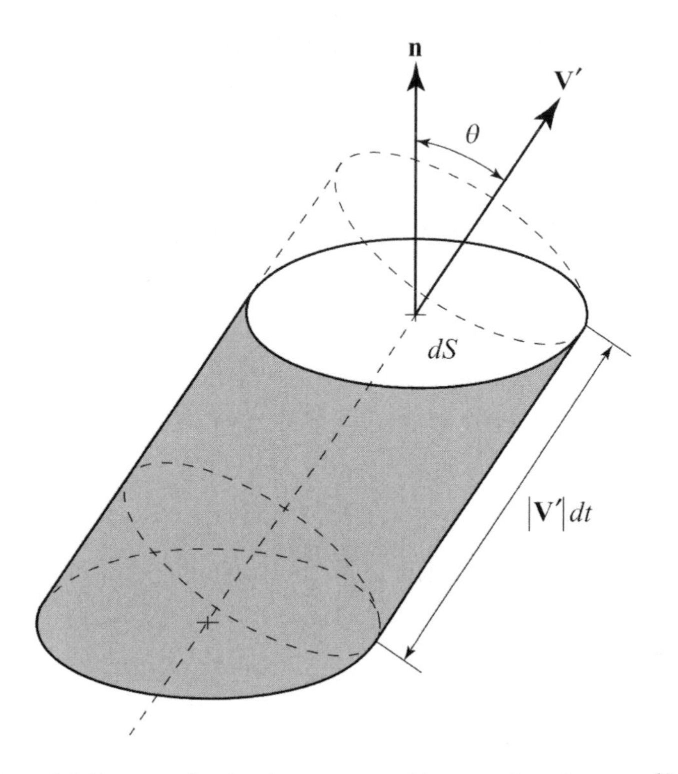

Figure 1-1. Passage of molecules across an arbitrary surface element, dS .

the direction from the negative side to the positive side. The velocity, $\mathbf{V}'$, of a molecule relative to dS is equal to $\mathbf{v}-\mathbf{u}'$, or $\mathbf{V}+\mathbf{u}-\mathbf{u}'$. In the Kinetic Theory of Gases, a standard analysis is usually employed to calculate the fluxes of the various molecular properties. First, the contribution to the flux across the surface element, dS, is determined for each separate velocity group of molecules. Then, the net flux is calculated by integrating over all velocity groups.

Consider the molecules of one velocity group in $d\tilde{\mathbf{r}}$ whose peculiar velocities lie in the range, $d\mathbf{V}$. Let a molecule cross the element dS in a time dt which is so short that the possibility of molecular encounters may be ignored. In this case, the molecule must lie somewhere inside the region having dS as the base and having a length and direction determined by $-\mathbf{V}'\,dt$. Thus, if $d\tilde{\mathbf{r}}$ denotes the volume of this region, the number of molecules crossing dS during dt is $fd\mathbf{V}d\tilde{\mathbf{r}}$.

From the above description, it follows that $d\tilde{\mathbf{r}}=\pm|\mathbf{V}'|\cos(\theta)dSdt$ where θ is the angle between $\mathbf{V}'$ and $\mathbf{n}$ and the sign, plus or minus ($\pm$), is chosen so as to make the expression for $d\tilde{\mathbf{r}}$ positive. Thus, the flux of $\phi(\mathbf{V})$ for this molecular group is expressed, both in magnitude and in sign, by the following relation:

$$\phi(\mathbf{V})V_n' f(\mathbf{V})d\mathbf{V}dSdt \ . \tag{1-4a}$$

The net flux is then the integral over all of the velocity groups and is written as:

$$dSdt\int\phi(\mathbf{V})V_n' f(\mathbf{V})d\mathbf{V} \ . \tag{1-4b}$$

The rate of flow of property ϕ per unit area is expressed in the form:

$$n\overline{V_n'\phi(\mathbf{V})} \ . \tag{1-4c}$$

This represents the n-component of the corresponding vector quantity:

$$n\overline{\mathbf{V}'\phi(\mathbf{V})} \ . \tag{1-5}$$

Substituting the definition of $\mathbf{V}'$ into Eq. (1-5), one can obtain:

$$n\overline{\mathbf{V}'\phi(\mathbf{V})} = n\overline{\mathbf{V}\phi(\mathbf{V})} + n(\mathbf{u}-\mathbf{u}')\overline{\phi(\mathbf{V})} \ . \tag{1-6}$$

The vector $n\overline{\mathbf{V}\phi(\mathbf{V})}$ is typically termed the flux density vector for the molecular property, $\phi(\mathbf{V})$, while the second term on the right in Eq. (1-6) is connected with the number flow of molecules across dS; each of which possesses the mean value of the property, $\phi(\mathbf{V})$.

6. THE PRESSURE TENSOR.

Let $\phi(\mathbf{V})$ be equal to some component of the molecular momentum, $m\mathbf{v}$. At the boundary of some containment vessel, every molecule that rebounds from the surface imparts momentum to it. These impacts simulate a continuous force on the surface equal to the rate at which momentum is imparted to the surface. The force per unit area on the surface is called the pressure which is a vector that is not necessarily normal to the surface.

Let an element of the surface, dS, move with a velocity, $\mathbf{u}'$, and let the direction pointing into the gas from the surface be taken as negative. The pressure on this boundary surface element, $\mathbf{P}_n$, is equal to the total rate of

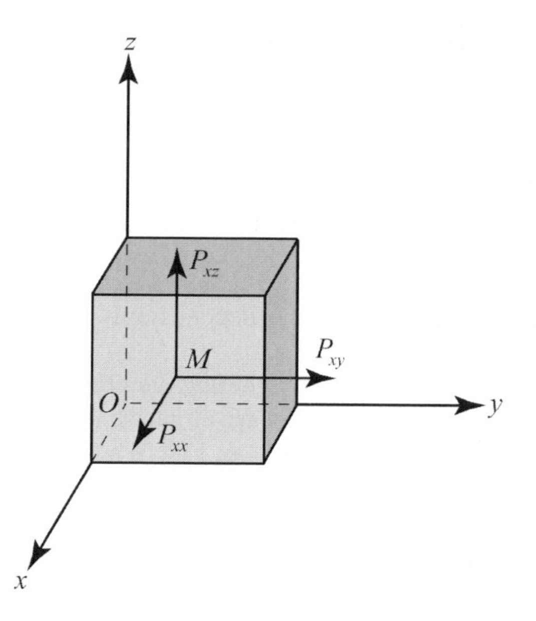

Figure 1-2. Components of the pressure tensor. Here, the point, M, is physically located at the center of the cube face nearest to the observer.

flow of the molecular momentum, $mn\mathbf{v}$, per unit area. Using Eq. (1-4c), one obtains [4]:

$$\mathbf{P}_n = \rho\overline{V_n'\mathbf{v}} \ . \tag{1-7}$$

If the surface is impenetrable to the gas molecules, then the mean velocity relative to the wall has no component normal to the wall. This may be expressed as:

$$\overline{V_n'} = 0 \ . \tag{1-8}$$

Using this condition, one can perform the following transformations:

$$\overline{V_n'\mathbf{v}} = \overline{V_n'(\mathbf{V}+\mathbf{u})} = \overline{\left[\mathbf{n}\cdot(\mathbf{V}+\mathbf{u}-\mathbf{u}')\right]\mathbf{V}} = \overline{(\mathbf{n}\cdot\mathbf{V})\mathbf{V}} \ ,$$

after which the following expression for $\mathbf{P}_n$ may be obtained:

$$\mathbf{P}_n = \rho\overline{V_n\mathbf{V}} = \mathbf{n}\cdot\rho\overline{\mathbf{V}\mathbf{V}} = \mathbf{n}\cdot\mathbf{P} \ . \tag{1-9}$$

Here, $\mathbf{P}$ is a symmetrical tensor defined by:

$$\mathbf{P} = \rho\overline{\mathbf{V}\mathbf{V}} = \rho\begin{Bmatrix} \overline{V_x^2} & \overline{V_xV_y} & \overline{V_xV_z} \\ \overline{V_yV_x} & \overline{V_y^2} & \overline{V_yV_z} \\ \overline{V_zV_x} & \overline{V_zV_y} & \overline{V_z^2} \end{Bmatrix} . \tag{1-10}$$

It is important to note that this tensor depends only upon the distribution of the peculiar velocities.

The same analysis may be used to define the pressure distribution at any point, M, within the gas. Let the surface element, dS, containing M share the mean motion of the gas so that $\mathbf{u}' = \mathbf{u}$, $\overline{\mathbf{V}}' = \overline{\mathbf{V}} = 0$ and, therefore, $\overline{V}_n' = 0$. Since the basic condition for deriving Eq. (1-9) is satisfied in this particular case, the pressure distribution across the surface element, dS, is defined by the same pressure tensor, $\mathbf{P}$, given by Eq. (1-10). The physical sense of each pressure tensor component is illustrated in Fig. 1-2.

7. THE HYDROSTATIC PRESSURE.

The sum of the normal pressures at the point, M, across the three planes parallel to the coordinate planes, is given by:

$$P_{xx} + P_{yy} + P_{zz} = \rho\overline{V^2} .$$

The mean of the normal pressures across these planes is called the hydrostatic pressure, p, or the pressure at M. Immediately from this definition, one can obtain:

$$p = \tfrac{1}{3} P_{ik}\delta_{ik} , \tag{1-11}$$

where δ_{ik} is the unit symmetrical tensor whose components are given by:

$$\delta_{ik} = \begin{cases} 1 ; & i = k , \\ 0 ; & i \neq k . \end{cases}$$

If $P_{ik} = 0$ when $i \neq k$ and the diagonal elements are equal, the pressure, p, may be expressed in the form:

$$p = P_{xx} = P_{yy} = P_{zz} \ , \tag{1-12a}$$

or:

$$P_{ik} = p\delta_{ik} \ . \tag{1-12b}$$

In this case, the pressure vector may be written as $(\mathbf{P}_n)_i = P_{ik}n_k = pn_i$ and, therefore, the pressure on any surface element through the point, M, is normal to the surface and is independent of the orientation of the surface.

8. THE AMOUNT OF HEAT.

The kinetic energy of translational motion of molecules in the volume element, $d\tilde{\mathbf{r}}$, at time, t, is:

$$(nd\tilde{\mathbf{r}})\tfrac{1}{2}m\overline{v^2} \ ,$$

which may be expressed in the form:

$$d\tilde{\mathbf{r}}\tfrac{1}{2}\rho\left(u^2 + \overline{V^2}\right) . \tag{1-13}$$

Here, the first term is the kinetic energy associated with the motion of the volume element which has a net mass. The second term is the kinetic energy associated with the peculiar motions of the molecules in the volume element and which is one component of the thermal energy of the gas in the volume element (and is the only component of the thermal energy for monatomic gases). This energy is communicable between molecules at encounters. It should be noted that there are also other forms of communicable energy that occur in polyatomic gases. In this case, the total heat energy, E, of a molecule is the sum of its peculiar kinetic energy and any of these other forms of communicable energy that may be present.

9. THE KINETIC TEMPERATURE.

In the Kinetic Theory of Gases, a temperature, T, is defined directly by the relationship:

$$\tfrac{1}{2}m\overline{V^2} = \tfrac{3}{2}kT \ , \tag{1-14}$$

where $k = 1.38066 \times 10^{-16}$ erg K^{-1} is Boltzmann's constant. It is important to note that the temperature definition given by Eq. (1-14) is more general than that employed in Thermodynamics and Statistical Mechanics where only equilibrium states are considered. Eq. (1-14) can be used for non-steady-states of a gas as well.

10. THE EQUATION OF STATE FOR A PERFECT GAS.

From the definitions of the hydrostatic pressure and the temperature given in Eqs. (1-11) and (1-14), one can obtain the following relationship between the pressure and the temperature:

$$p = nkT \ . \tag{1-15}$$

This equation is called the equation of state for a perfect gas.

11. THE THERMAL FLUX VECTOR.

The thermal flux vector, $\mathbf{Q}$, is an important quantity that expresses the rate of flow of heat energy across a unit surface element. This vector may be obtained from Eq. (1-6) by substituting the heat energy of a molecule (the energy which is communicable between molecules at encounters) in place of the generic molecular property, $\phi(\mathbf{V})$. After this substitution, one obtains:

$$\mathbf{Q} = n\overline{E\mathbf{V}} \ . \tag{1-16}$$

If a molecule possesses only kinetic energy, then this heat flux vector is defined by:

$$\mathbf{Q} = n\overline{\tfrac{1}{2}mV^2\mathbf{V}} \ . \tag{1-17}$$

12. SUMMARY.

Rarefied gases, since they are systems containing a huge number of molecules, can only be realistically described within a statistical framework. The probabilistic basis of this description is not readily apparent in many practical transport problems, however, because the accuracy of the results predicted by the Kinetic Theory of Gases is much greater than the realistically achievable accuracy in actual experimental measurements and because the relative deviations of the various macroscopic quantities from their mean values are extremely small.

Within this framework, the state of a rarefied gas is described by the distribution function, $f(\mathbf{v},\tilde{\mathbf{r}},t)$, so that the quantity, $fd\mathbf{v}d\tilde{\mathbf{r}}$, gives the probable number of molecules at time, t, which are located in the spatial region at $\tilde{\mathbf{r}}$ in $d\tilde{\mathbf{r}}$, and which have velocities of $\mathbf{v}$ in the velocity space interval, $d\mathbf{v}$. The description of a gas by the distribution function is excessively detailed while the gas, as a continuum, may be characterized by certain macroscopic quantities which are relatively few in number and which can be experimentally measured. These macroscopic quantities, and any other molecular property fluxes of interest, may be calculated as moments of the distribution function. The most significant of these macroscopic values for a monatomic gas include the number density, the mean velocity, the temperature, the pressure, and the thermal flux, and are given by:

$$n = \int f d\mathbf{v} \ , \tag{1-18}$$

$$\mathbf{u} = n^{-1} \int \mathbf{v} f d\mathbf{v} \ , \tag{1-19}$$

$$\tfrac{3}{2} kT = n^{-1} \int \tfrac{1}{2} m (\mathbf{v} - \mathbf{u})^2 f d\mathbf{v} \ , \tag{1-20}$$

$$P_{ik} = m \int (v_i - u_i)(v_k - u_k) f d\mathbf{v} \ , \tag{1-21}$$

and:

$$\mathbf{Q} = \tfrac{1}{2} m \int (\mathbf{v} - \mathbf{u})^2 (\mathbf{v} - \mathbf{u}) f d\mathbf{v} \ . \tag{1-22}$$

Thus, a complete knowledge of the distribution function can be used to fully describe the behavior of a gas.

PROBLEMS

1.1. Prove that the relative fluctuation of an additive quantity is proportional to $N^{-1/2}$, where N is the number of molecules of a gas.

Solution: Let L be an additive quantity for a gas that contains N molecules. The mean value of this quantity may be written as:

$$\bar{L} = \sum_{i=1}^{N} \bar{L}_i \ ,$$

and, therefore, this value is proportional to N. The relative fluctuation of L is given by:

$$\delta = \bar{L}^{-1} \sqrt{\overline{\left(\sum_i \Delta L_i\right)^2}} \ .$$

The radicand may be expressed in the form:

$$\overline{\left(\sum_i \Delta L_i\right)^2} = \sum_i \overline{(\Delta L_i)^2} + \sum_{\substack{i \\ i \neq k}} \sum_k \overline{\Delta L_i}\, \overline{\Delta L_k} \ .$$

The later term in this relation is equal to zero since separate molecules are statistically independent and, for one molecule, $\overline{\Delta L_i} = 0$. Hence, this sum containing N terms is proportional to N and one can obtain the following:

$$\delta \sim N^{-1} \sqrt{N} = N^{-1/2} \ .$$

1.2. Derive the condition of applicability of the point-mass hypothesis for a gas molecule.

Solution: A molecule in a gas may be considered as a point particle when the de Broglie wavelength of the molecule is much less than the mean distance between the gas molecules. This relationship may be expressed as:

$$\frac{2\pi\hbar}{m\bar{v}} << \left(\frac{V}{N}\right)^{1/3} = n^{-1/3} ,$$

where $\hbar$ is Planck's constant ($\hbar = 1.054 \times 10^{-27}$ erg sec) and $\bar{v}$ is the mean heat velocity of a molecule. The following expression may be written:

$$\frac{2\pi\hbar}{\sqrt{2mkT}} n^{1/3} << 1 .$$

1.3. Determine the temperature for which the de Broglie wavelength of a molecule is the same order of magnitude as the mean distance between gas molecules. Consider the case of hydrogen having a number density of $n = 2.69 \times 10^{19}$ cm^{-3}.

Solution:

$$T_0 = \frac{2\pi^2\hbar^2 n^{2/3}}{mk} ;\ T_0 = 0.43 \text{ K}.$$

1.4. Let the behavior of a mechanical system be described by exact equations of motion (N is the number of molecules, V is the volume, and $N >> 1$). How might a statistical description be introduced for such a system? How might the mean values and the distribution function be specified? What accuracy does the statistical description have?

Solution: Consider the state of a subsystem which is situated in a small volume element, $d\tilde{\mathbf{r}}$, surrounding the point, $\tilde{\mathbf{r}}$, during a time interval, dt. This interval is assumed to satisfy the following relations:

$$\frac{(d\tilde{\mathbf{r}})^{1/3}}{\bar{v}} << dt << \frac{V^{1/3}}{\bar{v}} ,$$

where $\bar{v}$ is the mean velocity of the particles. Let the time interval, dt, be divided into k smaller subintervals of width, $d\tau$, and the whole velocity space be divided into m smaller velocity ranges, $\Delta\mathbf{v}_j = \Delta v_{jx}\Delta v_{jy}\Delta v_{jz}$, so that $\sum_{j=1}^{m} \Delta\mathbf{v}_j$ is equal to the product of the maximum values of the velocity components. If the equations of motion are known, it is easy to calculate the number of molecules at time, $t + i\,d\tau$, which have velocities lying in the range, $\Delta\mathbf{v}_j$. Let this quantity be denoted by $n_{ij} = n\left(\tilde{\mathbf{r}}, \mathbf{v}_j, t + id\tau\right)$. The number density and the mean value of any molecular property may be defined, respectively, by:

$$n = \frac{1}{d\tilde{\mathbf{r}}} \sum_{j=1}^{m} \sum_{i=1}^{k} \frac{n_{ij}}{k} \text{ , and } n\bar{\phi} = \sum_{j=1}^{m} \sum_{i=1}^{k} \frac{\phi(\mathbf{v}_j) n_{ij}}{k} .$$

The distribution function may be written as:

$$f = \frac{1}{d\tilde{\mathbf{r}}\Delta\mathbf{v}_j} \sum_{i=1}^{k} \frac{n_{ij}}{k} ,$$

The relative fluctuations of additive quantities for such a system are given by $\delta \sim N^{-1/2}$.

REFERENCES

1. Landau, L.D. and Lifshitz, E.M., *Statistical Physics* (Pergamon, London; Addison Wesley, Reading, Mass, 1958).
2. Levich, V.G., *Theoretical Physics: An Advanced Text* (North-Holland, Amsterdam, 1970).
3. Huang, K., *Statistical Mechanics* (John Wiley and Sons, New York, 1963).
4. Chapman, S. and Cowling, T.G., *The Mathematical Theory of Non-Uniform Gases* (Cambridge University Press, Cambridge, U.K., 1990).
5. Ferziger, J.H. and Kaper, H.G., *Mathematical Theory of Transport Processes in Gases* (North-Holland, Amsterdam-London, 1972).
6. Kogan, M.N., *Rarefied Gas Dynamics* (Plenum Press, New York, 1969).

Chapter 2

THE BOLTZMANN EQUATION

1. DERIVATION OF THE BOLTZMANN EQUATION.

It has been established previously that knowledge of the distribution function gives all the necessary information for a gas. To obtain the basic equation for the distribution function, consider a balance of the number of molecules that are located in the element, $d\mathbf{v}d\tilde{\mathbf{r}}$, of the six-dimensional phase space for the time interval, dt. Consider a gas in which each molecule is subject to an external force, $m\mathbf{F}$, that is a function of $\tilde{\mathbf{r}}$ and t, but not a function of $\mathbf{v}$. For the time interval, dt, the position of this phase element will change from $(\mathbf{v},\tilde{\mathbf{r}})$ to $(\mathbf{v}+\mathbf{F}dt,\tilde{\mathbf{r}}+\mathbf{v}dt)$. It is important to note that the magnitude of this volume element is unaltered in accordance with the Liouville theorem of classical mechanics [1].

There are $f(\mathbf{v},\tilde{\mathbf{r}},t)d\mathbf{v}d\tilde{\mathbf{r}}$ molecules which, at the time, t, are situated in the volume element, $d\tilde{\mathbf{r}}$, and have velocities in the velocity range, $d\mathbf{v}$. These molecules are assumed to belong to the first molecular set. After the time interval, dt, there is another set in which the number of molecules is given by:

$$f(\mathbf{v}+\mathbf{F}dt,\tilde{\mathbf{r}}+\mathbf{v}dt,t+dt)d\mathbf{v}d\tilde{\mathbf{r}} \ .$$

The number of molecules in the second set will differ from that in the first set owing to molecular encounters. The net gain of molecules to the second set is proportional to $d\mathbf{v}d\tilde{\mathbf{r}}dt$ and will be denoted by $(\delta f/\delta t)d\mathbf{v}d\tilde{\mathbf{r}}dt$. Consequently, one can obtain:

$$\left[f\left(\mathbf{v}+\mathbf{F}dt,\tilde{\mathbf{r}}+\mathbf{v}dt,t+dt\right)-f\left(\mathbf{v},\tilde{\mathbf{r}},t\right)\right] d\mathbf{v}d\tilde{\mathbf{r}}=\left(\delta f/\delta t\right) d\mathbf{v}d\tilde{\mathbf{r}}dt \ . \qquad (2\text{-}1)$$

Now, if dt tends to zero, it is easy to obtain the Boltzmann equation which may be written in the form [2]:

$$\mathrm{D}f=\frac{\delta f}{\delta t} \ . \qquad (2\text{-}2a)$$

Here, the following notation is introduced:

$$\mathrm{D}f=\frac{\partial f}{\partial t}+v_i\frac{\partial f}{\partial \tilde{x}_i}+F_i\frac{\partial f}{\partial v_i} \ , \qquad (2\text{-}2b)$$

where $\tilde{x}_i$ are components of the position vector, $\tilde{\mathbf{r}}$. The quantity $\left(\delta f/\delta t\right)d\mathbf{v}d\tilde{\mathbf{r}}$ is the rate of change of the number of molecules in the phase element, $d\mathbf{v}d\tilde{\mathbf{r}}$, owing to molecular encounters.

2. THE MOMENT EQUATIONS.

Let $\phi=\phi\left(\mathbf{v},\tilde{\mathbf{r}},t\right)$ be any molecular property. After multiplying the Boltzmann equation by $\phi d\mathbf{v}$ and integrating throughout the velocity space, one can obtain the following integral relation:

$$\int \phi \mathrm{D}f d\mathbf{v}=n\Delta\overline{\phi} \ , \qquad (2\text{-}3)$$

where:

$$n\Delta\overline{\phi}=\int \phi\frac{\delta f}{\delta t}d\mathbf{v} \ .$$

In Eq. (2-3), all the integrals are assumed to be convergent and, moreover, products such as ϕf tend to zero if $\left|\mathbf{v}\right|\rightarrow\infty$. This equation is the general form of a moment equation in which $n\Delta\overline{\phi}$ is a moment of the Boltzmann collision integral.

The various terms on the left-hand-side of Eq. (2-3) may be transformed by means of the following relationships:

Table 2-1. Transformation of the various derivatives to new variables.

$\mathbf{v}, \tilde{\mathbf{r}}, t$	$\mathbf{V}, \tilde{\mathbf{r}}, t$
$\dfrac{\partial \phi}{\partial \mathbf{v}}$	$\dfrac{\partial \phi}{\partial \mathbf{V}}$
$\dfrac{\partial \phi}{\partial t}$	$\dfrac{\partial \phi}{\partial t} - \dfrac{\partial \phi}{\partial V_i}\dfrac{\partial u_i}{\partial t}$
$v_i \dfrac{\partial \phi}{\partial \tilde{x}_i}$	$v_i \dfrac{\partial \phi}{\partial \tilde{x}_i} - \dfrac{\partial \phi}{\partial V_k} v_i \dfrac{\partial u_k}{\partial \tilde{x}_i}$

$$\int \phi \frac{\partial f}{\partial t} d\mathbf{v} = \frac{\partial}{\partial t} \int \phi f d\mathbf{v} - \int f \frac{\partial \phi}{\partial t} d\mathbf{v} = \frac{\partial}{\partial t}\left(n\bar{\phi}\right) - n\overline{\frac{\partial \phi}{\partial t}} ,$$

$$\int \phi v_i \frac{\partial f}{\partial \tilde{x}_i} d\mathbf{v} = \frac{\partial}{\partial \tilde{x}_i} \int \phi v_i f d\mathbf{v} - \int f v_i \frac{\partial \phi}{\partial \tilde{x}_i} d\mathbf{v} = \frac{\partial}{\partial \tilde{x}_i}\left(n\overline{\phi v_i}\right) - n\overline{v_i \frac{\partial \phi}{\partial \tilde{x}_i}} ,$$

$$\int \phi \frac{\partial f}{\partial v_i} d\mathbf{v} = \int\int \phi f \Big|_{v_i=-\infty}^{v_i=\infty} dv_j dv_k - \int f \frac{\partial \phi}{\partial v_i} d\mathbf{v} = -n\overline{\frac{\partial \phi}{\partial v_i}} .$$

In the last of the above relationships, the double integral vanishes because $\phi f \big|_{v_i=-\infty}^{v_i=\infty} = 0$ in accordance with the notation made above. Substituting these expressions into Eq. (2-3), one can obtain [2]:

$$\frac{\partial}{\partial t}\left(n\bar{\phi}\right) = -\frac{\partial}{\partial \tilde{x}_i}\left(n\overline{\phi v_i}\right) + n\left\{\overline{\frac{\partial \phi}{\partial t}} + \overline{v_i \frac{\partial \phi}{\partial \tilde{x}_i}} + F_i \overline{\frac{\partial \phi}{\partial v_i}} + \Delta\bar{\phi}\right\} . \tag{2-4}$$

This equation is usually called the moment equation or the transport equation of the molecular property, ϕ.

3. ANOTHER FORM OF THE MOMENT EQUATIONS.

The basic moment relationship given in Eq. (2-4) is suitable for the laboratory frame of reference relative to which the gas has a mean velocity, $\mathbf{u}(\tilde{\mathbf{r}},t)$. To transform this equation into the variables, $(\mathbf{V},\tilde{\mathbf{r}},t)$, where $\mathbf{V}=\mathbf{v}-\mathbf{u}(\tilde{\mathbf{r}},t)$, one must consider that $\phi(\mathbf{V},\tilde{\mathbf{r}},t)$ depends upon $\tilde{\mathbf{r}}$ and t, both explicitly and implicitly through its dependence on $\mathbf{V}$. Table 2-1 expresses the transformation of the various derivatives in Eq. (2-4).

Substituting the derivatives from Table 2-1 into Eq. (2-4), one can obtain [2]:

$$\begin{gathered}\frac{D}{Dt}\left(n\overline{\phi}\right)+n\overline{\phi}\frac{\partial u_i}{\partial\tilde{x}_i}+\frac{\partial}{\partial\tilde{x}_i}\left(n\overline{\phi V_i}\right)\\-n\left\{\overline{\frac{D\phi}{Dt}}+\overline{V_i\frac{\partial\phi}{\partial\tilde{x}_i}}+\left(F_i-\frac{Du_i}{Dt}\right)\overline{\frac{\partial\phi}{\partial V_i}}-\overline{\frac{\partial\phi}{\partial V_k}V_i}\frac{\partial u_k}{\partial\tilde{x}_i}\right\}=n\Delta\overline{\phi}\ ,\end{gathered}\tag{2-5}$$

where $D/Dt=\partial/\partial t+u_i\left(\partial/\partial\tilde{x}_i\right)$ is the 'mobile operator' or time derivative with respect to the frame of reference moving with velocity, $\mathbf{u}$. By analogy, the Boltzmann equation for these variables may be expressed in the form:

$$\frac{Df}{Dt}+V_i\frac{\partial f}{\partial\tilde{x}_i}+\left(F_i-\frac{Du_i}{Dt}\right)\frac{\partial f}{\partial V_i}-\frac{\partial f}{\partial V_k}V_i\frac{\partial u_k}{\partial\tilde{x}_i}=\frac{\delta f}{\delta t}\ .\tag{2-6}$$

4. THE EQUATIONS FOR A CONTINUUM MEDIUM.

Some important moment equations can be obtained from Eq. (2-5) by using the molecular properties, ϕ, which are conserved during encounters. It is very important to note that these equations may be derived without knowledge of the collision operator because $\Delta\overline{\phi}=0$ for such molecular properties. The invariant quantities of molecular encounters in a monatomic gas are the number of molecules, momentum, and kinetic energy because the mutual potential energy of two molecules equals zero both before and after an encounter. Let the following notations be introduced with respect to these invariant quantities:

$$\psi^{(1)}=1\ ,\quad \psi^{(2)}=m\mathbf{V}\ ,\quad \psi^{(3)}=\tfrac{1}{2}mV^2\ .$$

The statement $n\overline{\Delta\psi^{(i)}} = 0$ (for $i = 1,2,3$, respectively) expresses the principles of conservation of number (of molecules), momentum, and energy during encounters. Substituting $\phi = \psi^{(i)}$ into Eq. (2-5), one can obtain:

$$\frac{Dn}{Dt} + n\frac{\partial u_i}{\partial \tilde{x}_i} = 0 , \tag{2-7}$$

$$\frac{\partial P_{ik}}{\partial \tilde{x}_k} - \rho\left(F_i - \frac{Du_i}{Dt}\right) = 0 , \tag{2-8}$$

$$\frac{DT}{Dt} = -\frac{2}{3nk}\left(P_{ij}\frac{\partial u_i}{\partial \tilde{x}_j} + \frac{\partial Q_i}{\partial \tilde{x}_i}\right) . \tag{2-9}$$

These equations are usually called the continuum equations. This system is not closed as it contains 13 unknown functions that cannot be specified from the five equations. This lack of closure is a general characteristic of all moment systems.

Since the distribution function is unknown, the moment transport equations may be considered to be the most general relations between the time and space derivatives of the macroscopic quantities because this system cannot be solved without definite suppositions or hypotheses about the distribution function which forms the basis of any moment method. The distribution function being used to close a moment system must satisfy certain general conditions. It must describe the characteristic features of the gas flow and must contain as many unknown parameters as the number of moment equations being employed. Then, all moments will depend on these parameters and the moment system becomes closed.

5. MOLECULAR ENCOUNTERS.

Knowledge of the nature of the molecular interaction during encounters is necessary to derive the expression for the Boltzmann collision integral. At present, there is not a theory that describes the details of molecular encounters exactly. Therefore, a definite model of the molecular interaction has to be introduced to complete this description.

One of the simplest molecular models is that of a smooth, rigid, and perfectly elastic sphere; commonly referred to as the rigid-sphere model for brevity. The force between two molecules at a collision appears only if these molecules are in close contact with each other. This model is approximate since the real force varies continuously with the distance between the molecules. This fact might be better interpreted if a molecule was considered as a point-center of force depending on the nature of the interacting molecules and on their separation distance. Molecules represented as either rigid spheres or as point-centers of force may be termed smooth molecules since their internal energy is not altered during encounters. In this book, only smooth molecules are considered and the analysis of encounters follows that described in [2].

Let the masses of the two molecules undergoing an encounter be m_1 and m_2. The interaction force is directed along the line joining their centers and it depends only upon the distance between molecules. The velocities before and after the encounter are denoted by $\mathbf{v}_1, \mathbf{v}_2$ and $\mathbf{v}'_1, \mathbf{v}'_2$, respectively. Some general relations between these velocities may be found without consideration of the details of the molecular encounter.

To start, one introduces the following notations:

$$m_0 = m_1 + m_2 \ , \quad M_1 = m_1/m_0 \ , \quad M_2 = m_2/m_0 \ .$$

Owing to the momentum conservation during encounters, the center of mass of the two molecules moves with a constant velocity, $\mathbf{G}$, that is given by:

$$m_0\mathbf{G} = m_1\mathbf{v}_1 + m_2\mathbf{v}_2 = m_1\mathbf{v}'_1 + m_2\mathbf{v}'_2 \ . \tag{2-10}$$

Let $\mathbf{g}_{21}$ and $\mathbf{g}'_{21}$ denote the relative velocities of the two molecules before and after an encounter:

$$\mathbf{g}_{21} = \mathbf{v}_2 - \mathbf{v}_1 \ , \quad \mathbf{g}'_{21} = \mathbf{v}'_2 - \mathbf{v}'_1 \ . \tag{2-11}$$

which have the following magnitudes:

$$|\mathbf{g}_{21}| = g \ , \quad |\mathbf{g}'_{21}| = g' \ .$$

From Eqs. (2-10) and (2-11), one can obtain:

$$\mathbf{v}_1 = \mathbf{G} - M_2\mathbf{g}_{21} \ , \quad \mathbf{v}'_1 = \mathbf{G} - M_2\mathbf{g}'_{21} \ , \tag{2-12a}$$

$$\mathbf{v}_2 = \mathbf{G} + M_1\mathbf{g}_{21} \ , \quad \mathbf{v}'_2 = \mathbf{G} + M_1\mathbf{g}'_{21} \ . \tag{2-12b}$$

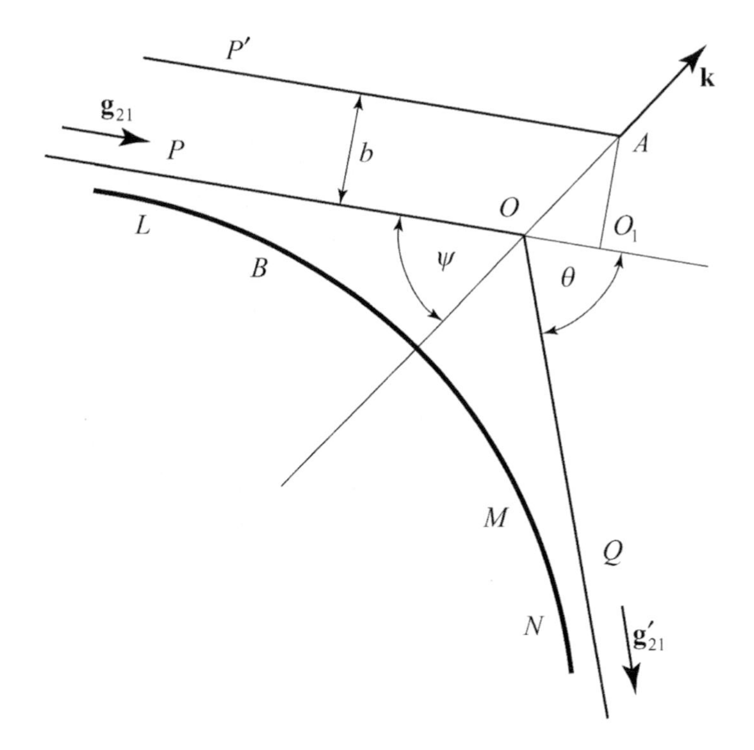

Figure 2-1. Geometry of a generic encounter.

The mutual potential energy of the two molecules equals zero both before and after an encounter and, hence, the energy conservation equation may be written in the following very simple form:

$$\tfrac{1}{2}\left(m_1 v_1^2 + m_2 v_2^2\right) = \tfrac{1}{2}\left(m_1 v_1'^2 + m_2 v_2'^2\right) \ .$$

Using this relation and Eqs. (2-12a) and (2-12b), one is able to determine that the relative velocity during an encounter is changed only in direction and not in magnitude, i.e.:

$$g = g' \ .$$

Consideration of momentum and energy conservation laws alone does not suffice to determine the direction of $\mathbf{g}'_{21}$ which depends not only upon the initial velocities but also upon two geometric variables characterizing the spatial orientation of the two interacting molecules under consideration.

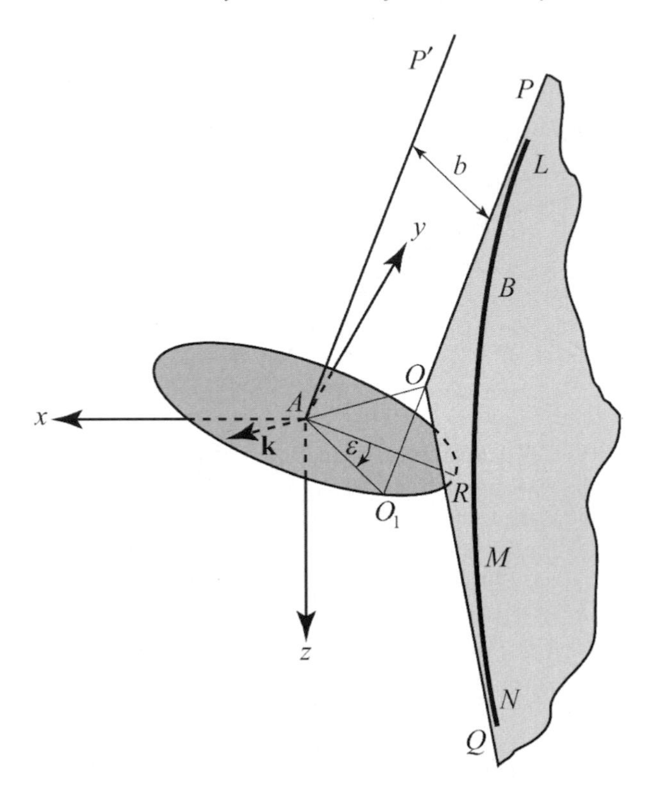

Figure 2-2. Parameters of a generic encounter.

Consider the motion of the center B of the second molecule relative to the center A of the first. This relative motion can formally be considered as the motion of an imaginary particle with a reduced mass, $\mu = m_1 m_2 / m_0$, in a central force field and, therefore, it will be confined to the plane through A [1] (the curve LMN in Fig. 2-1 is a trajectory of the second molecule). The asymptotes of this trajectory, PO and OQ, are in the directions of $\mathbf{g}_{21}$ and $\mathbf{g}'_{21}$. Let $P'A$ be a line parallel to PO so that the direction of $P'A$ is fixed by $\mathbf{g}_{21}$. The orientation of the plane LMN is specified by the angle ε between this plane and another plane containing both $P'A$ and the arbitrary fixed axis, Ax, as shown in Fig. 2-2. Let the plane RAO_1 (O_1 is the point of intersection of the line PO with the plane RAO_1) be perpendicular to $P'A$ and AR be the line of intersection of the plane RAO_1 with the plane through A which contains both $P'A$ and the axis, Ax. Let the polar coordinates of O_1 in this plane be b and ε. These geometric variables complete the specification of the encounter.

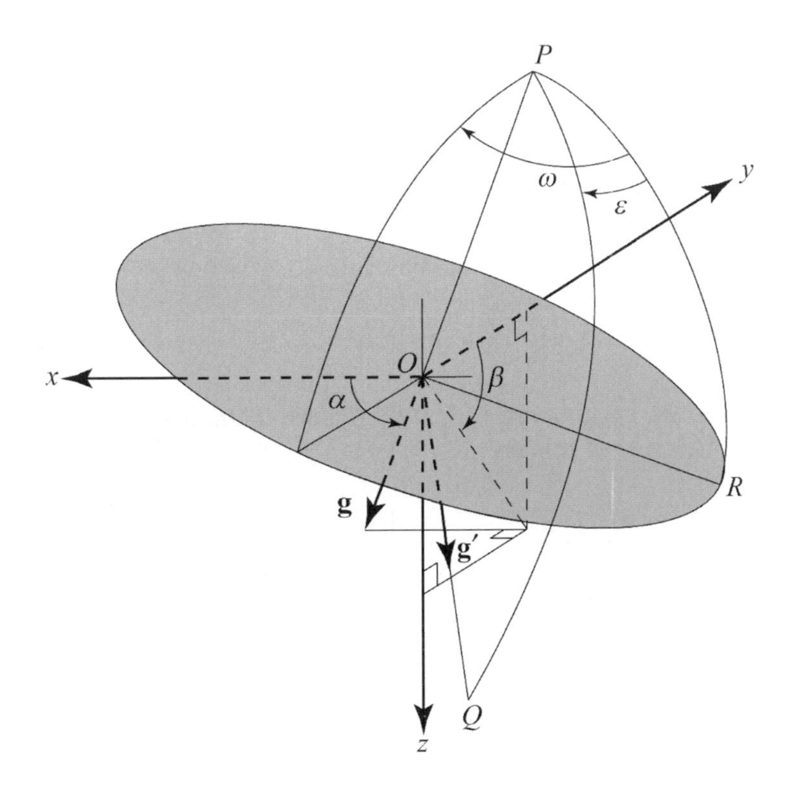

Figure 2-3. The spherical coordinates of the relative velocity.

The angle, θ, through which $\mathbf{g}_{21}$ is deflected, depends, in general, on the magnitude of g and on the parameter, b. To derive the geometrical connection between $\mathbf{g}_{21}$ and $\mathbf{g}'_{21}$, this angle is assumed to be known. Let the spherical coordinates of the vector $\mathbf{g}_{21} = \mathbf{g}$ be (g, α, β). Then, the Cartesian and spherical coordinates shown in Fig. 2-3 are related by the following:

$$g_x = g\cos(\alpha) \ , \tag{2-13a}$$

$$g_y = g\sin(\alpha)\cos(\beta) \ , \tag{2-13b}$$

$$g_z = g\sin(\alpha)\sin(\beta) \ . \tag{2-13c}$$

Let ε be the angle between the two planes, POQ and POR, as shown in Fig. 2-3. The vectors, $-\mathbf{g}$ and $\mathbf{g}'$, and $-\mathbf{g}$ and $-\mathbf{i}$, lie in the planes, POQ and POR, respectively, hence one can obtain [3]:

$$\cos(\varepsilon) = -\frac{(\mathbf{g}'\times\mathbf{g})\cdot(\mathbf{i}\times\mathbf{g})}{|\mathbf{g}'\times\mathbf{g}||\mathbf{i}\times\mathbf{g}|} = -\frac{g'_x - g_x\cos(\theta)}{g\sin(\alpha)\sin(\theta)} . \tag{2-14}$$

To simplify this vector expression, the following relation is employed:

$$(\mathbf{a}\times\mathbf{b})\cdot(\mathbf{c}\times\mathbf{b}) = (\mathbf{a}\cdot\mathbf{c})\mathbf{b}^2 - (\mathbf{a}\cdot\mathbf{b})(\mathbf{c}\cdot\mathbf{b}) .$$

Let ω be the angle between the two planes, POR and POy, shown in Fig. 2-3. Then, one can obtain:

$$\cos(\omega) = \frac{(\mathbf{i}\times\mathbf{g})\cdot(\mathbf{j}\times\mathbf{g})}{|\mathbf{i}\times\mathbf{g}||\mathbf{j}\times\mathbf{g}|} = -\frac{\cos(\alpha)\cos(\beta)}{\sqrt{1-\sin^2(\alpha)\cos^2(\beta)}} , \tag{2-15}$$

and:

$$\begin{aligned}\cos(\omega-\varepsilon) &= \frac{(\mathbf{j}\times\mathbf{g})\cdot(\mathbf{g}\times\mathbf{g}')}{|\mathbf{j}\times\mathbf{g}||\mathbf{g}\times\mathbf{g}'|} \\ &= -\frac{(g'_y/g) - \sin(\alpha)\cos(\beta)\cos(\theta)}{\sin(\theta)\sqrt{1-\sin^2(\alpha)\cos^2(\beta)}} .\end{aligned} \tag{2-16}$$

Using Eqs. (2-14)-(2-16) and analogous relations for g'_z, one can express the components of the relative velocity by:

$$g'_x = g\left[\cos(\alpha)\cos(\theta) - \sin(\alpha)\sin(\theta)\cos(\varepsilon)\right] , \tag{2-17a}$$

$$\begin{aligned}g'_y = g\big[&\sin(\alpha)\cos(\beta)\cos(\theta) \\ &+\cos(\alpha)\cos(\beta)\sin(\theta)\cos(\varepsilon) \\ &-\sin(\beta)\sin(\theta)\sin(\varepsilon)\big] ,\end{aligned} \tag{2-17b}$$

$$g'_z = g\big[\sin(\alpha)\sin(\beta)\cos(\theta) \\ +\cos(\alpha)\sin(\beta)\sin(\theta)\cos(\varepsilon) \\ +\cos(\beta)\sin(\theta)\sin(\varepsilon)\big] \ . \qquad (2\text{-}17c)$$

If the angle, θ, is known, these relations together with Eqs. (2-12a) and (2-12b) permit one to determine the velocity components after an encounter.

Some other relations for the molecular velocities before and after an encounter may be derived from Eqs. (12-12a) and (12-12b). The trajectory, LMN, of the second molecule relative to the first molecule is symmetrical about the line, OA (usually known as the apse-line), shown in Fig. 2-1. The vectors, $\mathbf{g}_{21}$ and $\mathbf{g}'_{21}$, differ by twice the component of $\mathbf{g}_{21}$ in the direction of $\mathbf{k}$. This difference is expressed as:

$$\mathbf{g}_{21} - \mathbf{g}'_{21} = 2(\mathbf{g}_{21} \cdot \mathbf{k})\mathbf{k} \ .$$

Using this relation and Eqs. (2-12a) and (2-12b), one can obtain:

$$\mathbf{v}'_1 - \mathbf{v}_1 = 2M_2(\mathbf{g}_{21} \cdot \mathbf{k})\mathbf{k} = -2M_2(\mathbf{g}'_{21} \cdot \mathbf{k})\mathbf{k} \ , \qquad (2\text{-}18a)$$

$$\mathbf{v}'_2 - \mathbf{v}_2 = -2M_1(\mathbf{g}_{21} \cdot \mathbf{k})\mathbf{k} = 2M_1(\mathbf{g}'_{21} \cdot \mathbf{k})\mathbf{k} \ . \qquad (2\text{-}18b)$$

These formulas form another set of relations between molecular velocities during an encounter.

6. THE RELATIVE MOTION OF TWO MOLECULES.

The relative motion of two molecules is equivalent to the motion of one particle with the reduced mass, $\mu = m_1 m_2 / m_0$, moving in a central force field, $\mathbf{F}(\tilde{r}) = -(\partial U/\partial \tilde{r})\tilde{\mathbf{r}}/\tilde{r}$, where $U(\tilde{r})$ is the mutual potential energy of the two molecules [1]. This problem can be solved in a general form by employing the energy and angular momentum conservation laws which, in polar coordinates (r,ψ) with the origin at the point, A, as shown in Fig. 2-1, may be written as [1,4]:

$$E = \tfrac{1}{2}\mu\left(\dot{\tilde{r}}^2 + \tilde{r}^2\dot{\psi}^2\right) + U(\tilde{r}) \ , \qquad (2\text{-}19)$$

$$M = \mu \tilde{r}^2 \dot{\psi} \ , \tag{2-20}$$

where E is the energy before an encounter and M is the constant angular momentum component which is normal to the motion plane. Using Eq. (2-19), one can obtain:

$$\dot{\tilde{r}} = \sqrt{\frac{2}{\mu}\left[E - U(\tilde{r})\right] - \frac{M^2}{\mu^2 \tilde{r}^2}} \ . \tag{2-21}$$

The polar angle, ψ, can be found from Eq. (2-20) which may be expressed in the form:

$$d\psi = \frac{M}{\mu \tilde{r}^2} dt \ .$$

From this relation and Eq. (2-21), after integrating with respect to $\tilde{r}$, one can obtain the following expression for ψ as a function of $\tilde{r}$:

$$\psi = \int \frac{\frac{M}{\tilde{r}^2} d\tilde{r}}{\sqrt{2\mu\left[E - U(\tilde{r})\right] - \frac{M^2}{\tilde{r}^2}}} + \text{const} \ . \tag{2-22}$$

After substituting $E = \frac{1}{2}\mu g^2$ and $M = \mu b g$, Eq. (2-22) becomes:

$$\psi_0 = b \int_{\tilde{r}_0}^{\infty} \frac{\frac{d\tilde{r}}{\tilde{r}^2}}{\sqrt{1 - \frac{b^2}{\tilde{r}^2} - \frac{2U(\tilde{r})}{\mu g^2}}} \ . \tag{2-23}$$

Here, the lower limit of integration, $\tilde{r}_0$, is defined as:

$$\tilde{r}_0^2 - b^2 = \frac{2\tilde{r}_0^2 U(\tilde{r}_0)}{\mu g^2} \ .$$

and the scattering angle, θ, can be found from:

$$\theta = \left|\pi - 2\psi_0\right| \ . \tag{2-24}$$

PROBLEMS

2.1. Prove that, if molecules move in accordance with Newton's equations of motion, the phase volume element is not altered for the time interval, dt.

Solution: For the time interval, dt, the position of the phase element, $d\Gamma = d\mathbf{v}d\tilde{\mathbf{r}}$, is changed from $(\mathbf{v},\tilde{\mathbf{r}})$ to $(\mathbf{v}' = \mathbf{v} + \mathbf{F}dt\,,\tilde{\mathbf{r}}' = \tilde{\mathbf{r}} + \mathbf{v}dt)$. Then:

$$d\Gamma' = \frac{\partial(\mathbf{v}',\tilde{\mathbf{r}}')}{\partial(\mathbf{v},\tilde{\mathbf{r}})}d\Gamma = \frac{\partial(\mathbf{v}',\tilde{\mathbf{r}}')}{\partial(\mathbf{v},\tilde{\mathbf{r}}')}\frac{\partial(\mathbf{v},\tilde{\mathbf{r}}')}{\partial(\mathbf{v},\tilde{\mathbf{r}})}d\Gamma$$

$$= \left.\frac{\partial(v_x',v_y',v_z')}{\partial(v_x,v_y,v_z)}\right|_{\tilde{\mathbf{r}}'=\mathrm{const}} \left.\frac{\partial(\tilde{x}',\tilde{y}',\tilde{z}')}{\partial(\tilde{x},\tilde{y},\tilde{z})}\right|_{\mathbf{v}=\mathrm{const}} d\Gamma = d\Gamma \ .$$

The external force, $\mathbf{F}$, is assumed to be independent of $\mathbf{v}$.

2.2. Prove that the relative motion of two interacting molecules with mutual potential energy depending only on the distance between the molecules may be considered as the motion of a single particle in a central force field.

Solution: First, introduce the vector: $\tilde{\mathbf{r}} = \tilde{\mathbf{r}}_2 - \tilde{\mathbf{r}}_1 = \mathbf{e}_r\tilde{r}$. The equations of motion of these molecules may be written in terms of this vector as $m_1\ddot{\tilde{\mathbf{r}}}_1 = -f(\tilde{r})\mathbf{e}_r$, and $m_2\ddot{\tilde{\mathbf{r}}}_2 = f(\tilde{r})\mathbf{e}_r$, where $f(\tilde{r}) = -\partial U(\tilde{r})/\partial\tilde{r}$ and $U(\tilde{r})$ is the mutual potential energy. From these equations, it is easy to find the relationship, $\mu\ddot{\tilde{\mathbf{r}}} = f(\tilde{r})\mathbf{e}_r$. This equation is the equation of motion of a single particle of mass, $\mu = m_1m_2/m_0$, in a central force field.

2.3. Prove that the relative motion of two interacting molecules is planar in nature.

Solution: The differential equation of the relative motion of two molecules is given by: $\mu\dot{\mathbf{g}} = f(\tilde{r})\mathbf{e}_r$, where $\mu = m_1m_2/m_0$, $\mathbf{e}_r = \tilde{\mathbf{r}}/\tilde{r}$, $\tilde{\mathbf{r}} = \tilde{\mathbf{r}}_2 - \tilde{\mathbf{r}}_1$, $\mathbf{g} = \mathbf{v}_2 - \mathbf{v}_1$, $f(\tilde{r}) = -\partial U(\tilde{r})/\partial\tilde{r}$ and $U(\tilde{r})$ is the mutual potential energy. The vector product of this equation and the position vector, $\tilde{\mathbf{r}}$, may be expressed as $\mu\tilde{\mathbf{r}}\times\dot{\mathbf{g}} = 0$ or $d\mathbf{M}/dt = 0$, where $\mathbf{M} = \tilde{\mathbf{r}}\times\mu\mathbf{g}$. This shows that $\mathbf{M} = \mathrm{const}$ and, consequently, that the position vector lies in the plane which is perpendicular to the constant vector, $\mathbf{M}$.

2.4. Determine the distribution function for a gas in which the magnitude of the velocity vector is the same for all of the molecules, in which all directions of velocity are equally probable, and where the number density of the gas is n.

Solution: The number of molecules situated in a phase element, $d\mathbf{v}d\tilde{\mathbf{r}}$, may be expressed as: $fd\mathbf{v}d\tilde{\mathbf{r}} = n\psi(v)\delta(v - v_0)d\mathbf{v}d\tilde{\mathbf{r}}$, where $\psi(v)$ is a function that depends only on the magnitude of the molecular velocity and $\delta(v - v_0)$ is the Dirac delta function. The function, $\psi(v)$, may be found from the condition:

$$\int f d\mathbf{v} = n \int \psi(v) \delta(v - v_0) v^2 \sin(\theta) dv d\theta d\phi = n \ .$$

Then, it is easy to obtain the following form for the distribution function:

$$f = \frac{n}{4\pi v^2} \delta(v - v_0) \ .$$

2.5. Using the conditions given in Problem 2.4, determine the distribution function for a two-dimensional gas (in which all molecular velocity vectors are parallel to the same plane) with a number density, n.

Answer:

$$f = \frac{n}{2\pi v} \delta(v - v_0) \ .$$

REFERENCES

1. Landau, L.D. and Lifshitz, E.M., *Mechanics* (Pergamon, Oxford, 1960).
2. Chapman, S. and Cowling, T.G., *The Mathematical Theory of Non-Uniform Gases* (Cambridge University Press, Cambridge, U.K., 1990).
3. Jeans, J., *Dynamical Theory of Gases* (Cambridge University Press, Cambridge, U.K., 1990).
4. Bird, G.A., *Molecular Gas Dynamics* (Clarendon Press, Oxford, 1976).

Chapter 3

THE COLLISION OPERATOR

1. THE DIFFERENTIAL AND TOTAL SCATTERING CROSS SECTIONS.

One of the most important concepts relating to the statistics of molecular collisions is that of the scattering cross section which is the basic quantity characteristic of molecular encounters. Consider a uniform flux of molecules that have a constant velocity, $\mathbf{g}$, towards a spherically symmetric force center, O. The parameters of importance to the scattering events are shown in Fig. 3-1. Let N_g be the g-component of the uniform molecular number flux and $d\Omega = \sin(\theta)d\theta d\varepsilon$ be the solid angle in which molecules are deflected after encounters. The differential scattering cross section, $\alpha(\theta, g)$, is defined by [1]:

$$\alpha(\theta,g)d\Omega = \frac{dN}{N_g}\ , \tag{3-1}$$

where dN is the number of molecules deflected per unit time by the central force field in $d\Omega$. Since the incident molecules scattered into $d\Omega$ must primarily pass through that portion of the annular differential element represented by $dS = bdbd\varepsilon$, dN may be expressed in the form:

$$dN = N_g bdbd\varepsilon\ . \tag{3-2}$$

From Eqs. (3-1) and (3-2), one can obtain:

$$\alpha(\theta,g)=\frac{b}{\sin(\theta)}\left|\frac{db}{d\theta}\right| . \tag{3-3}$$

The absolute value of $db/d\theta$ must be included in this relation since the signs of db and $d\theta$ are usually opposite. The integral or total scattering cross section is then defined as:

$$\alpha_{tot}(g)=2\pi\int\alpha(\theta,g)\sin(\theta)d\theta . \tag{3-4}$$

The differential scattering cross section has a very simple physical sense. The quantity, $\alpha(\theta,g)d\Omega=bdbd\varepsilon$, is that part of the annular differential area through which the molecular trajectory at infinity must pass in order for a molecule to be scattered into the solid angle, $d\Omega$, after an encounter. The total scattering cross section, $\alpha_{tot}(g)$, is the area of the circular region through which the incident molecular trajectory must pass for any scattering to occur. The range of scattering that would contribute to this total cross section would encompass all of the available scattering solid angle (4π) and any incident trajectory that were to lie outside of this circular region would be deemed to have not been scattered at all.

Now, for example, consider molecules which are rigid spheres with diameters, σ_1 and σ_2. The relation between the collision parameters shown in Fig. 3-2 may be expressed in the form:

$$b=\sigma_{12}\sin(\psi) , \tag{3-5}$$

where $\sigma_{12}=\frac{1}{2}(\sigma_1+\sigma_2)$ and $\psi=\frac{1}{2}(\pi-\theta)$. Since the relative motion of two molecules is equivalent to motion in a central force field, the differential scattering cross section specified by Eq. (3-3) is given by:

$$\alpha(\theta,g)=\tfrac{1}{4}\sigma_{12}^2 . \tag{3-6}$$

The total scattering cross section may be written as:

$$\alpha_{tot}(g)=\pi\sigma_{12}^2 . \tag{3-7}$$

It is very important to note that this molecular interaction model is unique in that $\alpha(\theta,g)$ and $\alpha_{tot}(g)$ are independent of the relative velocity of the two molecules, $\mathbf{g}$. This significant property of this model allows one to facilitate many calculations. For other models of intermolecular

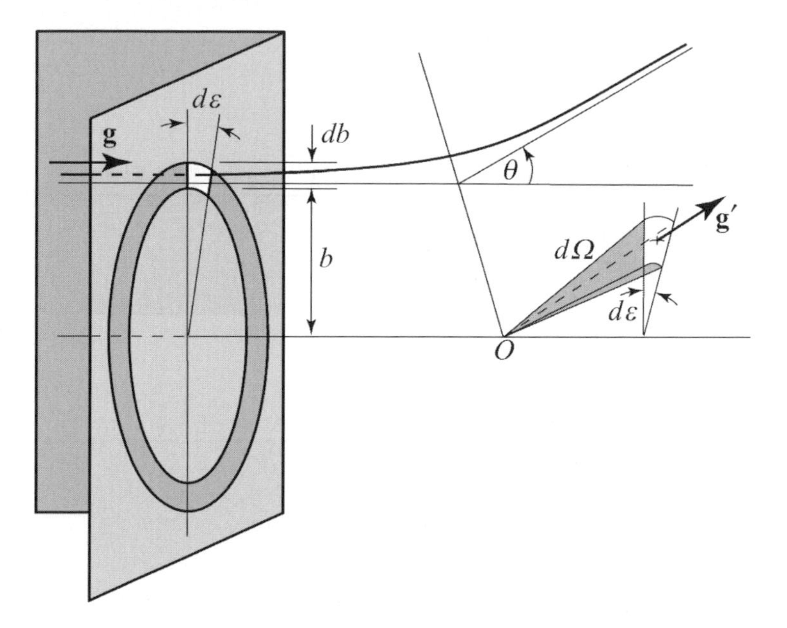

Figure 3-1. The scattering solid angle.

interaction that involve forces acting across infinite distances, the integral given by Eq. (3-4) is divergent since molecules are scattered even if the impact parameter, b, tends to infinity. Consequently, in this case, definite assumptions must be made regarding a cutoff of this parameter for which, when $b > b_{\max}$, no encounter occurs.

2. THE STATISTICS OF MOLECULAR ENCOUNTERS.

Consider encounters between two different molecules with masses, m_1 and m_2, situated in the volume element, $d\tilde{\mathbf{r}}$. To derive the collision operator, one should calculate the rate of change of the number of m_1-molecules having velocities in the range, $d\mathbf{v}_1$, due to collisions with all molecules of the second kind. First, encounters between the two molecular sets, $d\mathbf{v}_1$ and $d\mathbf{v}_2$, in the volume element, $d\tilde{\mathbf{r}}$, must be analyzed. Then, this analysis must be generalized for all molecules of the second kind.

Encounters in which more than two molecules take part are assumed to be negligible in number compared with binary encounters and, moreover, molecules of the two velocity sets are assumed to be distributed at random without any correlation between velocity and position in the neighborhood of the point, $\tilde{\mathbf{r}}$. These two assumptions are fundamental to the classical

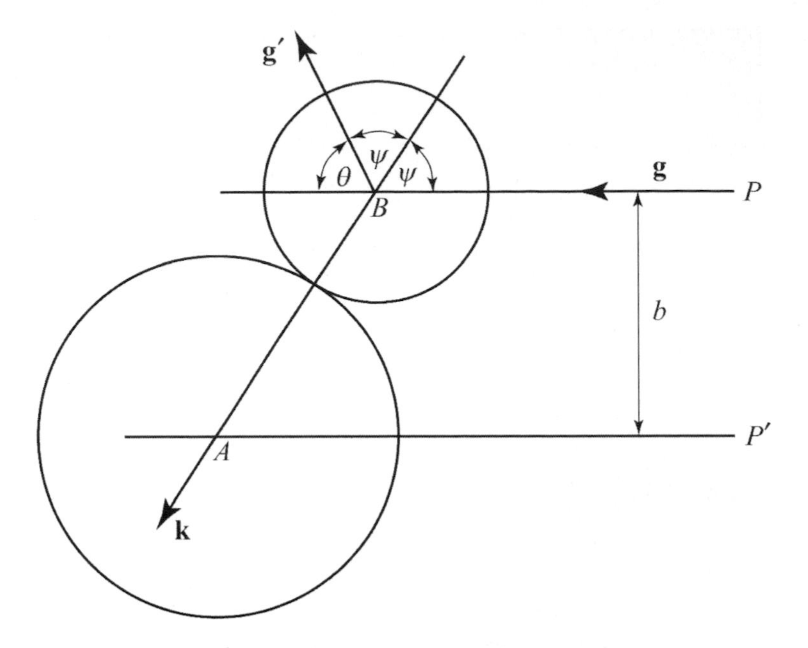

Figure 3-2. Encounters of rigid-sphere molecules.

kinetic theory. It is necessary to note that the physical sense of the second assumption cannot be understood within the framework of the one-particle distribution function and, therefore, it may be considered to be a hypothesis the applicability of which may only be confirmed by using a more general theory [2-4] or by comparison with experiment.

Examine the scattering of molecules of the second set by a molecule of the first in $d\tilde{\mathbf{r}}$. The molecular number flux in the direction of the relative velocity, may be expressed in the form:

$$gf_2\left(\mathbf{v}_2,\tilde{\mathbf{r}},t\right)d\mathbf{v}_2 \ . \tag{3-8}$$

Then, in accordance with Eq. (3-1), the number of molecules of the second set deflected into the solid angle, $d\Omega$, during the time interval, dt, is given by:

$$g\alpha_{12}\left(\theta,g\right)f_2\left(\mathbf{v}_2,\tilde{\mathbf{r}},t\right)d\Omega d\mathbf{v}_2 dt \ , \tag{3-9}$$

where the time interval, dt, is assumed to be short compared to the scale of the time variation of the macroscopic properties of the gas, but large compared to the duration of an encounter. The appropriate number of $\mathbf{v}_2$-

molecules reflected by all $\mathbf{v}_1$-molecules in $d\mathbf{r}$ and $d\mathbf{v}_1$ may be written as [5]:

$$g\alpha_{12}(\theta,g) f_1 f_2 d\Omega d\mathbf{v}_1 d\mathbf{v}_2 d\tilde{\mathbf{r}} dt \ . \tag{3-10}$$

In Eq. (3-10) it is assumed that the number of $(\mathbf{v}_1, \mathbf{v}_2)$-collisions is proportional to $f_1 f_2$ such that any potential correlation between velocity and position of molecules in $d\tilde{\mathbf{r}}$ has been ignored. Moreover, this expression holds true only if binary encounters are exclusively considered.

The total number of encounters of the first set molecules with all molecules of the second kind in $d\tilde{\mathbf{r}}$ during dt may be found by integration over all values of Ω and $\mathbf{v}_2$. Then, one can obtain:

$$N_{12}^{-} d\mathbf{v}_1 d\tilde{\mathbf{r}} dt = \left[\int\int f_1 f_2 g\alpha_{12}(\theta,g) d\Omega d\mathbf{v}_2\right] d\mathbf{v}_1 d\tilde{\mathbf{r}} dt \ . \tag{3-11}$$

Such encounters change the velocity of the first set molecules and thus the molecules are lost to the first set. As a result, N_{12}^{-} may be interpreted as the rate of loss of molecules from the first set per unit volume and velocity interval caused by collisions with molecules of the second kind.

In the same way, one can obtain the number of m_1-molecules that enter the velocity set, $d\mathbf{v}_1$. For this purpose, the m_1-molecules whose velocities after encounters lie in the range, $d\mathbf{v}_1$, must be considered. In such inverse collisions, the molecular velocities are changed from $\mathbf{v}_1', \mathbf{v}_2'$ to $\mathbf{v}_1, \mathbf{v}_2$. Using the above analysis, one can obtain the total gain of m_1-molecules to the first velocity set, $d\mathbf{v}_1$, for the time interval, dt, in the volume element, $d\tilde{\mathbf{r}}$, owing to inverse encounters. This is given by:

$$N_{12}^{+} d\mathbf{v}_1 d\tilde{\mathbf{r}} dt = \left[\int\int f_1' f_2' g'\alpha_{12}(\theta,g') d\Omega d\mathbf{v}_2'\right] d\mathbf{v}_1' d\tilde{\mathbf{r}} dt \ , \tag{3-12}$$

where f_1' and f_2' stand for $f_1(\mathbf{v}_1', \tilde{\mathbf{r}}, t)$ and , $f_2(\mathbf{v}_2', \tilde{\mathbf{r}}, t)$ respectively.

Now, one transforms the integrand in Eq. (3-12). First, since $g = g'$, one can obtain:

$$g'\alpha_{12}(\theta,g') = g\alpha_{12}(\theta,g) \ . \tag{3-13}$$

Next, one considers two subsystems of molecules the velocities of which, before an encounter, lie in the ranges, $d\mathbf{v}_1'$ and $d\mathbf{v}_2'$. If the collision parameters of these molecules lie in the ranges, db and $d\varepsilon$, then their velocities after encounters will lie in the ranges, $d\mathbf{v}_1$ and $d\mathbf{v}_2$. In accordance with Liouville's theorem [1], the phase elements of these subsystems are connected by the relationship:

$$d\mathbf{v}_1' d\tilde{\mathbf{r}}_1' d\mathbf{v}_2' d\tilde{\mathbf{r}}_2' = d\mathbf{v}_1 d\tilde{\mathbf{r}}_1 d\mathbf{v}_2 d\tilde{\mathbf{r}}_2 \ . \tag{3-14}$$

Since the collisions between the two molecular sets occur in the same volume element, $d\mathbf{r}$, Eq. (3-14) may be written as:

$$d\mathbf{v}_1' d\mathbf{v}_2' = d\mathbf{v}_1 d\mathbf{v}_2 \ . \tag{3-15}$$

Taking into account Eqs. (3-13) and (3-15), one can express Eq. (3-12) in the form:

$$N_{12}^{+} d\mathbf{v}_1 d\tilde{\mathbf{r}} dt = \left[\int\int f_1' f_2' g \alpha_{12}(\theta, g) d\Omega d\mathbf{v}_2 \right] d\mathbf{v}_1 d\tilde{\mathbf{r}} dt \ , \tag{3-16}$$

where N_{12}^{+} is equal to the rate of gain of the first set molecules, per unit volume and velocity interval, due to encounters.

Now, it is straightforward to obtain the collision operator for encounters between molecules of the first set and of the second kind:

$$\left(\frac{\delta f_1}{\delta t} \right)_2 = N_{12}^{+} - N_{12}^{-} = \int\int (f_1' f_2' - f_1 f_2) g \alpha_{12}(\theta, g) d\Omega d\mathbf{v}_2 \ . \tag{3-17}$$

This form of the collision operator is usually called the Boltzmann collision integral.

For a simple gas $(m_1 = m_2 = m, \mathbf{v}_1 \to \mathbf{v}, \mathbf{v}_2 \to \mathbf{v}_1)$, Eq. (3-17) may be written as:

$$\frac{\delta f}{\delta t} = \int\int (f' f_1' - f f_1) g \alpha(\theta, g) d\Omega d\mathbf{v}_1 \ , \tag{3-18}$$

where $\delta f / \delta t$ denotes the rate of variation of the distribution function due to encounters at a fixed point, $\tilde{\mathbf{r}}$. The rate of variation of any molecular property, $\phi(\mathbf{v}, \tilde{\mathbf{r}}, t)$, per unit volume, due to encounters, is defined by:

$$n \Delta \bar{\phi} = \int\int\int \phi (f' f_1' - f f_1) g \alpha(\theta, g) d\Omega d\mathbf{v}_1 d\mathbf{v} \ . \tag{3-19}$$

This very important quantity may be called the moment of the collision integral associated with the property, ϕ.

3. THE TRANSFORMATION OF SOME INTEGRALS.

Let the variables of integration in Eq. (3-19) be changed from $\mathbf{v}, \mathbf{v}_1$ to $\mathbf{v}', \mathbf{v}_1'$. Then, this integral may be written as:

$$n\Delta\overline{\phi'} = \iiint \phi'(ff_1 - f'f_1')g'\alpha(\theta, g')d\Omega d\mathbf{v}'d\mathbf{v}_1' \ . \qquad (3\text{-}20)$$

Since the integration over all values of $\mathbf{v}', \mathbf{v}_1'$ is equivalent to integration over all encounters, one can conclude that $n\Delta\overline{\phi'} = n\Delta\overline{\phi}$. Taking into account Eqs. (3-15), (3-19) and (3-20), one can obtain:

$$n\Delta\overline{\phi} = \tfrac{1}{2}\iiint(\phi - \phi')(f'f_1' - ff_1)g\alpha(\theta, g)d\Omega d\mathbf{v}d\mathbf{v}_1 \ . \qquad (3\text{-}21)$$

Since the variables, $\mathbf{v}_1$ and $\mathbf{v}$, both refer to the same molecules, an interchange of them does not alter the value of this integral. After making this interchange of variables, Eq. (3-21) may be expressed in the form:

$$n\Delta\overline{\phi} = \tfrac{1}{2}\iiint(\phi_1 - \phi_1')(f'f_1' - ff_1)g\alpha(\theta, g)d\Omega d\mathbf{v}d\mathbf{v}_1 \ . \qquad (3\text{-}22)$$

Then, combining Eqs. (3-21) and (3-22), one can obtain the following integral relation:

$$n\Delta\overline{\phi} = \tfrac{1}{4}\iiint(\phi + \phi_1 - \phi' - \phi_1')(f'f_1' - ff_1)g\alpha(\theta, g)d\Omega d\mathbf{v}d\mathbf{v}_1 \ . \qquad (3\text{-}23)$$

PROBLEMS

3.1. Evaluate the time duration of an encounter between two molecules.

Solution: Let $\tilde{r}_0$ be the radius of action of the molecular forces and let $\overline{V}$ be the mean velocity of a molecule. The time duration of an encounter, τ_{coll}, is then given by $\tau_{coll} \sim \tilde{r}_0/\overline{V}$. For rigid-sphere molecules, this may be written as $\tau_{coll} \sim \sigma/\overline{V}$ where σ is a molecular diameter.

3.2. Derive the condition of applicability of the Boltzmann equation if n is the gas number density.

Solution: The mean distance between molecules is proportional to $n^{-1/3}$. Let $\tilde{r}_0$ be the radius of action of the molecular forces. The binary interaction hypothesis holds true if the molecular force action sphere contains not more than one molecule. This condition may be expressed: $\tilde{r}_0^3 n \approx 1$.

3.3. Prove by direct calculation that $d\mathbf{v}_1' d\mathbf{v}_2' = d\mathbf{v}_1 d\mathbf{v}_2$ during molecular encounters.

Solution: First, derive the expressions; $d\mathbf{v}_1 d\mathbf{v}_2 = d\mathbf{G} d\mathbf{g}_{21}$ and $d\mathbf{v}_1' d\mathbf{v}_2' = d\mathbf{G} d\mathbf{g}_{21}'$ in which $d\mathbf{g}_{21} = g^2 dg d\Omega$. The desired proof follows directly if one takes into account the relationship $g = g'$.

3.4. For a rigid-sphere gas in which all of the molecules have velocities of the same magnitude, determine the molecular mean free path assuming that all velocity directions are equally probable (number density, n, molecular diameter, σ).

Solution: The total number of collisions occurring between molecules per unit volume and time is given by:

$$N_{11} = \iiint f f_1 g \alpha(\theta, g) d\Omega d\mathbf{v}_1 d\mathbf{v} \ , \tag{P-1}$$

where $g = \left(v^2 + v_1^2 - 2 v v_1 \cos(\chi)\right)^{1/2}$ and χ is the angle between $\mathbf{v}$ and $\mathbf{v}_1$. The distribution function (Problem 2.4) may be expressed in the form:

$$f = \frac{n}{4\pi v_0^3 c^2} \delta(c-1) \ , \tag{P-2}$$

where $\mathbf{c} = \mathbf{v}/v_0$ and $c = |\mathbf{c}|$. Substituting Eq. (P-2) into Eq. (P-1), one can obtain:

$$N_{11} = \frac{\pi 2^{1/2} \sigma^2 n^2 v_0}{(4\pi)^2} \iint \left(1 - \cos(\chi)\right)^{1/2} d\omega_1 d\omega \ , \tag{P-3}$$

where $d\omega = \sin(\chi) d\chi d\beta$ and $d\omega_1 = \sin(\chi_1) d\chi_1 d\beta_1$. After integration over all directions of $\mathbf{c}$ and $\mathbf{c}_1$, Eq. (P-3) becomes $N_{11} = \frac{4}{3}\pi\sigma^2 n^2 v_0$. The mean free path is then given by:

$$\lambda = v_0 \frac{n}{N_{11}} = \left(\tfrac{4}{3}\pi\sigma^2 n\right)^{-1} .$$

3.5. For a two-dimensional rigid-sphere gas (number density, n, molecular diameter, σ), determine the molecular mean free path using the conditions in Problem 3.4.

Solution: All molecular velocity vectors for this gas lie in the same plane. Therefore, it is convenient to use polar coordinates in the velocity space. The distribution function, in dimensionless variables, is given by (Problem 2.5):

$$f = \frac{n}{2\pi v_0^2 c}\delta(c-1) \ .$$

The total number of molecular collisions per unit volume and time is then:

$$N_{11} = \frac{\pi 2^{1/2}\sigma^2 n^2 v_0}{(2\pi)^2}\int_0^{2\pi}\left(1-\cos(\phi)\right)^{1/2} d\phi \int_0^{2\pi} d\phi_1 = 4\sigma^2 n^2 v_0 \ ,$$

where ϕ is the angle between $\mathbf{v}$ and $\mathbf{v}_1$, and ϕ_1 is the polar coordinate of $\mathbf{v}_1$. The mean free path may then be expressed as:

$$\lambda = v_0 \frac{n}{N_{11}} = \left(4\sigma^2 n\right)^{-1} \ .$$

REFERENCES

1. Landau, L.D. and Lifshitz, E.M., *Mechanics* (Pergamon, Oxford, 1960).
2. Bogoliubov, N.N., *Problems of a Dynamical Theory in Statistical Physics* (State Technical Press, Moscow, 1946). English translation by Gora, E.K. in *Studies in Statistical Mechanics*, Vol. 1, Eds. de Boer, J. and Uhlenbeck, G.E. (Wiley, New York, 1964).
3. Ferziger, J.H. and Kaper, H.G., *Mathematical Theory of Transport Processes in Gases* (North-Holland, Amsterdam-London, 1972).
4. Lifshitz, E.M. and Pitaevskii, L.P., *Physical Kinetics* (Pergamon, Oxford, 1960).
5. Anselm, A.I., *Principles of Statistical Physics and Thermodynamics* (in Russian) (Nauka, Moscow, 1973).

Chapter 4

THE UNIFORM STEADY-STATE OF A GAS

1. THE BOLTZMANN H-THEOREM.

Consider a simple gas whose molecules possess only energy of translation, and are subject to no external forces. Let the state of the gas be uniform so that the distribution function, f, is independent of $\tilde{\mathbf{r}}$. For this case, the Boltzmann equation may be expressed in the following simple form:

$$\frac{\partial f}{\partial t} = \int\int\left(f'f_1' - ff_1\right)g\alpha(\theta,g)\,d\Omega d\mathbf{v}_1 \ . \tag{4-1}$$

Let the following H-function be introduced:

$$H = \int f\ln(f)\,d\mathbf{v} \ . \tag{4-2}$$

The time derivative of the H-function may be written as:

$$\begin{aligned}\frac{\partial H}{\partial t} &= \int\left(1+\ln(f)\right)\frac{\partial f}{\partial t}\,d\mathbf{v} \\ &= \int\int\int\left(1+\ln(f)\right)\left(f'f_1' - ff_1\right)g\alpha(\theta,g)\,d\Omega d\mathbf{v}_1 d\mathbf{v} \ .\end{aligned} \tag{4-3}$$

Now, taking into account Eq. (3.23), one can obtain:

$$\frac{\partial H}{\partial t} = \tfrac{1}{4} \iiint \ln\left(ff_1/f'f_1'\right)\left(f'f_1' - ff_1\right) g\alpha(\theta, g)\, d\Omega d\mathbf{v}_1 d\mathbf{v} \; . \tag{4-4}$$

In this relation, $\ln\left(ff_1/f'f_1'\right)$ is always opposite in sign to $\left(f'f_1' - ff_1\right)$ and, therefore, the integral is either negative or zero. As a result of this:

$$\frac{\partial H}{\partial t} \leq 0 \; . \tag{4-5}$$

The latter differential relation indicates that the H-function can never increase. This fundamental inference is known as Boltzmann's H-theorem and gives the time direction of processes for the uniform state of a gas [1-7].

2. THE MAXWELLIAN VELOCITY DISTRIBUTION.

A very important question arises regarding whether there exists a limiting state for a gas after it has been disturbed from an initial uniform state. To show that there is, in fact, a limiting state, it is necessary to prove that the H-function cannot decrease indefinitely (i.e., that the H-function is bounded below). It is clear that $H \to -\infty$ only if the integral given by Eq. (4-2) is divergent. If $|\mathbf{v}| \to \infty$, then the distribution function tends to zero and $\ln(f) \to -\infty$. Since the integral, $\int \frac{1}{2} m v^2 f d\mathbf{v}$, expressing the energy of translation of the molecules, must be convergent, the H-function would be divergent only if $-\ln(f)$ tended to infinity more rapidly than c^2, where $\mathbf{c} = \left(m/2kT\right)^{1/2} \mathbf{v}$ is the dimensionless velocity. In this case, however, f must tend to zero more rapidly than $\exp\left(-c^2\right)$ and, if this occurs in the integral given by Eq. (4-2), then the integral is convergent and the H-function is bounded below.

Since it is bounded below, the H-function cannot decrease indefinitely but must tend to some limit, corresponding to a state of the gas, in which $\partial H/\partial t = 0$. From Eq. (4-4), one may then conclude that the distribution function for the gas in this state may be found from the relation, $f'f_1' = ff_1$. This equation is equivalent to:

$$\ln\left(f'\right) + \ln\left(f_1'\right) = \ln\left(f\right) + \ln\left(f_1\right) \; . \tag{4-6}$$

If the distribution function satisfies Eq. (4-6), then, from Eq.(4-1), one can obtain that $\partial f/\partial t = 0$ also, so that such a state of the gas is steady as well as uniform.

Now, consider the form of this distribution function. Eq. (4-6) shows that $\ln(f)$ is a summational invariant of encounters and, therefore, must be a linear combination of the three summational invariants. It may be expressed in the form [1,2]:

$$\ln(f) = \alpha^{(1)} + \mathbf{a}^{(2)} \cdot m\mathbf{v} - \alpha^{(3)} \tfrac{1}{2} m v^2 , \tag{4-7}$$

or:

$$f = \alpha^{(0)} \exp\left(-\alpha^{(3)} \tfrac{1}{2} m (V')^2\right) ; \quad \mathbf{V}' = \mathbf{v} - \frac{\mathbf{a}^{(2)}}{\alpha^{(3)}} . \tag{4-8}$$

The constants, $\alpha^{(0)}$, $\mathbf{a}^{(2)}$, and $\alpha^{(3)}$, can be expressed in terms of the number density, n, the mean velocity, $\mathbf{u}$, and the temperature, T [1,2]. Using definitions of these quantities given by Eqs. (1-18)-(1-20), one can obtain:

$$n = \alpha^{(0)} \left(\frac{2\pi}{m\alpha^{(3)}}\right)^{3/2} , \quad \mathbf{u} = \frac{\mathbf{a}^{(2)}}{\alpha^{(3)}} , \quad \alpha^{(3)} = \frac{1}{kT} .$$

Then, the uniform steady-state of a gas is described by:

$$f = n \left(\frac{m}{2\pi kT}\right)^{3/2} \exp\left(-\frac{m(\mathbf{v}-\mathbf{u})^2}{2kT}\right) . \tag{4-9}$$

This function is usually called the Maxwellian distribution function.

Some mean values for the Maxwellian state of a gas may be calculated quite easily as they involve only simple integrations. For example, the mean value of the magnitude of the peculiar molecular velocity is given by:

$$\overline{V} = \frac{4}{\pi^{1/2}} \left(\frac{2kT}{m}\right)^{1/2} \int_0^{\infty} C^3 \exp\left(-C^2\right) dC = \left(\frac{8kT}{\pi m}\right)^{1/2} . \tag{4-10}$$

The number of molecules per unit time crossing a unit area surface element that is normal to the axis Ox and which moves with the same mean velocity as the gas is given by:

$$N_x = \int_{+} V_x f d\mathbf{v} , \tag{4-11}$$

where the notation, $(+)$, signifies that the range of integration is over all values of $\mathbf{v}$ for which V_x is positive. After integration, one can obtain:

$$N_x = \tfrac{1}{4} n \bar{V} \ . \tag{4-12}$$

The last expression is very frequently used in the elementary kinetic analysis of transport phenomena [8-10].

3. THE MEAN FREE PATH OF A MOLECULE.

Consider collisions between pairs of molecules with masses, m_1 and m_2, in a binary gas mixture having a uniform steady-state. The number of collisions per unit volume and time between the two molecular sets having collision parameters in the ranges, db and $d\varepsilon$, is given by:

$$dN_{12} = f_1 f_2 g b db d\varepsilon d\mathbf{v}_1 d\mathbf{v}_2 = f_1 f_2 g \alpha_{12}(\theta, g) d\Omega d\mathbf{v}_1 d\mathbf{v}_2 \ . \tag{4-13}$$

The total number of collisions per unit volume and time, N_{12}, is obtained by integrating Eq. (4-13) over all values of the collision parameters and the molecular velocities. For a gas having a uniform steady-state, the Maxwellian distribution function is used in the integrand. Then, N_{12} for rigid-sphere molecules may be written as [1]:

$$N_{12} = \frac{\pi n_1 n_2 (m_1 m_2)^{3/2} \sigma_{12}^2}{(2\pi kT)^3} \int\int \exp\left(-\frac{m_1 v_1^2 + m_2 v_2^2}{2kT} \right) g d\mathbf{v}_1 d\mathbf{v}_2 \ . \tag{4-14}$$

Let the variables of integration be changed from $\mathbf{v}_1, \mathbf{v}_2$ to $\mathbf{G}, \mathbf{g}\ (\mathbf{g} = \mathbf{g}_{21})$ as introduced in Section 2-5. Due to the following relationships between Jacobians:

$$\frac{\partial(\mathbf{G},\mathbf{g})}{\partial(\mathbf{v}_1,\mathbf{v}_2)} = \frac{\partial(\mathbf{v}_1 + M_2\mathbf{g},\mathbf{g})}{\partial(\mathbf{v}_1,\mathbf{v}_2)} = \frac{\partial(\mathbf{v}_1,\mathbf{g})}{\partial(\mathbf{v}_1,\mathbf{v}_2)}$$
$$= \frac{\partial(\mathbf{v}_1,\mathbf{v}_2 - \mathbf{v}_1)}{\partial(\mathbf{v}_1,\mathbf{v}_2)} = \frac{\partial(\mathbf{v}_1,\mathbf{v}_2)}{\partial(\mathbf{v}_1,\mathbf{v}_2)} = 1 \ ,$$

the velocity space element, $d\mathbf{v}_1 d\mathbf{v}_2$, may be replaced by $d\mathbf{G} d\mathbf{g}$. Then, Eq. (4-14) becomes:

$$N_{12} = \frac{\pi n_1 n_2 \left(m_1 m_2\right)^{3/2} \sigma_{12}^2}{\left(2\pi kT\right)^3} \times \int\int \exp\left(-\frac{m_0\left(G^2 + M_1 M_2 g^2\right)}{2kT}\right) g d\mathbf{G} d\mathbf{g} \ . \tag{4-15}$$

Having performed the integration in spherical coordinates for $\mathbf{G}$ and $\mathbf{g}$, one obtains:

$$N_{12} = 2 n_1 n_2 \sigma_{12}^2 \left(\frac{2\pi kT m_0}{m_1 m_2}\right)^{1/2} \ . \tag{4-16}$$

The number of encounters solely between molecules of the first-kind is given by:

$$N_{11} = 4 n_1^2 \sigma_1^2 \left(\frac{\pi kT}{m_1}\right)^{1/2} \ . \tag{4-17}$$

The average number of collisions of a molecule of the first-kind per unit volume and time is usually called the collision frequency and may be written as:

$$\nu_1 = \frac{N_{11} + N_{12}}{n_1} \ . \tag{4-18}$$

The mean time between collisions (the collision interval) is then defined by:

$$\tau_1 = \frac{1}{\nu_1} = \frac{n_1}{N_{11} + N_{12}} \ . \tag{4-19}$$

The mean distance between collisions is called the mean free path. This quantity is:

$$\lambda_1 = \bar{V}_1 \tau_1 = \left[2^{1/2} \pi n_1 \sigma_1^2 + \pi n_2 \sigma_{12}^2 \left(1 + \frac{m_1}{m_2}\right)^{1/2}\right]^{-1} \ . \tag{4-20}$$

For a simple gas $(m_1 = m_2 = m, \sigma_1 = \sigma_2 = \sigma, n_1 = n, n_2 = 0)$, the mean free path can be expressed in the form:

$$\lambda = \left[2^{1/2}\pi n\sigma^2\right]^{-1} . \tag{4-21}$$

Using Eq. (4-21) the mean free path in a simple gas can now be estimated for various conditions. Assuming that one has nitrogen under conditions of Standard Temperature and Pressure (STP; 0 °C and 1 atm), the appropriate density and molecular diameter are $n \sim 2.7\times10^{19}$ cm^{-3} and $\sigma \sim 3\times10^{-8}$ cm, respectively. Using these values in Eq. (4-21), the calculated mean free path for nitrogen at STP is then approximately 10^{-5} cm.

PROBLEMS

4.1. For the Maxwellian gas state, determine the number of molecules per unit volume with velocities in the ranges, dv_z and dv_ρ, where v_z is the molecular velocity component along the axis, OZ, and v_ρ is that along any perpendicular axis.

Solution: In cylindrical coordinates, the distribution function is given by:

$$f = n\left(\frac{m}{2\pi kT}\right)^{3/2} \exp\left(-\frac{mv^2}{2kT}\right) v_\rho .$$

The number of molecules per unit volume in dv_ρ and dv_z may be written as:

$$dn(v_\rho, v_z) = \left[\int_0^{2\pi} f d\beta\right] dv_\rho dv_z = 2\pi n\left(\frac{m}{2\pi kT}\right)^{3/2} \exp\left(-\frac{mv^2}{2kT}\right) v_\rho dv_\rho dv_z .$$

4.2. For the Maxwellian gas state, determine the number of molecules per unit volume, $n(v_x \geq v_{0x})$, having an x-velocity component in the interval, $v_x \geq v_{0x}$.

Solution: The number of molecules in the above range may be determined from:

$$n\left(v_x \geq v_{0x}\right) = n\left(\frac{m}{2\pi kT}\right)^{3/2} \int_{-\infty}^{\infty} dv_y \int_{-\infty}^{\infty} dv_z \int_{v_{0x}}^{\infty} \exp\left(-\frac{mv^2}{2kT}\right) dv_x$$

$$= \frac{n}{\sqrt{\pi}} \int_{c_{0x}}^{\infty} \exp\left(-c_x^2\right) dc_x = \tfrac{1}{2} n\left[1 - \operatorname{erf}\left(c_{0x}\right)\right] = \tfrac{1}{2} n \operatorname{cerf}\left(c_{0x}\right) .$$

4.3. For the Maxwellian gas state, determine the distribution function of the kinetic energy of translation.

Solution: Starting with the Maxwellian distribution function for the magnitude of the molecular velocity use the relationship $\frac{1}{2}mv^2 = E$ to transform the differential elements of the relationship $f(v)dv = f(E)dE$. Substitute this transformation and the original energy relationship into the molecular velocity distribution to obtain the following distribution for the kinetic energy of translation:

$$f(E) = \frac{2}{\sqrt{\pi}} n(kT)^{-3/2} \exp\left(-\frac{E}{kT}\right)\sqrt{E} .$$

4.4. For the Maxwellian gas state, determine the number of molecules impinging upon a unit surface element per unit time and lying in the angular range, $d\theta$, where θ is the angle with respect to the surface element normal.

Solution: From the Maxwellian distribution function in spherical coordinates, after integration over all velocities and the azimuthal angle, one can obtain:

$$dN_\theta = n\left(\frac{2kT}{\pi m}\right)^{1/2} \cos(\theta)\sin(\theta)d\theta .$$

4.5. For the Maxwellian gas state, determine the number of molecules impinging on a unit surface element per unit time and having molecular velocity magnitudes in the range, dv.

Solution: This flux of molecules is given by:

$$dN_v = 2\pi n\left(\frac{m}{2\pi kT}\right)^{3/2} \exp\left(-\frac{mv^2}{2kT}\right) v^3 dv \int_0^{\pi/2} \cos(\theta)\sin(\theta)d\theta$$

$$= \pi n\left(\frac{m}{2\pi kT}\right)^{3/2} \exp\left(-\frac{mv^2}{2kT}\right) v^3 dv .$$

4.6. A narrow molecular beam exits a system of slots into a vacuum. Determine the mean velocity and the root mean square (rms) velocity of the beam molecules.

Solution: The distribution function of the beam molecules may be expressed in the form:

$$f(v)dv = n\frac{dN_v}{N} \quad ,$$

where dN_v has been found in Problem 4.5 and N is defined by Eq. (4-12). Then, any mean value for the beam molecules may be determined by means of this distribution function with the following results:

$$\bar{V} = \left(\frac{9\pi kT}{8m}\right)^{1/2} \text{ and } \sqrt{\overline{V^2}} = \left(\frac{4kT}{m}\right)^{1/2} .$$

4.7. For an equilibrium gas, prove that the ratio of the mean velocity of molecules crossing a surface element to the mean velocity of other molecules is $\frac{3}{8}\pi$.

Solution: Using Eq. (4-10) and the result from Problem 4.6 yields:

$$\frac{\bar{V}_S}{\bar{V}} = \left(\frac{9\pi kT}{8m}\right)^{1/2} \Big/ \left(\frac{8kT}{\pi m}\right)^{1/2} = \tfrac{3}{8}\pi .$$

4.8. Consider molecules reflecting from a differential surface element, dS', on one surface and subsequently passing through a differential surface element, dS, on another surface. Assume that the reflected molecules at the first surface are being diffusely reflected and that the gas has a number density of n' and a temperature of T'. Moreover, assume that no molecular collisions occur while a molecule is transiting the space between the two differential surface elements. Determine the rate at which molecules reflected from dS' cross dS. Use the geometry in Fig. 4-1.

Solution: The number of molecules crossing dS per second in a given direction may be expressed in the form (see Problem 4.4):

$$dN_\theta = n'\left(\frac{2kT}{\pi m}\right)^{1/2} dS\cos(\theta)\sin(\theta)d\theta = \frac{n'\bar{V}'dS\cos(\theta)}{4\pi}d\Omega ,$$

where $d\Omega$ is the solid angle subtended by dS' at dS. Taking into account that $d\Omega = dS'\cos(\theta')/\tilde{r}^2$, one can obtain:

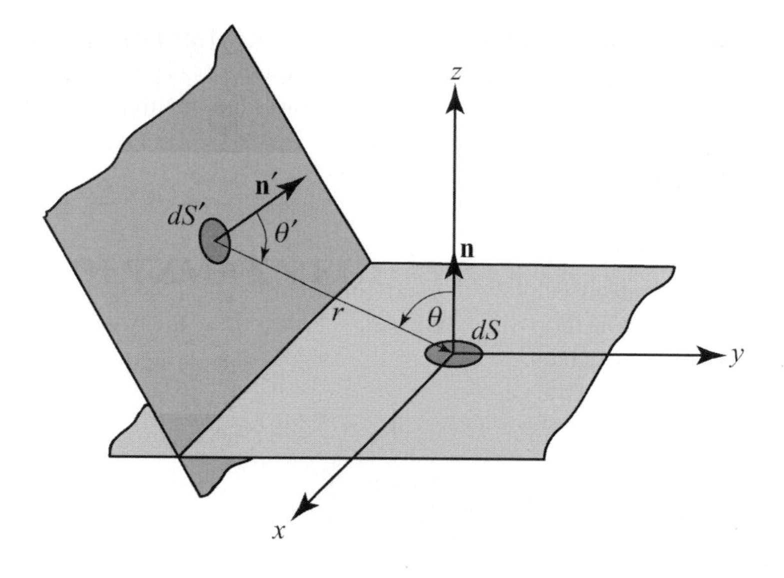

Figure 4-1. The geometry to be used in Problem 4.8 in determining the rate at which molecules reflected from one differential surface element cross another.

$$dN_\theta = \frac{n'\overline{V}'dSdS'\cos(\theta)\cos(\theta')}{4\pi\tilde{r}^2}, \ \overline{V}' = \left(\frac{8kT'}{\pi m}\right)^{1/2}.$$

4.9. Two vessels containing the same gas at the pressures and temperatures, p_1, T_1 and p_2, T_2, respectively, are connected by a small slot possessing a typical dimension of, L, which is much less than the mean free path of the gas molecules in either chamber. Derive the condition of kinetic equilibrium.

Solution: If $\lambda >> L$, molecules are leaving the vessels independently of each other. The condition of kinetic equilibrium requires that the net number of molecules crossing the area of the slot per second be equal to zero. This condition may be expressed as $\frac{1}{4}n_1\overline{V}_1 - \frac{1}{4}n_2\overline{V}_2 = 0$. From this relationship, one can obtain:

$$\frac{p_1}{\sqrt{T_1}} = \frac{p_2}{\sqrt{T_2}}.$$

4.10. A binary gas mixture exits a chamber through a small slot by means of effusion. The escaping gas is collected in another chamber without loss. Calculate the ratio of the gas densities in the collection chamber.

Solution: Let n_1, m_1 and n_2, m_2 be the number densities and molecular masses, respectively, of the gases in the first chamber. The total number of molecules of each constituent in the collection chamber may be expressed as:

$$N_i' = \tfrac{1}{4} n_i \overline{V_i} S \Delta t \ ,$$

where $\overline{V_i}$ is the mean velocity of each constituent, S is the area of the slot, and Δt is the collection time interval. From this, the ratio of the gas densities can be written as:

$$\frac{n_1'}{n_2'} = \frac{N_1'}{N_2'} = \frac{n_1}{n_2} \frac{\overline{V_1}}{\overline{V_2}} \ .$$

Equation (4-10) then yields:

$$\frac{n_1'}{n_2'} = \frac{n_1}{n_2} \left(\frac{m_2}{m_1} \right)^{1/2} \ .$$

4.11. Electrons evaporating from a hot wire form an electron gas with a number density, n. These electrons exit the chamber in which they are generated by means of a system of slots which serve to direct the electrons in one specific direction forming a beam. While exiting the chamber, this electron beam must pass through a hindering potential, φ, which stops some of the electrons. Determine the flux density of the electrons that are able to overcome this hindering potential.

Solution: The flux density of electrons having velocities in the interval, dv, is given by (Problem 4.5):

$$dN_v = \pi n \left(\frac{m}{2\pi kT} \right)^{3/2} \exp\left(-\frac{mv^2}{2kT} \right) v^3 dv \ .$$

The flux density of electrons overcoming the hindering potential may be written as:

$$N = \int_{v_0}^{\infty} dN_v = \frac{n}{\sqrt{2\pi}} \left(\frac{kT}{m} \right)^{1/2} \left(\frac{e\varphi}{kT} + 1 \right) \exp\left(-\frac{e\varphi}{kT} \right) .$$

where $\frac{1}{2}mv_0^2 = e\varphi$.

4.12. The work function of a metal is W. Determine the current density of the emitted electrons if the temperature is T and the electron number density inside the metal is n.

Solution: Let OX be the axis normal to the surface of the metal. The current density along this axis is given by:

$$j_x = ne\left(\frac{m}{2\pi kT}\right)^{1/2}\int_{v_{0x}}^{\infty} v_x \exp\left(-\frac{mv_x^2}{2kT}\right)dv_x = \tfrac{1}{4}ne\bar{V}\exp\left(-\frac{W}{kT}\right),$$

where $\frac{1}{2}mv_{0x}^2 = W$. This is equivalent to the classical Richardson thermionic emission formula.

4.13. Determine the number density variation for the steady-state of a gas that is situated in a force field described by the potential, $\varphi(\tilde{\mathbf{r}})$.

Solution: Let the distribution function be represented in the form:

$$f = n(\tilde{\mathbf{r}})\left(\frac{m}{2\pi kT(\tilde{\mathbf{r}})}\right)^{3/2}\exp\left(-\frac{mv^2}{2kT(\tilde{\mathbf{r}})}\right).$$

Substituting this function into the Boltzmann equation, one obtains:

$$\left(\mathbf{v}\cdot\frac{\partial}{\partial\tilde{\mathbf{r}}} + \mathbf{F}\cdot\frac{\partial}{\partial\mathbf{v}}\right)\ln(f) = 0 .$$

This expression may be reduced to:

$$\mathbf{v}\cdot\left\{\frac{\partial}{\partial\tilde{\mathbf{r}}}\ln\left(\frac{n}{T^{3/2}}\right) + \frac{mv^2}{2kT^2}\frac{\partial T}{\partial\tilde{\mathbf{r}}} - \frac{m}{kT}\mathbf{F}\right\} = 0 .$$

which is an identity in $\mathbf{v}$. As such, the following system of equations may be obtained:

$$\frac{\partial T}{\partial\tilde{\mathbf{r}}} = 0 \text{ and } \frac{\partial}{\partial\tilde{\mathbf{r}}}\left[\ln\left(\frac{n}{T^{3/2}}\right) + \frac{m}{kT}\varphi(\tilde{\mathbf{r}})\right] = 0 .$$

From the above system of equations, it is easy to obtain:

$$n(\tilde{\mathbf{r}}) = n_0 \exp\left(-\frac{m\varphi(\tilde{\mathbf{r}})}{kT}\right) .$$

where n_0 is the number density at the point for which $\varphi(\tilde{\mathbf{r}}) = 0$ and T is the constant temperature.

4.14. A gas rotates around a fixed axis with an angular speed, ω. Determine the number density distribution in the gas.

Solution: The rotation of a gas is equivalent to an external force field having the potential: $\varphi(\tilde{\mathbf{r}}) = -\frac{1}{2}\omega^2\tilde{r}^2$, where $\tilde{r}$ is the distance from the rotational axis. From Problem 4.13, the variation in the number density is known to be:

$$n(\tilde{\mathbf{r}}) = n_0 \exp\left(-\frac{m\varphi(\tilde{\mathbf{r}})}{kT}\right) .$$

Substituting the rotational potential into this expression yields:

$$n(\tilde{\mathbf{r}}) = n_0 \exp\left(\frac{m\omega^2\tilde{r}^2}{2kT}\right) .$$

4.15. Determine the number density distribution, $n(\tilde{\mathbf{r}})$, of a gas contained in a cylindrical vessel of radius, R, and length, L, that is rotating about its axis with an angular velocity, ω. The total number of molecules in the vessel is N.

Solution: Using the number distribution obtained in Problem 4.14, one can obtain:

$$n(\tilde{\mathbf{r}}) = \frac{Nm\omega^2}{2\pi L} \exp\left(\frac{m\omega^2\tilde{r}^2}{2kT}\right)\left[\exp\left(\frac{m\omega^2 R^2}{2kT}\right) - 1\right]^{-1} .$$

4.16. Evaluate the mass of the Earth's atmosphere assuming that almost all of the mass is contained in the thin layer near the Earth's surface.

Solution: The potential energy of a molecule near the Earth's surface is given by: $U(h) = mgh$. The total mass of the atmosphere is then approximated by:

$$M \approx 4\pi R_0^2 mn \int_0^\infty \exp\left(-\frac{mgh}{kT}\right) dh = 4\pi R_0^2 nkTg^{-1} ,$$

where R_0 is the radius of the Earth. Using typical values for the quantities involved yields $M = 5.3 \times 10^{18}$ kg.

REFERENCES

1. Chapman, S. and Cowling, T.G., *The Mathematical Theory of Non-Uniform Gases* (Cambridge University Press, Cambridge, U.K., 1990).
2. Ferziger, J.H. and Kaper, H.G., *Mathematical Theory of Transport Processes in Gases* (North-Holland, Amsterdam-London, 1972).
3. Huang, K., *Statistical Mechanics* (John Wiley and Sons, New York, 1963).
4. Cercignani, C., *Theory and Application of the Boltzmann Equation* (Scottish Academic Press, Edinburgh, U.K., 1975).
5. Brush, S.G., *Selected Readings in Physics. Vol. 2. Irreversible Processes* (Pergamon Press, New York, 1966).
6. Klein, M.J., *Paul Ehrenfest* (North-Holland, Amsterdam and Elsevier, New York, 1970).
7. Uhlenbeck, G.E. and Ford, G.W., *Lectures in Statistical Mechanics* (American Mathematical Society, Providence, RI, 1963).
8. Kennard, E.H. , *Kinetic Theory of Gases* (McGraw-Hill, New York and London, 1938).
9. Knudsen, M. , *The Kinetic Theory of Gases*, third edition (Methuen & Co., London, and Wiley, New York, reprinted 1952).
10. Loeb, L.B. , *The Kinetic Theory of Gases* (Dover, New York, 1961).

Chapter 5

THE NON-UNIFORM STATE FOR A SIMPLE GAS

1. EXPANSION IN POWERS OF A SMALL PARAMETER.

Consider the unsteady, non-uniform state of an infinite gas. For this general case, the Boltzmann equation has the form [1]:

$$\mathrm{D}f = -J\left(ff_1\right) , \tag{5-1}$$

where:

$$\mathrm{D}f = \frac{\partial f}{\partial t} + \mathbf{v} \cdot \frac{\partial f}{\partial \tilde{\mathbf{r}}} + \mathbf{F} \cdot \frac{\partial f}{\partial \mathbf{v}} ,$$

and:

$$J\left(ff_1\right) = \int\int\int\left(ff_1 - f'f_1'\right) gbdbd\varepsilon d\mathbf{v}_1 .$$

The Boltzmann equation may be readily converted into a dimensionless form by introducing the following quantities:

$$\mathbf{v} = \left(\frac{2kT}{m}\right)^{1/2} \mathbf{c} \ , \ f = n\left(\frac{m}{2kT}\right)^{3/2} f^* \ ,$$

$$t = \tau t^* \ , \ \tilde{\mathbf{r}} = L\mathbf{r} \ , \ \mathbf{F} = w\mathbf{F}^* \ , \ b = \sigma 2^{1/4} b^* \ , \ \varepsilon = \pi\varepsilon^* \ ,$$

where τ and L are typical values for the time and length variations of the macroscopic quantities of a gas.

In terms of the above dimensionless quantities, Eq. (5-1) becomes:

$$\begin{aligned} \mathrm{D}^* f^* &= \frac{L}{\tau}\left(\frac{m}{2kT}\right)^{1/2} \frac{\partial f^*}{\partial t^*} + \mathbf{c}\cdot\frac{\partial f^*}{\partial \mathbf{r}} + \frac{wmL}{2kT}\mathbf{F}^*\cdot\frac{\partial f^*}{\partial \mathbf{c}} \\ &= -\frac{1}{\in} J^*\left(f^*\, f_1^*\right) , \end{aligned} \tag{5-2}$$

where $\in = \lambda/L$. The parameter, $\in$, is usually called the typical parameter of non-uniformity of a gas but, since L is not necessarily the typical dimension of a boundary problem, $\in$ is not the Knudsen number [2].

Let the macroscopic values vary slightly in the mean free path length, so that the following relations are satisfied:

$$\lambda\left|\nabla \ln(n)\right| \ll 1 \ , \ \lambda\left|\nabla \ln(T)\right| \ll 1 \ , \ \lambda\left|\nabla \ln(u)\right| \ll 1 \ .$$

Moreover, the typical values τ and w are assumed to be confined by the conditions $\tau \geq L(m/2kT)^{1/2}$ and $w \leq 2kT/mL$. Then, the basic dimensionless parameter, $\in \ll 1$. This makes it quite natural to seek a solution to Eq. (5-2) in the form of a power series in $\in$. Taking into account only the linear terms of this expansion, the distribution function may be approximated by:

$$f^* = f^{*(0)} + \in f^{*(1)} \ , \tag{5-3a}$$

or:

$$f = f^{(0)} + f^{(1)} \ , \tag{5-3b}$$

in ordinary variables.

Substituting Eq. (5-3a) into Eq. (5-2), and retaining only the main terms in $\in$, one obtains:

$$\mathrm{D}^* f^{*(0)} = -\frac{1}{\in} J^*\left(f^{*(0)} f_1^{*(0)}\right) - \left\{J^*\left(f^{*(0)} f_1^{*(1)}\right) + J^*\left(f^{*(1)} f_1^{*(0)}\right)\right\} \ . \tag{5-4}$$

After equating coefficients of like powers of $\in$, the following relationships may be written:

$$J^*\left(f^{*(0)} f_1^{*(0)}\right) = 0 \ , \tag{5-5}$$

$$\mathrm{D}^* f^{*(0)} = -\left\{J^*\left(f^{*(0)} f_1^{*(1)}\right) + J^*\left(f^{*(1)} f_1^{*(0)}\right)\right\} \ . \tag{5-6}$$

These integral equations are the basic ones used to obtain analytical expressions for the first and second-order correction terms to the distribution function.

2. THE FIRST APPROXIMATION.

In the usual variables, Eq. (5-5) has the form:

$$\int\!\!\int\left(f^{(0)\prime} f_1^{(0)\prime} - f^{(0)} f_1^{(0)}\right) gbdbd\varepsilon d\mathbf{v}_1 = 0 \ .$$

The general solution of this integral equation is defined by [1]:

$$f^{(0)\prime} f_1^{(0)\prime} = f^{(0)} f_1^{(0)} \ ,$$

and therefore, $\ln\left(f^{(0)}\right)$ is a linear combination of the summational invariants of molecular encounters. By a simple transformation one can obtain:

$$f^{(0)} = n\left(\frac{m}{2\pi kT}\right)^{3/2} \exp\left(-\frac{m\left(\mathbf{v}-\mathbf{u}\right)^2}{2kT}\right) , \tag{5-7}$$

where n, $\mathbf{u}$, and T denote arbitrary values related to each of the summational invariants. These values can be identified with number density, mean velocity, and temperature at a point, $\left(\tilde{\mathbf{r}},t\right)$. This means that the Maxwellian distribution function in which $n = n\left(\tilde{\mathbf{r}},t\right)$, $T = T\left(\tilde{\mathbf{r}},t\right)$, and $\mathbf{u} = \mathbf{u}\left(\tilde{\mathbf{r}},t\right)$ is chosen as the first approximation to f. This function is usually called the local Maxwellian distribution function, which, for the general, non-uniform, non-steady-state of a gas, may be written as:

$$f^{(0)}\left(\tilde{\mathbf{r}},t\right) = n\left(\tilde{\mathbf{r}},t\right)\left(\frac{m}{2\pi kT\left(\tilde{\mathbf{r}},t\right)}\right)^{3/2} \exp\left(-\frac{m\left(\mathbf{v}-\mathbf{u}\left(\tilde{\mathbf{r}},t\right)\right)^2}{2kT\left(\tilde{\mathbf{r}},t\right)}\right) . \tag{5-8}$$

As a consequence of the identification of n, $\mathbf{u}$, and T with the density, mean velocity, and temperature at the point, $(\tilde{\mathbf{r}},t)$, the function $f^{(1)}$ must satisfy the following relations:

$$\int f^{(1)}\psi^{(i)}d\mathbf{v}=0 \ , \tag{5-9}$$

where $\psi^{(i)}=1, m\mathbf{V}, \frac{1}{2}mV^2$. These conditions must be satisfied in the formulation of any expressions for the second-order correction term to the distribution function.

3. A GENERAL FORMAL SOLUTION FOR THE SECOND CORRECTION.

Eq. (5-6) in ordinary variables may be written in the form:

$$J\left(f^{(0)}f_1^{(1)}\right)+J\left(f^{(1)}f_1^{(0)}\right)=-\mathrm{D}f^{(0)} \ . \tag{5-10}$$

This is the linear non-homogeneous integral equation for the second-order correction to the distribution function, $f^{(1)}$. Let $F^{(1)}$ and $\chi^{(1)}$ be the particular and general solutions of the non-homogeneous and homogeneous equations, respectively. Then, the general solution can be represented by:

$$f^{(1)}=F^{(1)}+\chi^{(1)} \ . \tag{5-11}$$

Substituting this expression into Eq. (5-10), one can obtain the following form of the homogeneous integral equation:

$$I\left(\varphi^{(1)}\right)=0 \ . \tag{5-12}$$

Here, the correction $\varphi^{(1)}$ is given by the relation, $\chi^{(1)}=f^{(0)}\varphi^{(1)}$ and $I\left(\varphi^{(1)}\right)$ is the standard linearized collision operator which may be expressed as:

$$I(F)=n^{-2}\int\int f^{(0)}f_1^{(0)}\left(F+F_1-F'-F_1'\right)gbdbd\varepsilon d\mathbf{v}_1 \ . \tag{5-13}$$

Eq. (5-12) means that $\varphi^{(1)}$ must be the summational collision invariant which may be written as:

$$\varphi^{(1)} = \alpha^{(1,1)} + \boldsymbol{\alpha}^{(2,1)} \cdot m\mathbf{v} + \alpha^{(3,1)} \tfrac{1}{2} m v^2 \ , \tag{5-14}$$

where $\alpha^{(i,1)}$ are arbitrary functions of $\mathbf{r}$ and t. These functions may be found from the conditions given by Eq. (5-9).

Now, to find the necessary conditions of solubility for Eq. (5-10), one uses the integral theorem [1] given by:

$$\int \psi^{(i)} \left[J\left(f^{(0)} f_1^{(1)}\right) + J\left(f^{(1)} f_1^{(0)}\right) \right] d\mathbf{v} = 0 \ .$$

Using this theorem in Eq. (5-10), one can obtain the following conditions:

$$\int \psi^{(i)} \mathrm{D} f^{(0)} d\mathbf{v} = 0 \ . \tag{5-15}$$

These relations, when $i = 1,2,3$, mean that the values n, $\mathbf{u}$, and T contained in $f^{(0)}$ must satisfy the continuum equations obtained by means of the use of the first-order distribution function [1].

Let $F^{(1)}$ be represented in the form, $F^{(1)} = f^{(0)} \Phi^{(1)}$. Then, Eq. (5-10) gives the following basic equation for this correction:

$$n^2 I\left(\Phi^{(1)}\right) = -\mathrm{D} f^{(0)} \ . \tag{5-16}$$

From general conditions given by Eq. (5-9), one can conclude that the function, $\Phi^{(1)}$, must satisfy the following additional conditions:

$$\int \psi^{(i)} f^{(0)} \left(\Phi^{(1)} + \varphi^{(1)} \right) d\mathbf{v} = 0 \ . \tag{5-17}$$

4. THE TRANSFORMATION OF THE NON-HOMOGENEOUS TERM.

If the variables $\mathbf{V}, \tilde{\mathbf{r}}, t$ are used, the right-hand-side of Eq. (5-16) may be expressed in a form analogous to that given in Eq. (2.6):

$$\mathrm{D} f^{(0)} = f^{(0)} \ln\left(f^{(0)}\right) \times \left\{ \frac{D}{Dt} + \mathbf{V} \cdot \frac{\partial}{\partial \tilde{\mathbf{r}}} + \left(\mathbf{F} - \frac{D\mathbf{u}}{Dt} \right) \cdot \frac{\partial}{\partial \mathbf{V}} - \frac{\partial}{\partial \mathbf{V}} \mathbf{V} : \frac{\partial \mathbf{u}}{\partial \tilde{\mathbf{r}}} \right\} , \tag{5-18}$$

where:

$$\frac{D}{Dt} = \frac{\partial}{\partial t} + \mathbf{u} \cdot \frac{\partial}{\partial \tilde{\mathbf{r}}} \ .$$

The space and time derivatives of n, $\mathbf{u}$, and T are connected by the equations of the continuous medium in which the pressure tensor and the heat flux must be calculated by use of the local Maxwellian distribution function given in Eq. (5-8). This results in the following expressions:

$$P_{ik}^{(0)} = nkT\delta_{ik} = p\delta_{ik} \ , \ \mathbf{q}^{(0)} = 0 \ . \tag{5-19}$$

Substituting these quantities into Eqs. (2.7)-(2.9), one obtains:

$$\frac{Dn}{Dt} = -n\frac{\partial}{\partial \tilde{\mathbf{r}}} \cdot \mathbf{u} \ , \tag{5-20}$$

$$\frac{D\mathbf{u}}{Dt} = \mathbf{F} - \frac{1}{\rho}\frac{\partial p}{\partial \tilde{\mathbf{r}}} \ , \tag{5-21}$$

$$\frac{DT}{Dt} = -\tfrac{2}{3}T\frac{\partial}{\partial \tilde{\mathbf{r}}} \cdot \mathbf{u} \ . \tag{5-22}$$

Eqs. (5-20) and (5-22) result in:

$$\frac{D}{Dt}\ln\left(nT^{-3/2}\right) = 0 \ , \ \ln\left(nT^{-3/2}\right) = \text{const} \ . \tag{5-23}$$

Using the second relationship in Eq. (5-23), one can write:

$$\ln\left(f^{(0)}\right) = \text{const} - \frac{mV^2}{2kT} \ .$$

Taking into account Eq. (5-22), the first term in Eq. (5-18) may be expressed in the form:

$$\frac{D}{Dt}\ln\left(f^{(0)}\right)=\tfrac{1}{2}\frac{mV^2}{kT^2}\frac{DT}{Dt}=-\tfrac{1}{3}\frac{mV^2}{kT}\frac{\partial}{\partial\tilde{\mathbf{r}}}\cdot\mathbf{u}=-\tfrac{1}{3}\frac{mV^2}{kT}\delta_{ik}\frac{\partial u_i}{\partial\tilde{x}_k}\ .$$

The last term in Eq. (5-18) may be written as:

$$-\frac{\partial\ln\left(f^{(0)}\right)}{\partial\mathbf{V}}\mathbf{V}:\frac{\partial\mathbf{u}}{\partial\tilde{\mathbf{r}}}=\frac{m}{kT}V_iV_k\frac{\partial u_i}{\partial\tilde{x}_k}\ .$$

The sum of the first and last terms of Eq. (5-18) is then:

$$\frac{m}{kT}\left(V_iV_k-\tfrac{1}{3}V^2\delta_{ik}\right)\frac{\partial u_i}{\partial\tilde{x}_k}=\frac{m}{kT}\overset{\circ}{\mathbf{V}\mathbf{V}}:\frac{\partial\mathbf{u}}{\partial\tilde{\mathbf{r}}}\ . \tag{5-24}$$

The sum of the two middle terms of Eq. (5-18) is given by:

$$\begin{aligned}&\mathbf{V}\cdot\left(\frac{\partial}{\partial\tilde{\mathbf{r}}}\ln\left(f^{(0)}\right)-\frac{\partial}{\partial\tilde{\mathbf{r}}}\ln\left(nkT\right)\right)=\mathbf{V}\cdot\frac{\partial}{\partial\tilde{\mathbf{r}}}\ln\left(\frac{f^{(0)}}{nkT}\right)\\&=\mathbf{V}\cdot\left(\frac{\partial}{\partial\tilde{\mathbf{r}}}\ln\left(T^{-5/2}\right)+\tfrac{1}{2}\frac{mV^2}{kT^2}\frac{\partial T}{\partial\tilde{\mathbf{r}}}\right)=\left(\tfrac{1}{2}\frac{mV^2}{kT}-\tfrac{5}{2}\right)\mathbf{V}\cdot\frac{\partial\ln(T)}{\partial\tilde{\mathbf{r}}}\ .\end{aligned} \tag{5-25}$$

Using Eqs. (5-24) and (5-25), Eq. (5-18) can be represented as:

$$\mathrm{D}f^{(0)}=f^{(0)}\left\{\left(C^2-\tfrac{5}{2}\right)\mathbf{V}\cdot\frac{\partial\ln(T)}{\partial\tilde{\mathbf{r}}}+2\overset{\circ}{\mathbf{C}\mathbf{C}}:\frac{\partial\mathbf{u}}{\partial\tilde{\mathbf{r}}}\right\}\ . \tag{5-26}$$

The basic relation expressed by Eq. (5-16) can now be written in the form:

$$n^2I\left(\Phi^{(1)}\right)=-f^{(0)}\left\{\left(C^2-\tfrac{5}{2}\right)\mathbf{V}\cdot\frac{\partial\ln(T)}{\partial\tilde{\mathbf{r}}}+2\overset{\circ}{\mathbf{C}\mathbf{C}}:\frac{\partial\mathbf{u}}{\partial\tilde{\mathbf{r}}}\right\}\ . \tag{5-27}$$

5. THE SECOND APPROXIMATION.

Since Eq. (5-27) is a linear integral equation and the right-hand-side of this equation is linear in $\partial\ln(T)/\partial\tilde{\mathbf{r}}$ and $\partial\mathbf{u}/\partial\tilde{\mathbf{r}}$, one may seek a solution of this equation in the form of a sum of the two terms each of which is proportional to the appropriate space non-uniformity. Thus, one can write [1,3-5]:

$$\Phi^{(1)} = -\mathbf{A} \cdot \frac{\partial \ln(T)}{\partial \tilde{\mathbf{r}}} - 2\mathbf{B} : \frac{\partial \mathbf{u}}{\partial \tilde{\mathbf{r}}} , \tag{5-28}$$

where $\mathbf{A}$ is a vector and $\mathbf{B}$ is a tensor. The quantities, $\mathbf{A}$ and $\mathbf{B}$, introduced here are identical to those obtained from the Chapman-Enskog theory for multiple gas mixtures in the limiting case of a simple gas [1].

Substituting Eq. (5-28) into Eq. (5-27), one can separate out equations for the unknown functions, $\mathbf{A}$ and $\mathbf{B}$. These functions are then particular solutions for the following equations:

$$n^2 I(\mathbf{A}) = f^{(0)}\left(C^2 - \tfrac{5}{2}\right)\mathbf{V} , \tag{5-29}$$

$$n^2 I(\mathbf{B}) = f^{(0)} \overset{\circ}{\mathbf{C}\mathbf{C}} , \tag{5-30}$$

which, in the coordinate form, can be written as:

$$n^2 I(A_i) = f^{(0)}\left(C^2 - \tfrac{5}{2}\right)V_i , \tag{5-31a}$$

$$n^2 I(B_{ik}) = f^{(0)}\left(C_i C_k - \tfrac{1}{3} C^2 \delta_{ik}\right) . \tag{5-31b}$$

One now assumes that the solutions of Eqs. (5-31a) and (5-31b) are of the following form:

$$\mathbf{A} = A(C)\mathbf{C} , \tag{5-32}$$

$$\mathbf{B} = B(C)\overset{\circ}{\mathbf{C}\mathbf{C}} , \tag{5-33}$$

where $A(C)$ and $B(C)$ are two new unknown functions of n, T, and $C = |\mathbf{C}|$. The general formal solution of Eq. (5-27) then has the form:

$$f^{(1)} = f^{(0)}\left\{-A(C)\mathbf{C}\cdot\frac{\partial\ln(T)}{\partial\tilde{\mathbf{r}}} - 2B(C)\overset{\circ}{\mathbf{CC}}:\frac{\partial\mathbf{u}}{\partial\tilde{\mathbf{r}}}\right.$$
$$\left. + \alpha^{(1,1)} + \boldsymbol{\alpha}^{(2,1)}\cdot m\mathbf{V} + \alpha^{(3,1)}\tfrac{1}{2}mV^2\right\}d\mathbf{v}\ . \qquad (5\text{-}34)$$

where the constants $\alpha^{(i,1)}$ in Eq. (5-34) are chosen such that the function $f^{(1)}$ satisfies the following equations:

$$\int f^{(1)}\psi^{(i)}d\mathbf{v} = 0\ ;\ i = 1,2,3\ . \qquad (5\text{-}35)$$

Neglecting vanishing integrals, these equations have the form:

$$\int f^{(0)}\left(\alpha^{(1,1)} + \alpha^{(3,1)}\tfrac{1}{2}mV^2\right)d\mathbf{v} = 0\ ,$$

$$\int f^{(0)}\left(-A(C)\frac{\partial\ln(T)}{\partial\tilde{\mathbf{r}}} + (2mkT)^{1/2}\boldsymbol{\alpha}^{(2,1)}\right)\cdot C^2 d\mathbf{v} = 0\ ,$$

$$\int f^{(0)}\left(\alpha^{(1,1)} + \alpha^{(3,1)}\tfrac{1}{2}mV^2\right)\tfrac{1}{2}mV^2 d\mathbf{v} = 0\ .$$

The first and third of the above relations show that $\alpha^{(1,1)} = \alpha^{(3,1)} = 0$. The second shows that $\boldsymbol{\alpha}^{(2,1)} \sim \partial\ln(T)/\partial\tilde{\mathbf{r}}$ and, hence, that this term may be absorbed into the first term. In order that $\boldsymbol{\alpha}^{(2,1)} = 0$ also, the function $A(C)$ must satisfy the additional relation:

$$\int A(C)C^2 f^{(0)}d\mathbf{v} = 0\ . \qquad (5\text{-}36)$$

Now, the second approximation to the distribution function is given by:

$$f^{(1)} = f^{(0)}\left(\Phi^{(1)} + \varphi^{(1)}\right)$$
$$= f^{(0)}\left\{-A(C)\mathbf{C}\cdot\frac{\partial\ln(T)}{\partial\tilde{\mathbf{r}}} - 2B(C)\overset{\circ}{\mathbf{CC}}:\frac{\partial\mathbf{u}}{\partial\tilde{\mathbf{r}}}\right\}\ . \qquad (5\text{-}37)$$

6. THE FIRST-ORDER CHAPMAN-ENSKOG SOLUTION FOR THERMAL CONDUCTION.

Suppose that the function $\mathbf{A}$ can be expressed as a convergent series of the form:

$$\mathbf{A} = \sum_{r=0}^{\infty} a_r \mathbf{a}^{(r)} \ , \quad \mathbf{a}^{(r)} = S_{3/2}^{(r)}\left(C^2\right)\mathbf{C} \ , \tag{5-38}$$

where $S_{3/2}^{(r)}\left(C^2\right)$ are the Sonine polynomials. The polynomial $S_m^{(n)}(x)$ is defined by:

$$S_m^{(n)}(x) = \sum_{p=0}^{n} (-x)^p \frac{(m+n)_{n-p}}{p!(n-p)!} \ ,$$

where:

$$(m+n)_{n-p} = (m+n)(m+n-1)\ldots\left[(m+n)-(n-p-1)\right] \ ,$$

and $(m+n)_0 = 1$. Two special cases exist where:

$$S_m^{(0)}(x) = 1 \ , \quad S_m^{(1)}(x) = m+1-x \ .$$

These polynomials satisfy the following integral relations:

$$\int_0^\infty \exp(-x) S_m^{(p)}(x) S_m^{(q)}(x) x^m dx = \begin{cases} 0 & ; \ p \neq q \ , \\ \dfrac{\Gamma(m+p+1)}{p!} & ; \ p = q \ , \end{cases}$$

where $\Gamma(k)$ is the gamma function.

The additional condition expressed by Eq. (5-36) gives:

$$\int \exp\left(-C^2\right) C^2 A(C) d\mathbf{C}$$
$$= 4\pi \int_0^\infty C^4 \exp\left(-C^2\right) \left\{ \sum_{p=0}^\infty a_p S_{3/2}^{(p)}\left(C^2\right) \right\} dC$$
$$= 2\pi \sum_{p=0}^\infty a_p \int_0^\infty \exp\left(-C^2\right) S_{3/2}^{(0)}\left(C^2\right) S_{3/2}^{(p)}\left(C^2\right) C^3 d\left(C^2\right)$$
$$= 2\pi a_0 \Gamma\left(\tfrac{5}{2}\right) = 0 ,$$

and therefore $a_0 = 0$. To get acceptable accuracy [1], only one term of the series in Eq. (5-38) need to be taken into account, i.e.:

$$\mathbf{A} = a_1^{(1)} \mathbf{C} S_{3/2}^{(1)}\left(C^2\right) .$$

Substituting this expression into Eq. (5-29), one obtains:

$$n^2 a_1^{(1)} I\left(\mathbf{C} S_{3/2}^{(1)}\left(C^2\right)\right) = f^{(0)}\left(C^2 - \tfrac{5}{2}\right)\mathbf{V} = -f^{(0)} \mathbf{V} S_{3/2}^{(1)}\left(C^2\right) . \tag{5-39}$$

After multiplying this equation by $\mathbf{C} S_{3/2}^{(1)}\left(C^2\right)$ and integrating over all values of $\mathbf{v}$, Eq. (5-39) can be expressed in the form:

$$n^2 a_1^{(1)} \int \mathbf{C} S_{3/2}^{(1)}\left(C^2\right) I\left(\mathbf{C} S_{3/2}^{(1)}\left(C^2\right)\right) d\mathbf{v} = \alpha_1 ,$$

or $a_1^{(1)} a_{11} = \alpha_1$ where:

$$a_{11} = \left[\mathbf{C} S_{3/2}^{(1)}\left(C^2\right), \mathbf{C} S_{3/2}^{(1)}\left(C^2\right)\right] ,$$

$$\alpha_1 = -n^{-2} \int f^{(0)} \mathbf{V} S_{3/2}^{(1)}\left(C^2\right) \cdot \mathbf{C} S_{3/2}^{(1)}\left(C^2\right) d\mathbf{v} = -\tfrac{15}{4} n^{-1} \left(\frac{2kT}{m}\right)^{1/2} .$$

The expression for the bracket integral a_{11} is calculated in Appendix A. Using Eq. (A-9), the following solution of Eq. (5-39) can be obtained for rigid-sphere molecules:

$$a_1^{(1)} = -\tfrac{15}{4} n^{-1} \left(\frac{2kT}{m} \right)^{1/2} \frac{1}{a_{11}} = -\tfrac{15}{16} \sqrt{\pi} \lambda \ . \tag{5-40}$$

The vector, $\mathbf{A}$, for the first-order Chapman-Enskog solution is given by:

$$\mathbf{A} = -\tfrac{15}{16} \sqrt{\pi} \lambda \mathbf{C} S_{3/2}^{(1)} \left(C^2 \right) \ . \tag{5-41}$$

7. THE FIRST-ORDER CHAPMAN-ENSKOG SOLUTION FOR VISCOSITY.

To simplify the following calculations, consider the viscosity equation for only one component of the tensor, $\mathbf{B}$. The equation for B_{xy} may be written as:

$$n^2 I\left(C_x C_y B(C) \right) = f^{(0)} C_x C_y \ . \tag{5-42}$$

Since the tensor, $\mathbf{B}$, is a non-divergent tensor of molecular velocity components, it can be expressed in the general form [1]:

$$\mathbf{B} = \overset{\circ}{\mathbf{C}\mathbf{C}} B(C) = \overset{\circ}{\mathbf{C}\mathbf{C}} \sum_{r=1}^{\infty} b_r S_{5/2}^{(r-1)} \left(C^2 \right) \ , \tag{5-43}$$

where the coefficients, b_r, are constants to be determined.

When only one tensor component, B_{xy}, is being considered, it can be expressed as:

$$B_{xy} = C_x C_y \sum_{r=1}^{\infty} b_r S_{5/2}^{(r-1)} \left(C^2 \right) \ . \tag{5-44}$$

To obtain the first-order Chapman-Enskog solution (retaining only one term of the expansion), this tensor is given by:

$$B_{xy} = b_1^{(1)} C_x C_y \ . \tag{5-45}$$

Substituting Eq. (5-45) into Eq. (5-42) one can obtain:

$$n^2 b_1^{(1)} I\left(C_x C_y\right) = f^{(0)} C_x C_y \ . \tag{5-46}$$

After multiplying both sides of Eq. (5-46) by $C_x C_y d\mathbf{v}$ and integrating over all velocities one obtains:

$$b_1^{(1)} b'_{11} = \beta'_1 \ , \tag{5-47}$$

where:

$$b'_{11} = \left[C_x C_y \, , C_x C_y\right] \, ,$$

$$\beta'_1 = n^{-2} \int C_x^2 C_y^2 f^{(0)} d\mathbf{v} = \tfrac{1}{4} n^{-1} \ .$$

It is important to note that the values of b'_{11} and β'_1 introduced here differ from the corresponding values in [1] by a factor equal to $\frac{1}{10}$. Using Eq. (A-12) for b'_{11}, one obtains:

$$b_1^{(1)} = \frac{1}{4 b'_{11} n} = \tfrac{5}{8} \sqrt{\pi} \lambda \left(\frac{m}{2kT}\right)^{1/2} \ . \tag{5-48}$$

Both Eqs. (5-40) and (5-48) have been obtained for rigid-sphere molecules.

From the preceding analysis, in general, the components of the tensor for the first-order Chapman-Enskog solution, B_{ik}, have the form:

$$B_{ik} = \tfrac{5}{8} \sqrt{\pi} \lambda \left(\frac{m}{2kT}\right)^{1/2} \left(C_i C_k - \tfrac{1}{3} \delta_{ik} C^2\right) \ . \tag{5-49}$$

Using Eqs. (5-41) and (5-49) and introducing the mean free path, the first-order Chapman-Enskog distribution function is then given by:

$$f = f^{(0)} \left\{ 1 + \tfrac{15}{16} \lambda \sqrt{\pi} S_{3/2}^{(1)}\left(C^2\right) C_i \frac{\partial \ln(T)}{\partial \tilde{x}_i} - \tfrac{5}{4} \lambda \sqrt{\pi} \left(\frac{m}{2kT}\right)^{1/2} \left(C_i C_k - \tfrac{1}{3} C^2 \delta_{ik}\right) \frac{\partial u_i}{\partial \tilde{x}_i} \right\} \ . \tag{5-50}$$

Here, molecules have been assumed to act as rigid spheres.

8. THE THERMAL CONDUCTIVITY AND VISCOSITY COEFFICIENTS.

Since the distribution function is known, the coefficients of viscosity and thermal conductivity may be readily calculated. The thermal conductivity coefficient can be obtained from the relation:

$$q_i = \tfrac{1}{2} m \int V^2 V_i f d\mathbf{V} = -\kappa \frac{\partial T}{\partial \tilde{x}_i} \ . \tag{5-51}$$

Having performed the very simple integration in Eq. (5-51), one obtains the following expression for the thermal conductivity coefficient of a monatomic gas:

$$\kappa = \tfrac{75}{64} \frac{k}{\sigma^2} \left(\frac{kT}{\pi m} \right)^{1/2} \ . \tag{5-52}$$

The molecules here have been assumed to behave as rigid spheres.

To derive a similar expression for the viscosity coefficient, the simplest case of $\mathbf{u} = \{0, u, 0\}$ and $u = u(\tilde{x})$ is considered. In this case:

$$P_{xy} = m \int V_x V_y f d\mathbf{V} = -\mu \frac{\partial u}{\partial \tilde{x}} \ .$$

Using Eq. (5-50) in this expression and integrating yields:

$$\mu = \tfrac{5}{16} \sigma^{-2} \left(\frac{mkT}{\pi} \right)^{1/2} \ . \tag{5-53}$$

The general expression for the pressure tensor [1] is given by:

$$P_{ik} = p\delta_{ik} - \mu \left[\left(\frac{\partial u_i}{\partial \tilde{x}_k} + \frac{\partial u_k}{\partial \tilde{x}_i} \right) - \tfrac{2}{3} \delta_{ik} \nabla \cdot \mathbf{u} \right] \ . \tag{5-54}$$

9. THE FIRST-ORDER APPROXIMATION FOR ARBITRARY INTERMOLECULAR POTENTIAL.

In the first approximation, the thermal conductivity and viscosity coefficients depend only on the bracket integrals, a_{11} and b_{11}, respectively. These bracket integrals for arbitrary intermolecular potentials may be expressed in the form [1]:

$$a_{11} = \left[\mathbf{C}S_{3/2}^{(1)}\left(C^2\right), \mathbf{C}S_{3/2}^{(1)}\left(C^2\right) \right] = 4\Omega^{(2,2)} , \tag{5-55}$$

and:

$$b_{11} = \left[\overset{\circ}{\mathbf{C}\mathbf{C}}, \overset{\circ}{\mathbf{C}\mathbf{C}} \right] = 4\Omega^{(2,2)} . \tag{5-56}$$

where the b_{11} given here is in the same form as that used by Chapman and Cowling [1].

The $\Omega^{(i,j)}$-integrals can be evaluated only if the form of the intermolecular interaction is specified. All potential functions may be characterized by two parameters, ε and σ, and it is convenient to represent the mutual potential energy of two molecules in the form:

$$\varphi(r) = \varepsilon f(r/\sigma) , \tag{5-57}$$

where $f(r/\sigma)$ is the same function for all gases described by the specific interaction model.

The Lennard-Jones (6-12) potential model has found the widest use in applied transport problems. This model is described by:

$$\varphi(r) = 4\varepsilon\left[\left(\frac{\sigma}{r}\right)^{12} - \left(\frac{\sigma}{r}\right)^{6}\right] , \tag{5-58}$$

where ε is the depth of the potential well and $\sigma = r$ at the point where $\varphi(r) = 0$. This parameter may be called an effective molecular diameter. The two parameters of this model, ε/k (K) and σ, have been tabulated for many gases by a comparison of the theory with experimental data.

All of the Ω-integrals are expressible in the form [3,5]:

$$\Omega^{(i,j)} = \left[\Omega^{(i,j)} \right]_{r.s.} \Omega^{(i,j)\text{å}} \ , \tag{5-59}$$

where $\left[\Omega^{(i,j)} \right]_{r.s.}$ is the integral for rigid-sphere molecules that is given by:

$$\left[\Omega^{(i,j)} \right]_{r.s.} = \pi\sigma^2 \left(\frac{kT}{\pi m} \right)^{1/2} \frac{(j+1)!}{2} \left[1 - \frac{1+(-1)^i}{2(i+1)} \right] . \tag{5-60}$$

The reduced integral, $\Omega^{(i,j)\text{å}}$, depends only on the reduced temperature which is defined as:

$$T^* = \frac{kT}{\varepsilon} . \tag{5-61}$$

The numerical values of the reduced Ω-integrals for various models of the intermolecular potential may be found in [3,5,6].

Taking into account Eq. (5-59), one can express the first approximation to the transport coefficients for a monatomic gas as follows:

$$[\mu]_1 = \tfrac{5}{16} \frac{(\pi mkT)^{1/2}}{\pi\sigma^2 \Omega^{(2,2)\text{å}}\left(T^*\right)} , \tag{5-62}$$

$$[\kappa]_1 = \tfrac{25}{32} \frac{(\pi mkT)^{1/2}}{\pi\sigma^2 \Omega^{(2,2)\text{å}}\left(T^*\right)} c_V , \tag{5-63}$$

where c_V is the specific heat per unit mass at constant volume which is (for a monatomic gas, for example) $c_V = 3k/2m$.

10. THE SECOND-ORDER APPROXIMATION FOR ARBITRARY INTERMOLECULAR POTENTIAL.

The parameters of the second-order Chapman-Enskog solutions may be found from the following algebraic systems [1]:

$$a_1^{(2)} a_{11} + a_2^{(2)} a_{21} = -\tfrac{15}{4} n^{-1} \left(\frac{2kT}{m} \right)^{1/2} ,$$

$$a_1^{(2)} a_{12} + a_2^{(2)} a_{22} = 0 ,$$

and:

$$b_1^{(2)} b_{11} + b_2^{(2)} b_{21} = \tfrac{5}{2} n^{-1} ,$$

$$b_1^{(2)} b_{12} + b_2^{(2)} b_{22} = 0 ,$$

All of the parameters are expressible in terms of the bracket integrals defined in [1]. In the second approximation, the thermal conductivity and viscosity coefficients can be shown to be:

$$[\kappa]_2 = [\kappa]_1 f_\kappa^{(2)}\left(T^*\right) , \tag{5-64}$$

and:

$$[\mu]_2 = [\mu]_1 f_\mu^{(2)}\left(T^*\right) , \tag{5-65}$$

where the functions, $f_\kappa^{(2)}\left(T^*\right)$ and $f_\mu^{(2)}\left(T^*\right)$, are given by:

$$f_\kappa^{(2)}\left(T^*\right) = \frac{77 - 112v + 80t}{28 - 64v^2 + 80t} , \tag{5-66}$$

and:

$$f_\mu^{(2)}\left(T^*\right) = \frac{301 - 336v + 240t}{154 + 48t} . \tag{5-67}$$

Here, the following notations have been introduced:

$$v = \Omega^{(2,3)å}\left(T^*\right) / \Omega^{(2,2)å}\left(T^*\right) , \tag{5-68}$$

Table 5-1. Functions for calculating the second-order transport coefficients.

T^*	$f_\kappa^{(2)}(T^*)$	$f_\mu^{(2)}(T^*)$	$a_2^{(2)*}(T^*)$	$b_2^{(2)*}(T^*)$
0.30	1.00189	1.01030	0.02680	0.01724
0.50	1.00009	1.01020	0.00588	0.00376
0.75	1.00004	1.02181	–0.00369	–0.00237
1.00	1.00000	1.02775	–0.00030	–0.00019
1.25	1.00012	1.03074	0.00680	0.00441
1.50	1.00055	1.03149	0.01419	0.00922
2.00	1.00214	1.02832	0.02801	0.01825
2.50	1.00380	1.02580	0.03719	0.02426
3.00	1.00516	1.02448	0.04325	0.02824
4.00	1.00737	1.02105	0.05158	0.03370
5.00	1.00867	1.01948	0.05590	0.03653
10.0	1.01096	1.01582	0.06279	0.04103
50.0	1.01160	1.01419	0.06465	0.04222
100.0	1.01167	1.01410	0.06483	0.04233
400.0	1.01170	1.01383	0.06492	0.04239

$$t = \Omega^{(2,4)å}\left(T^*\right)\Big/\Omega^{(2,2)å}\left(T^*\right) . \tag{5-69}$$

The second coefficients of the Chapman-Enskog solutions are found to be:

$$a_2^{(2)*} = \frac{32v - 28}{77 - 112v + 80t} , \tag{5-70}$$

and:

$$b_2^{(2)*} = \frac{96v - 84}{301 - 336v + 240t} . \tag{5-71}$$

where $a_2^{(2)*} = a_2^{(2)}\big/a_1^{(2)}$ and $b_2^{(2)*} = b_2^{(2)}\big/b_1^{(2)}$. For the Lennard-Jones (6-12) potential, the functions, $f_\kappa^{(2)}\left(T^*\right)$, $f_\mu^{(2)}\left(T^*\right)$, $a_2^{(2)*}\left(T^*\right)$, and $b_2^{(2)*}\left(T^*\right)$, are given in Table 5.1.

PROBLEMS

5.1. Show that if:

$$f^{(0)} = n(\mathbf{r},t)\left(\frac{m}{2\pi kT(\mathbf{r},t)}\right)^{3/2} \exp\left(-\frac{m\left(\mathbf{v}-\mathbf{u}(\mathbf{r},t)\right)^2}{2kT(\mathbf{r},t)}\right) ,$$

then the collision operator associated with this function is equal to zero.

Solution: Very simple transformations result in the following expression which occurs in the collision operator:

$$f^{(0)} f_1^{(0)} - f^{(0)\prime} f_1^{(0)\prime} = A(\mathbf{r},t) \left\{ \exp\left(-\frac{m\left(v^2 + v_1^2\right)}{2kT} + \frac{m}{kT}(\mathbf{v} + \mathbf{v}_1) \cdot \mathbf{u} \right) \right.$$

$$\left. - \exp\left(-\frac{m\left(v'^2 + v_1'^2\right)}{2kT} + \frac{m}{kT}(\mathbf{v}' + \mathbf{v}_1') \cdot \mathbf{u} \right) \right\} = 0 \ .$$

That this is equal to zero (and hence the collision operator is equal to zero) follows from the conservation of energy and momentum during molecular encounters.

5.2. Obtain the second-order Chapman-Enskog solution for the thermal conduction problem by using the following polynomial expansion:

$$\mathbf{A} = a_1^{(2)} \mathbf{C} \left\{ S_{3/2}^{(1)}\left(C^2\right) + a_2^{(2)*} S_{3/2}^{(2)}\left(C^2\right) \right\} .$$

Assume that the molecules in this problem are rigid spheres.

Solution: Multiplying Eq. (5-29) alternately by the terms, $\mathbf{C} S_{3/2}^{(1)}\left(C^2\right)$ and $\mathbf{C} S_{3/2}^{(2)}\left(C^2\right)$, one can obtain the following algebraic system:

$$a_1^{(2)} a_{11} + a_2^{(2)} a_{21} = -\tfrac{15}{4} n^{-1} \left(\frac{2kT}{m} \right)^{1/2} , \quad a_1^{(2)} a_{12} + a_2^{(2)} a_{22} = 0 ,$$

where $a_{12} = a_{21} = -\frac{1}{4} a_{11}$ and $a_{22} = \frac{45}{16} a_{11}$ for rigid-sphere molecules [1]. The solution of this system yields:

$$a_1^{(2)} = \tfrac{45}{44} a_1^{(1)} , \quad a_2^{(2)} = \tfrac{1}{11} a_1^{(1)} , \quad a_2^{(2)*} = \tfrac{4}{45} .$$

5.3. Derive the expression for the thermal conductivity coefficient for the second-order Chapman-Enskog solution for rigid-sphere molecules.

Solution: From Eqs. (5-37) and (5-51), one can obtain:

$$q_i = -\tfrac{1}{2} mn\pi^{-3/2}\left(\frac{2kT}{m}\right)^{3/2} \frac{1}{T}\frac{\partial T}{\partial \tilde{x}_i} \int C^2 C_i^2 A(C)\exp\left(-C^2\right) d\mathbf{C}$$

$$= \tfrac{1}{6} mn\pi^{-3/2}\left(\frac{2kT}{m}\right)^{3/2} \frac{1}{T}\frac{\partial T}{\partial \tilde{x}_i} \int \left(\tfrac{5}{2} - C^2\right) C^2 A(C)\exp\left(-C^2\right) d\mathbf{C}$$

$$= \tfrac{1}{3} mn\pi^{-1/2}\left(\frac{2kT}{m}\right)^{3/2} \frac{1}{T}\frac{\partial T}{\partial \tilde{x}_i} \int_0^\infty x^{3/2} S_{3/2}^{(1)}(x)\left(\sum_{p=1}^\infty a_p S_{3/2}^{(p)}(x)\right)\exp(-x)\,dx$$

$$= \tfrac{5}{4}\left(\frac{2kT}{m}\right)^{1/2} kna_1 \frac{\partial T}{\partial \tilde{x}_i} = -\kappa \frac{\partial T}{\partial \tilde{x}_i} \ .$$

This implies that the thermal conductivity coefficient is proportional to a_1 only and may be written as:

$$\kappa = -\tfrac{5}{4}\left(\frac{2kT}{m}\right)^{1/2} kna_1 \ .$$

For the second-order Chapman-Enskog solution (Problem 5.2), one can obtain:

$$\kappa^{(2)} = -\tfrac{5}{4}\left(\frac{2kT}{m}\right)^{1/2} kna_1^{(2)} = \tfrac{45}{44}\kappa^{(1)} \ ,$$

where the molecules have been assumed to be rigid spheres.

5.4. Obtain the second-order Chapman-Enskog solution for the viscosity problem by using the following polynomial expansion:

$$B_{xy} = b_1^{(2)} C_x C_y \left\{1 + b_2^{(2)*} S_{5/2}^{(1)}\left(C^2\right)\right\} \ .$$

Assume that the molecules in this problem are rigid spheres.

Solution: Multiplying Eq. (5-30) alternately by the terms, $C_x C_y$ and $C_x C_y S_{5/2}^{(1)}\left(C^2\right)$, one can obtain the following algebraic system:

$$b_1^{(2)} b_{11}' + b_2^{(2)} b_{21}' = (4n)^{-1} \ ,$$

$$b_1^{(2)} b_{12}' + b_2^{(2)} b_{22}' = 0 \ ,$$

where $b'_{12} = b'_{21} = -\frac{1}{4}b'_{11}$ and $b'_{22} = \frac{205}{48}b'_{11}$ for rigid-sphere molecules [1]. The solution of this system yields:

$$b_1^{(2)} = \tfrac{205}{202} b_1^{(1)} \ , \ b_2^{(2)} = \tfrac{6}{101} b_1^{(1)} \ , \ b_2^{(2)*} = \tfrac{12}{205} \ .$$

5.5. Derive the expression for the viscosity coefficient for the second-order Chapman-Enskog solution for rigid-sphere molecules.

Solution: Taking into account the definition of the pressure tensor and Eq. (5-37), one can obtain:

$$\begin{aligned} P_{xy} &= -4\pi^{-3/2} nT \frac{\partial u}{\partial \tilde{x}} \int C_x^2 C_y^2 B(C) \exp\left(-C^2\right) d\mathbf{C} \\ &= -\tfrac{16}{15}\pi^{-1/2} nkT \frac{\partial u}{\partial \tilde{x}} \int_0^\infty C^6 B(C) \exp\left(-C^2\right) dC \\ &= -\tfrac{8}{15}\pi^{-1/2} nkT \frac{\partial u}{\partial \tilde{x}} \int_0^\infty x^{5/2} S_{5/2}^{(0)}(x) \left(\sum_{p=1}^{\infty} b_p S_{5/2}^{(p-1)}(x) \right) \exp(-x)\, dx \\ &= -nkTb_1 \frac{\partial u}{\partial \tilde{x}} \ . \end{aligned}$$

This implies that the viscosity coefficient is proportional to b_1 only and does not contain any other coefficients from the polynomial expansion. Then the viscosity coefficient can be written as $\mu = nkTb_1$. For the second-order Chapman-Enskog solution (Problem 5.4), one can obtain:

$$\mu^{(2)} = kTb_1^{(2)} = \tfrac{205}{202}\mu^{(1)} \ .$$

REFERENCES

1. Chapman, S. and Cowling, T.G., *The Mathematical Theory of Non-Uniform Gases* (Cambridge University Press, Cambridge, U.K., 1990).
2. Uhlenbeck, G.E. and Ford, G.W., *Lectures in Statistical Mechanics* (American Mathematical Society, Providence, RI, 1963).
3. Hirschfelder, J.O., Curtiss, C.F. and Bird, R.B., *Molecular Theory of Gases and Liquids* (John Wiley and Sons, New York, 1954).
4. Kogan, M.N., *Rarefied Gas Dynamics* (Plenum Press, New York, 1969).
5. Ferziger, J.H. and Kaper, H.G., *Mathematical Theory of Transport Processes in Gases* (North-Holland, Amsterdam-London, 1972).
6. Maitland, G.C., Rigby, M.A., Smith, E.B., and Wakeham, W.A., *Intermolecular Forces* (Oxford University Press, Oxford, 1981).

Chapter 6

REGIMES OF RAREFIED GAS FLOWS

1. THE KNUDSEN NUMBER.

Consider the steady-state flow of an infinite stream of a rarefied gas over a body having a characteristic dimension, R, in the absence of external forces. In this case, the Boltzmann equation may be expressed in the form:

$$\mathbf{v}\cdot\frac{\partial f}{\partial \tilde{\mathbf{r}}}=\int\int\int\left(f'f_1'-ff_1\right)gbdbd\varepsilon d\mathbf{v}_1 \ . \tag{6-1}$$

Let this equation be transformed into dimensionless variables by means of the following:

$$\mathbf{v}=\left(\frac{2kT}{m}\right)^{1/2}\mathbf{c}\ ,\ f=n\left(\frac{m}{2kT}\right)^{3/2}f^*\ ,\ \tilde{\mathbf{r}}=R\mathbf{r}\ ,\ b=2^{1/4}\sigma b^*\ ,\ \varepsilon=\pi\varepsilon^*\ .$$

In these variables, Eq. (6-1) becomes:

$$\mathbf{c}\cdot\frac{\partial f^*}{\partial \mathbf{r}}=-\frac{R}{\lambda}J^*\left(f^*f_1^*\right)=-Kn^{-1}J^*\left(f^*f_1^*\right)\ , \tag{6-2}$$

where $Kn=\lambda/R$ is the Knudsen number and $J^*\left(f^*f_1^*\right)$ is the standard dimensionless collision operator. The Knudsen number is the natural parameter characterizing the degree of rarefaction of a gas for a given boundary problem. The Knudsen number is defined as the ratio of the mean free path of a molecule to a characteristic external geometrical dimension of the boundary problem. The Knudsen number is not a property of the gas alone as it explicitly involves the system dimension, R. It is, however, the basic parameter for any given boundary value transport problem.

2. A GENERAL ANALYSIS OF THE DIFFERENT GAS FLOW REGIMES.

The problem of the steady-state flow of a rarefied gas past a body has a characteristic dimensionless parameter. This parameter is the Knudsen number described in Section 6.1. Large or small values of the Knudsen number permit simplification or modification of the Boltzmann equation for each particular case. Corresponding to certain ranges of the Knudsen number, there are four regimes in the Kinetic Theory of Gases. This classification has a schematic character.

For conditions when $Kn \to 0$ ($Kn < 0.01$, the continuum regime), one may use the Navier-Stokes equations with the usual hydrodynamic boundary conditions (Eqs. (2-7)-(2-9) together with Eqs. (5-51) and (5-54)). If $Kn \ll 1$ ($0.01 < Kn < 0.1$, the slip-flow regime), the gas may be described by the Navier-Stokes equations but the boundary conditions should be modified by the introduction of slip conditions on the surface of the body. The slip-flow regime will be described in detail in Chapter 10.

The other regions are those in which the Knudsen number is greater than unity. Two sub-regions can be distinguished here. The limiting regime, where $Kn \to \infty$, is usually called the free-molecular regime. If the dimensions of the body are insignificant compared with the mean free path, one says that the flow past the body is of the free-molecular type. For such flows the presence of the body does not disturb the distribution function of the incident molecules of the gas in the neighborhood of the body. In this case, the molecules that rebound from the surface of the body collide with other molecules at distances on the order of the mean free path from the body. Therefore, in a free-molecular flow, the collisions between the molecules that rebound and those coming from the gas occur at a great enough distance from the body that the influence of these collisions can be neglected. The free-molecular distribution function is constant along the molecular trajectories. On the surface of the body the distribution function of incident molecules is the same as that in the surrounding volume of gas far from the body. The distribution function of molecules rebounded from the body can be found from the boundary conditions which will be considered in Section 6.3.

The region where $Kn \gg 1$ is usually called the near free-molecular regime. For flow in this regime, Knudsen iteration [1] is a well known approximate solution of the Boltzmann equation. The distribution function is assumed to have the form:

$$f = f^{(0)} + Kn^{-1} f^{(1)} \quad . \tag{6-3}$$

The function, $f^{(1)}$, may be found from the non-homogeneous linear partial differential equation that contains the free-molecular distribution function inside the collision integral. The solutions of several specific problems of interest in this regime may be found in [2-9].

The region where $Kn \sim 1$ is the most difficult to describe theoretically. In this region the flow has features typical of both the free-molecular and continuum regimes. In this transition region, it is often convenient to use the various moment methods that will be discussed in Chapter 11 in order to find the solutions to boundary problems of interest.

3. THE BOUNDARY CONDITIONS.

Both the mathematical formulation and the solution of all boundary value transport problems, by both the direct and the moment methods, requires knowledge of the applicable microscopic boundary conditions. These conditions on the surface of a body can be formulated for the distribution function of the reflected molecules by means of a dispersion kernel [10-15]. Within this framework only a general description may be provided at present, which cannot be used to solve practical boundary problems because no theory currently exists from which the necessary kernels can be derived. The absence of a strict theory for the molecular interaction of a gas with a solid surface results in the need to use definite models to postulate the distribution function for the reflected molecules. This function contains a certain number of free parameters which are usually called accommodation coefficients. These coefficients may be determined by a comparison of the theory constructed from a given model with an appropriate experiment.

Maxwell was the first to propose the boundary model that has been widely used in various modified forms. At present, this model still appears to be the most convenient and correct formulation for the various boundary value transport problems; particularly those in which one assumes a non-condensable gas for which the boundary surface is impenetrable. The Maxwellian boundary model is constructed on the assumption that some fraction $(1-\alpha_{\tau})$ of the incident gas molecules are reflected from the surface specularly, while the remaining fraction, α_{τ}, are reflected diffusely with a Maxwellian distribution. This supposition can be expressed as:

$$f^{+}\left(\mathbf{v},\tilde{\mathbf{r}}_S,t\right)=\alpha_\tau f_r+\left(1-\alpha_\tau\right)f^{-}\left(\mathbf{v}',\tilde{\mathbf{r}}_S,t\right) , \tag{6-4}$$

where $f^{\pm}$ are the distribution functions of the molecules reflected and incident on the surface at a point, $\tilde{\mathbf{r}}_S$, $\mathbf{v}'=\mathbf{v}-2(\mathbf{v}\cdot\mathbf{n})\mathbf{n}$, $\mathbf{n}$ is the unit normal vector to the surface at the given point, and f_r is the Maxwellian distribution function given by:

$$f_r=n_r\left(\tilde{\mathbf{r}}_S,t\right)\left(\frac{m}{2\pi kT_r\left(\tilde{\mathbf{r}}_S,t\right)}\right)^{3/2}\exp\left(-\frac{mv^2}{2kT_r\left(\tilde{\mathbf{r}}_S,t\right)}\right) . \tag{6-5}$$

There are two parameters in this distribution function, n_r and T_r, which correspond to the number density and temperature of some fictitious gas emitted by the surface. Also, in this form of the reflected distribution function, the surface element is assumed to be stationary.

Now, consider the physical meaning of the coefficient, α_τ. The tangential momentum transferred to a unit surface element per second by incident molecules is given by:

$$P_\tau^{-}=-m\int_{(\mathbf{v}\cdot\mathbf{n})<0} f^{-}v_\tau\left(\mathbf{v}\cdot\mathbf{n}\right)d\mathbf{v} , \tag{6-6}$$

and that carried away by the reflected molecules may be expressed as:

$$P_\tau^{+}=m\int_{(\mathbf{v}\cdot\mathbf{n})>0} f^{+}v_\tau\left(\mathbf{v}\cdot\mathbf{n}\right)d\mathbf{v}=\left(1-\alpha_\tau\right)P_\tau^{-} . \tag{6-7}$$

From Eqs. (6-6) and (6-7), the following relation can be written:

$$\alpha_\tau P_\tau^{-}=P_\tau^{-}-P_\tau^{+} \; ; \quad 0\le\alpha_\tau\le 1 .$$

Therefore, α_τ gives the fraction of the tangential momentum of incident molecules transmitted to the surface by all molecules. This parameter is usually called the tangential momentum accommodation coefficient.

The unknown parameters, n_r and T_r, in Eq. (6-5) may be found if the impenetrability condition for the surface and the energy accommodation coefficient, α_T, are introduced by means of the relations:

$$\left|N^{-}\right|-N^{+}=0 , \tag{6-8}$$

and:

$$\alpha_T = \frac{|Q^-| - Q^+}{|Q^-| - Q_{eq}^+} \; ; \quad 0 \le \alpha_T \le 1 \; , \tag{6-9}$$

where $N^{\mp}$ and $Q^{\mp}$ are the fluxes of the number of molecules and energy for the incident and reflected molecules across a unit element of the surface at the point $\tilde{\mathbf{r}}_S$ per unit time. The value Q_{eq}^+ is the energy flux which would be carried away by the reflected molecules if the gas were in equilibrium with the surface, i.e., when $T_r = T_w$, where T_w is the temperature of the surface at the point $\tilde{\mathbf{r}}_S$. The energy accommodation coefficient can be interpreted as the fraction of the equilibrium heat flux transmitted to the surface by the gas molecules. The details of the calculational technique associated with this formulation of the boundary problem are given in detail in Problem 6.3.

The accommodation coefficients, α_τ and α_T, may be found via experiment. The classical Millikan experiments [16], for instance, have shown that for most cases $\alpha_\tau \cong 1$. This would seem to indicate that the model of diffuse reflection is generally the most appropriate model to employ in practice.

The boundary conditions needed for a condensable gas are significantly different from those considered above. On a liquid surface, for instance, the evaporation (condensation) coefficient, α_m, should be introduced for the gas (the vapor phase of the liquid comprising the surface) via the relationship:

$$\alpha_m = \frac{|N^-| - N^+}{|N^-| - N_S{}^+} \, , \tag{6-10}$$

where $N_S{}^+$ is the reflected flux of molecules which have the parameters of a saturated vapor at the applicable surface temperature.

This coefficient should be interpreted to mean that a fraction of the incident flux of molecules at the surface, α_m, is reflected from the surface having the parameters of a saturated vapor and that the remaining fraction of the incident molecules, $(1-\alpha_m)$, leaves the surface without having undergone a phase transition. This latter fraction forms what is, in essence, a non-condensable gas for which the parameters may be specified by an impenetrability condition for the surface and an energy accommodation coefficient, α_T, as was done above for the case of a purely non-condensable gas. One must keep in mind, however, that the use of an energy accommodation coefficient in such a scenario makes sense only for the non-condensable part of the gas. The evaporating molecules must have the equilibrium parameters of a saturated vapor.

To be consistent with the definition of the evaporation coefficient, the reflected distribution function can be expressed in the form:

$$
\begin{aligned}
& f^{+}\left(\mathbf{v}, \tilde{\mathbf{r}}_S, t\right) \\
& = \alpha_m n_S\left(T_w\left(\tilde{\mathbf{r}}_S, t\right)\right)\left(\frac{m}{2\pi k T_w\left(\tilde{\mathbf{r}}_S, t\right)}\right)^{3/2} \exp\left(-\frac{m v^2}{2 k T_w\left(\tilde{\mathbf{r}}_S, t\right)}\right) \\
& + \left(1-\alpha_m\right) n_r\left(\tilde{\mathbf{r}}_S, t\right)\left(\frac{m}{2\pi k T_r\left(\tilde{\mathbf{r}}_S, t\right)}\right)^{3/2} \exp\left(-\frac{m v^2}{2 k T_r\left(\tilde{\mathbf{r}}_S, t\right)}\right),
\end{aligned}
\tag{6-11}
$$

where $n_S\left(T_w\left(\tilde{\mathbf{r}}_S, t\right)\right)$ is the saturated vapor density at the given surface temperature $T_w\left(\tilde{\mathbf{r}}_S, t\right)$ which, in the general case, is a function of the specific location on the surface, $\tilde{\mathbf{r}}_S$, and the time, t. The unknown parameters associated with the non-condensable fraction of the reflected molecules may be found using the standard impenetrability and energy accommodation conditions. A detailed description of this technique is given in Problem 6.4.

4. THE BOUNDARY DISPERSION KERNEL.

The Maxwellian model of the boundary conditions is frequently used to analyze boundary value transport problems. This model gives a simple expression for the distribution function of molecules reflected from the surface of the wall. For a non-condensable gas, the diffusely reflected part of this distribution function contains two unknown parameters which may be defined by introducing the energy accommodation coefficient and the impenetrability condition for the surface. Note that the expression for the energy accommodation coefficient contains an additional unknown parameter connected with the number density of the molecules that are reflected and which are in thermal equilibrium with the wall which is assumed to be located at $\tilde{x}=0$ and to have a constant temperature, T_0. Hence, three additional integral conditions must be used to obtain the unknown parameters. For the linearized case, if evaporation (condensation) is absent at the wall, a general form of the boundary condition may be derived in which these additional relations are taken into account by the dispersion kernel, $W\left(\mathbf{c}, \mathbf{c}_1\right)$:

$$\Phi_f^+(\mathbf{c},0) = \int\limits_{-} W(\mathbf{c},\mathbf{c}_1)\Phi_f^-(\mathbf{c}_1,0)d\mathbf{c}_1 \ , \qquad (6\text{-}12)$$

where $f^{\pm}(\mathbf{c},x) = f^{(0)}\left(1+\Phi_f^{\pm}(\mathbf{c},x)\right)$ and $\Phi_f^{\pm}(\mathbf{c},x)$ are the full corrections to the distribution function which, in general, contain the Chapman-Enskog distributions.

For the correction to the reflected distribution function, the Maxwellian boundary condition may be written as:

$$\Phi_f^+(\mathbf{c},0) = (1-\alpha_\tau)\Phi_f^-(\mathbf{c}',0) + \alpha_\tau\left[\nu_r + \left(c^2 - \tfrac{3}{2}\right)\tau_r\right] , \qquad (6\text{-}13)$$

where:

$$\mathbf{c}' = \mathbf{c} - \left(2(\mathbf{c}\cdot\mathbf{n})\mathbf{n}\right)$$

Using Eq. (6-13) and Eqs. (6-8) and (6-9), one can obtain:

$$\nu_r = -\pi^{-1}\int\limits_{-} c_x \exp\left\{-c^2\right\}\left\{2 + \tfrac{1}{2}(1-\alpha_T)\left(2-c^2\right)\right\}\Phi_f^- d\mathbf{c} \ ,$$

$$\tau_r = (1-\alpha_T)\pi^{-1}\int\limits_{-} c_x\left(2-c^2\right)\exp\left\{-c^2\right\}\Phi_f^- d\mathbf{c} \ .$$

Then, Eq. (6-13) may be written as:

$$\begin{aligned}\Phi_f^+(\mathbf{c},0) = (1-\alpha_\tau)\Phi_f^-(\mathbf{c}',0) - \alpha_\tau\pi^{-1}\int\limits_{-} c_{1x}\exp\left\{-c_1^2\right\} \\ \times\left\{2 + (1-\alpha_T)\left(2-c^2\right)\left(2-c_1^2\right)\right\}\Phi_f^-(\mathbf{c}_1,0)d\mathbf{c}_1 \ ,\end{aligned} \qquad (6\text{-}14)$$

Allowing for Eq. (6-12), one can express the dispersion kernel in the form:

$$\begin{aligned}W(\mathbf{c},\mathbf{c}_1) = (1-\alpha_\tau)\delta\left[\mathbf{c}_1 - \left(\mathbf{c} - 2(\mathbf{c}\cdot\mathbf{n})\mathbf{n}\right)\right] \\ -\pi^{-1}\alpha_\tau c_{1x}\exp\left\{-c_1^2\right\}\left\{2 + (1-\alpha_T)\left(2-c^2\right)\left(2-c_1^2\right)\right\} ,\end{aligned} \qquad (6\text{-}15)$$

where $\delta(x)$ is the Dirac delta function [17].

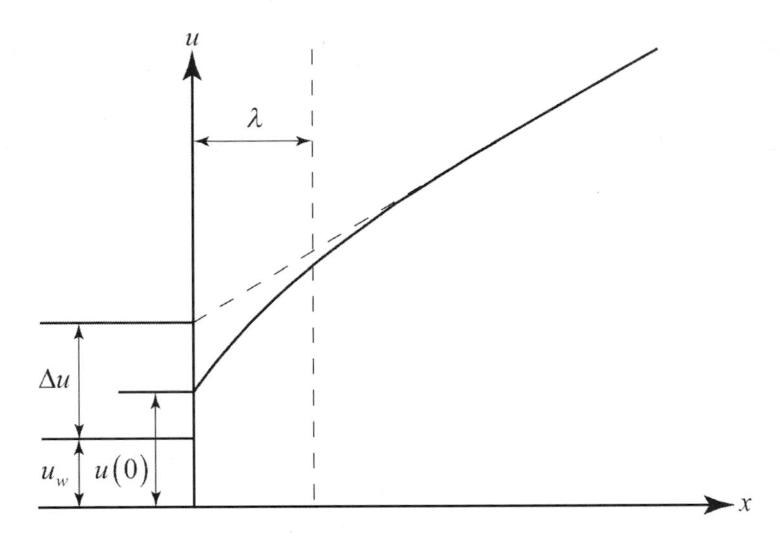

Figure 6-1. The variation of the mean velocity of a gas near a moving wall. Here, u_w is the velocity of the wall, Δu is the slip velocity, and $u(0)$ is the real mean velocity at the wall.

5. FEATURES OF THE BOUNDARY CONDITIONS FOR SMALL KNUDSEN NUMBER.

Consider the gas flow along a plane wall under conditions in which the mean velocity has only a tangential component that is a linear function of the normal coordinate beyond the wall. This linear dependence may be obtained from the solution of both the Boltzmann equation and Navier-Stokes equations. Since the Navier-Stokes distribution function is a solution of the Boltzmann equation only at large distances from the wall, the gas state in the Knudsen layer must be investigated in order to establish the boundary conditions for Navier-Stokes equations [18-26].

Let the solid curve (Fig. 6-1) be the variation of the mean velocity of a gas that has been obtained by the solution of the Boltzmann equation. Then, the linear extrapolation of the asymptotic solution (outside of the Knudsen layer) of this equation gives a fictitious value of mean velocity at the wall, $u_w + \Delta u$.

If one uses this fictitious velocity, $u_w + \Delta u$, on the wall as a boundary condition for the Navier-Stokes equations, then the same velocity profile may be obtained as from the solution of the Boltzmann equation outside of the Knudsen layer. The difference between the fictitious velocity and the actual wall velocity is called the slip velocity. The most reliable expression for the slip velocity has been obtained by Ivchenko, Loyalka and Tompson [27]:

$$\Delta u = c_m \lambda_\mu \left(\frac{\partial u}{\partial \tilde{x}} \right)_\infty , \tag{6-16a}$$

where:

$$c_m = \frac{2-\alpha_\tau}{\alpha_\tau} \tfrac{5}{8} \sqrt{\pi} \left[\frac{(0.6690)+(0.1775)\alpha_\tau}{(0.7549)+(0.09714)\alpha_\tau} \right], \tag{6-16b}$$

is the isothermal-creep coefficient, the molecules have been assumed to be rigid spheres, and $\lambda_\mu = \frac{205}{202}\lambda$ for the second-order Chapman-Enskog solution. In the very important case when $\alpha_\tau = 1$, one obtains:

$$\Delta u = (1.1006)\lambda_\mu \left(\frac{\partial u}{\partial \tilde{x}} \right)_\infty . \tag{6-17}$$

The temperature-jump may be introduced in the same way (Fig. 6-2). The most reliable analytical expression for the temperature-jump has been obtained by Loyalka [18]. Loyalka's formula, if the tangential momentum accommodation coefficient is taken into account, can be expressed as [28]:

$$\Delta T = c_T \lambda_\kappa \left(\frac{\partial T}{\partial \tilde{x}} \right)_\infty , \tag{6-18a}$$

where:

$$c_T = \tfrac{75}{128} \pi \frac{2-\alpha_\tau \alpha_T}{\alpha_\tau \alpha_T} \left[1 + \frac{\alpha_\tau \alpha_T (2-\alpha_\tau)}{(2-\alpha_\tau \alpha_T)} \left(\tfrac{52}{25} \frac{\varepsilon_T}{\pi} - \tfrac{1}{2} \right) \right], \tag{6-18b}$$

is the temperature-jump coefficient, $\lambda_\kappa = \frac{45}{44}\lambda$, and $\varepsilon_T = (0.9378)$ for rigid-sphere molecules if the second-order Chapman-Enskog solution is employed.

Consider a gas having a tangential temperature gradient located over a non-uniform heated plane surface (Fig. 6-3). Under these conditions, the gas has a mean velocity in the direction of the temperature gradient, i.e. the gas

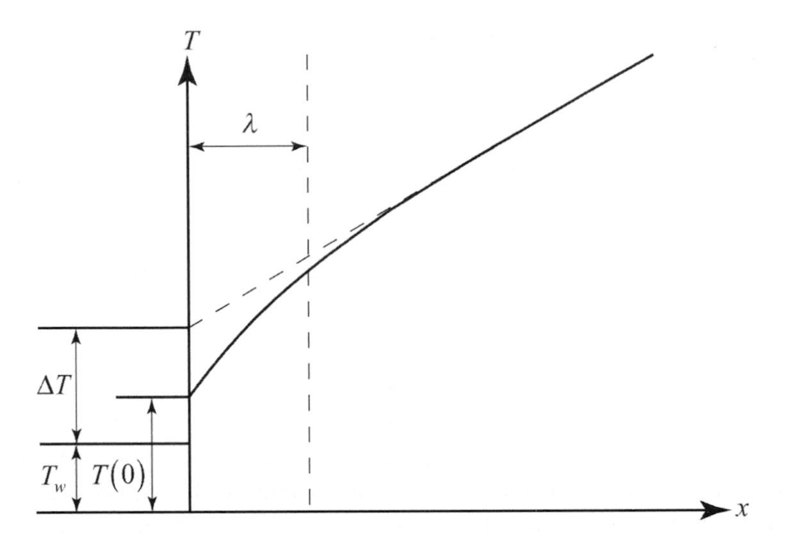

Figure 6-2. The variation of the gas temperature near a wall. Here, T_w is the wall temperature, ΔT is the temperature-jump, and $T(0)$ is the real temperature of the gas at the wall

slips along the wall. The gas velocity beyond the wall is the thermal-creep velocity or the thermal-creep velocity.

The most accurate analytical expression for the thermal-creep velocity has been obtained by Ivchenko, Loyalka and Tompson [28,29]:

$$u_{sl} = c_{Tsl} \nu^* \frac{\partial \ln(T)}{\partial \tilde{y}}, \tag{6-19}$$

where:

$$c_{Tsl} = \tfrac{3}{2} \frac{(0.4354) + (0.2179)\alpha_\tau}{(0.8518) + (0.1096)\alpha_\tau},$$

is the thermal-creep coefficient, $\nu^* = (\lambda_\kappa / \lambda_\mu)\nu$, $\nu = \mu/\rho$ is the kinematic viscosity, and molecules have been assumed to be rigid spheres. For two specific cases of interest, this coefficient may be written as:

$$c_{Tsl} = \begin{cases} 1.0193 & ; \alpha_\tau = 1, \\ 0.7667 & ; \alpha_\tau = 0. \end{cases} \tag{6-20}$$

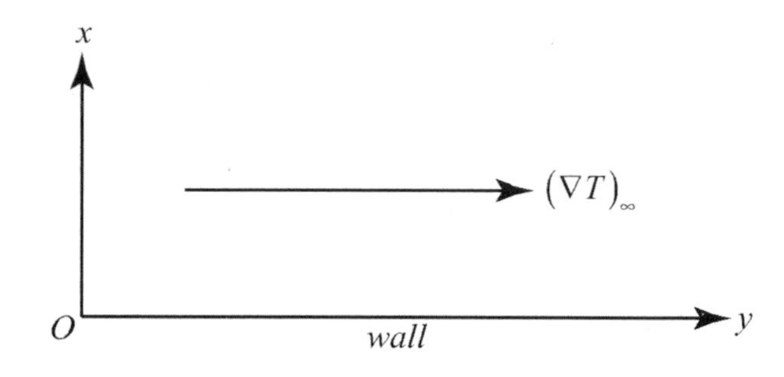

Figure 6-3. The thermal-creep geometry.

The slip boundary conditions may be obtained only from the solution of the Boltzmann equation with microscopic boundary conditions for the distribution function. These problems will be considered in Chapter 8.

PROBLEMS

6.1. Using the Maxwellian boundary model, determine the form of the reflected distribution function if the boundary surface is moving and $\mathbf{u}'$ is the wall velocity.

Solution: The mean velocity of the diffusely reflected molecules is equal to $\mathbf{u}'$. Therefore, the reflected distribution function may be written as:

$$f^+(\mathbf{v},\tilde{\mathbf{r}}_S) = \alpha_\tau n_r(\tilde{\mathbf{r}}_S)\left(\frac{m}{2\pi k T_r(\tilde{\mathbf{r}}_S)}\right)^{3/2} \exp\left(-\frac{m(\mathbf{v}-\mathbf{u}')^2}{2kT_r(\tilde{\mathbf{r}}_S)}\right)$$
$$+(1-\alpha_\tau) f^-\left(-(v_x - u'_x), v_y, v_z, \tilde{\mathbf{r}}_S\right) ,$$

where x is a normal coordinate.

6.2. Determine the form of the dispersion kernel which is specified by:

$$f^+(\mathbf{v},\mathbf{r}_S) = \int_{(\mathbf{v}_1\cdot\mathbf{n})<0} W(\mathbf{v}_1,\mathbf{v}) f^-(\mathbf{v}_1,\tilde{\mathbf{r}}_S) d\mathbf{v}_1 ,$$

for the Maxwellian boundary model.

Solution: The distribution function of reflected molecules is given by Eqs. (6-4) and (6-5). The condition of wall impenetrability gives:

$$n_r(\tilde{\mathbf{r}}_S) = -2\sqrt{\pi}\left(\frac{m}{2kT_r(\tilde{\mathbf{r}}_S)}\right)^{1/2} \int_{(\mathbf{v}_1\cdot\mathbf{n})<0} (\mathbf{v}_1\cdot\mathbf{n}) f^-(\mathbf{v}_1,\tilde{\mathbf{r}}_S)\, d\mathbf{v}_1 \ .$$

Using this expression, one can obtain:

$$W(\mathbf{v}_1,\mathbf{v}) = -\frac{\alpha_\tau}{2\pi}\left(\frac{m}{kT_r(\tilde{\mathbf{r}}_S)}\right)^2 \exp\left(-\frac{m\mathbf{v}^2}{2kT_r(\tilde{\mathbf{r}}_S)}\right)(\mathbf{v}_1\cdot\mathbf{n})$$
$$+(1-\alpha_\tau)\delta\left[\mathbf{v}_1 - \mathbf{v} + 2(\mathbf{v}\cdot\mathbf{n})\mathbf{n}\right] .$$

6.3. Write down the full system of boundary conditions at the solid impenetrable wall if the wall temperature is T_0.

Solution: The number density, n_r, and the temperature, T_r, in Eqs. (6-4) and (6-5) are defined by $N^+ + N^- = 0$ and:

$$\alpha_T = \frac{\left|Q^-\right| - Q^+}{\left|Q^-\right| - Q^+_{eq}} ,$$

where $N^\pm$ and $Q^\pm$ are normal components of the relative fluxes. The flux, Q^+_{eq}, is determined by means of the reflected distribution function in which $n_r = n_{eq}$ and $T_r = T_0$. The unknown equilibrium number density, n_{eq}, may be found from $N^+_{eq} + N^- = 0$.

6.4. At a liquid surface, write down the boundary conditions for reflected molecules of the same liquid vapor if the temperature of the surface is slightly different from the equilibrium temperature T_0, $\left(T_w = T_0\left(1+\tau_w\right)\right)$.

Solution: The reflected distribution function may be presented in the form:

$$f^+(\mathbf{v},0) = \alpha_m n_S(T_w)\left(\frac{m}{2\pi kT_w}\right)^{3/2} \exp\left(-\frac{m\mathbf{v}^2}{2kT_w}\right)$$
$$+(1-\alpha_m)n_r\left(\frac{m}{2\pi kT_r}\right)^{3/2} \exp\left(-\frac{m\mathbf{v}^2}{2kT_r}\right) ,$$

where the first term describes molecules evaporated from the liquid surface while the second term is related to molecules reflected from the surface without a phase transition. It should be noted that, for this latter group of molecules, it is appropriate to treat the liquid surface as if it were impenetrable and to employ an appropriate surface condition. The unknown parameters n_r and T_r, may be determined from the relations $(1-\alpha_m)N^- + N_r^+ = 0$ and:

$$\alpha_T = \frac{(1-\alpha_m)Q^- + Q_r^+}{(1-\alpha_m)Q^- + Q_{req}^+} ,$$

where Q_{req}^+ contains an additional unknown parameter related to the density of a fictitious gas reaching equilibrium with the surface if $T_r = T_w$. This parameter may be found from $(1-\alpha_m)N^- + N_{req}^+ = 0$. The fluxes, N_r^+, N_{req}^+, Q_r^+, and Q_{req}^+, are calculated by using the second term of the reflected distribution function. The number density of the saturated vapor at the temperature, T_w, may be found by using the Clausius-Clapeyron equation [30] which yields $n_S(T_w) = n_S(T_0)\left[1 + (\zeta - 1)\tau_w\right]$ where $\zeta = h_{fg}/kT_0$ and h_{fg} is the latent heat of condensation for a vapor molecule.

6.5. Generalize Eq. (6-14) to obtain an expression for the distribution function of the rebounding molecules for a case when the equilibrium temperature of the wall is given by $T_w = T_0 + \Delta T$ where $\Delta T \ll T_0$.

Solution: The boundary condition for a correction to the distribution function may be written as:

$$\Phi_f^+(\mathbf{c},0) = (1-\alpha_\tau)\Phi_f^-(-c_x, c_y, c_z, 0) + \alpha_\tau\left[\nu_r + \left(c^2 - \tfrac{3}{2}\right)\tau_r\right] ,$$

where the two boundary quantities, ν_r and τ_r, may be found by introducing the energy accommodation coefficient, α_T, together with the condition of impenetrability of the wall for molecules of the gas. This yields:

$$\Phi_f^+(\mathbf{c},0) = (1-\alpha_\tau)\Phi_f^-(-c_x, c_y, c_z, 0)$$
$$+\alpha_\tau\left\{(2-c^2)\left[-\alpha_T\tau_w + (1-\alpha_T)\left(\frac{\delta_2}{\pi} - \frac{2\delta_1}{\pi}\right)\right] - \frac{2\delta_1}{\pi}\right\} ,$$

where:

$$\tau_w = \frac{\Delta T}{T_0} ,$$

$$\delta_1 = \int_- \Phi_f^-(\mathbf{c},0)c_x \exp(-c^2)d\mathbf{c} ,$$

$$\delta_2 = \int_- \Phi_f^-(\mathbf{c},0)c_x c^2 \exp(-c^2)d\mathbf{c} .$$

6.6. Generalize Eq. (6-14) to obtain an expression for the distribution function of the rebounding molecules for a case when the evaporation/condensation processes at the wall are taken into account and the equilibrium temperature of the surface is given by $T_w = T_0 + \Delta T$ where $\Delta T \ll T_0$ and T_0 is the equilibrium vapor temperature.

Solution:

$$\Phi_f^+(\mathbf{c},0) = \alpha_m\left[\nu_w + \left(c^2 - \tfrac{3}{2}\right)\tau_w\right] + \left(1-\alpha_m\right)\left[\nu_r + \left(c^2 - \tfrac{3}{2}\right)\tau_r\right],$$

where ν_w is the correction to the number density of the saturated vapor at the temperature, $T_w = T_0\left(1+\tau_w\right)$, and ν_r and τ_r, may be found by introducing the energy accommodation coefficient, α_T, and using the impenetrability condition for molecules reflected without a phase transition. This yields:

$$\Phi_f^+(\mathbf{c},0) = \alpha_m\left[\nu_w + \left(c^2 - \tfrac{3}{2}\right)\tau_w\right]$$

$$+\left(1-\alpha_m\right)\left\{\left(2-c^2\right)\left[-\alpha_T\tau_w + \left(1-\alpha_T\right)\left(\frac{\delta_2}{\pi} - \frac{2\delta_1}{\pi}\right)\right] - \frac{2\delta_1}{\pi}\right\},$$

where:

$$\tau_w = \frac{\Delta T}{T_0},$$

$$\delta_1 = \int_{-} \Phi_f^-(\mathbf{c},0)\, c_x \exp\left(-c^2\right) d\mathbf{c},$$

$$\delta_2 = \int_{-} \Phi_f^-(\mathbf{c},0)\, c_x c^2 \exp\left(-c^2\right) d\mathbf{c}.$$

The correction, ν_w, may be found by using the Clausius-Clapeyron equation [30] which yields $n_S\left(T_w\right) = n_S\left(T_0\right)\left[1 + \left(\zeta - 1\right)\tau_w\right]$ where $\zeta = h_{fg}/kT_0$ and h_{fg} is the latent heat of condensation (evaporation) for a vapor molecule.

REFERENCES

1. Keller, J.B., "On the Solution of the Boltzmann Equation for Rarefied Gases," *Comm. Pure App. Math.* **1(3)**, 275-284 (1948).
2. Rose, M.H., "Drag on an Object in Nearly-Free-molecular Flow," *Phys. Fluids* **7(8)**, 1262-1269 (1964).

3. Liu, V.C., Pang, S.C., and Jew, H., "Sphere Drag in Flows of Almost Free Molecules," *Phys. Fluids* **8(5)**, 788-796 (1964).
4. Brock, J.R., "The Thermal Force in the Transition Region," *J. Colloid Interface Sci.* **23(3)**, 448-452 (1967).
5. Brock, J.R., "Highly Non-equilibrium Evaporation of Moving Particles in the Transition Region of Knudsen Number," *J. Colloid Interface Sci.* **24**, 344-351 (1967).
6. Brock, J.R., "The Diffusion Force in the Transition Region of Knudsen Number," *J. Colloid Interface Sci.* **27(1)**, 95-100 (1968).
7. Ivchenko, I.N. and Yalamov, Yu.I., "On the Thermophoresis of Aerosol Particles in the Almost-Free-molecular Regime," *Izv. AN SSSR, M. Zh. G.* **(3)**, 3 (1970).
8. Kelly, G.E. and Sengers, J.V., "Droplet Growth in a Dilute Vapor," *J. Chem. Phys.* **61(7)**, 2800-2807 (1974).
9. Barrett, J.C. and Shizgal, B., "Condensation and Evaporation of a Spherical Droplet in the Near Free Molecule Regime," in *Rarefied Gas Dynamics: Physical Phenomena,* Progress in Astronautics and Aeronautics Series, V. 117, edited by Muntz, E.P., Weaver, D.P., and Campbell, P.H. (AIAA, Washington, D.C., 1989) pp. 447-459.
10. Gross, E.P., Jackson, E.A., and Ziering, S., "Boundary Value Problems in Kinetic Theory of Gases," *Annals of Physics* **1(2)**, 141-167 (1957).
11. Cercignani, C., *Theory and Application of the Boltzmann Equation* (Scottish Academic Press, Edinburgh, U.K., 1975).
12. Kogan, M.N., *Rarefied Gas Dynamics* (Plenum Press, NY, 1969).
13. Barantsev, R.G., *Rarefied Gas Interactions with Streamlined Surfaces* (Nauka, Moscow, 1975).
14. Goodman, F.O. and Wachman, H.Y., *Dynamics of Gas-Surface Scattering* (Academic Press, New York, 1976).
15. Kuščer, I., "Phenomenology of Gas-Surface Accommodation," in *Proceedings of the Ninth International Symposium on Rarefied Gas Dynamics*, edited by Becker, M. and Fiebig, M. (DFVLR Press, Pozz-Wahn, 1974). Vol. 2, p. E.1-1.
16. Millikan, R.A., "Coefficients of Slip in Gases and the Law of Reflection from the Surfaces of Solids and Liquids," *Phys. Rev.* **21(1)**, 217-238 (1923).
17. Dirac, P.A.M., *The Principles of Quantum Mechanics*, 4-th edition revised (Clarendon Press, Oxford, reprinted 1993).
18. Ivchenko, I.N. and Yalamov, Yu.I., "The Hydrodynamic Method of Calculation of the Thermophoresis of Aerosol Particles," *J. Phys. Chem. (Russia)* **45(3)**, 577-582 (1971).
19. Loyalka, S.K., "Momentum and Temperature-Slip Coefficients with Arbitrary Accommodation at the Surface," *J. Chem. Phys.* **48(12)**, 5432-5436 (1968).
20. Hidy, G.M. and Brock, J.R., *Topics in Current Aerosol Research, part 2* (Pergamon Press, 1972).
21. Derjaguin, B.V., Ivchenko, I.N., and Yalamov, Yu.I., "About Construction of Solutions of the Boltzmann Equation in the Knudsen Layer," *Izv. AN SSSR, M. Zh. G.* **(4)**, 167-172 (1968).
22. Derjaguin, B.V. and Yalamov, Yu.I., "The Theory of Thermophoresis and Diffusiophoresis of Aerosol Particles and Their Experimental Testing," in *Topics in Current Aerosol Research, part 2*, edited by Hidy, G.M. and Brock, J.R. (Pergamon Press, 1972).
23. Loyalka, S.K. and Lang, H., "On Variational Principles in the Kinetic Theory," in *Proceedings of the Seventh International Symposium on Rarefied Gas Dynamics* (Editrice Tecnico Scientifica, Pisa, Italy, 1970).

24. Loyalka, S.K., "Slip and Jump Coefficients for Rarefied Gas Flows: Variational Results for Lennard-Jones and n(r)-6 Potentials," *Physica* **A163**, 813-821 (1990).
25. Rolduguin, V.I., *Application of the Non-Equilibrium Thermodynamics Method in Boundary Problems of the Kinetic Theory of Gases* (M.S. Thesis, Moscow, 1979).
26. Savkov, S.V., *The Slip-Flow Boundary Conditions for the Non-Uniform Binary Gas Mixture and an Application of Them in Aerosol Dynamics* (M.S. Thesis, Moscow, 1987).
27. Ivchenko, I.N., Loyalka, S.K. and Tompson, R.V., "The Precision of Boundary Models in the Gas Slip Problem," *High Temp.* **31(1)**, 127-129 (1993).
28. Ivchenko, I.N., Loyalka, S.K. and Tompson, R.V., "On the Use of Conservation Laws in Plane Slip Problems," *Teplofizika Vysokikh Temperatur* (in Russian), **33(1)**, 66-72 (1995).
29. Ivchenko, I.N., Loyalka, S.K. and Tompson, R.V., "On One Boundary Model for the Thermal Creep Problem," *Fluid Dynamics* **28(6)**, 876-888 (1993).
30. Landau, L.D. and Lifshitz, E.M., *Statistical Physics* (Pergamon, London, Addison Wesley, Reading, Mass., 1958).

Chapter 7

THE FREE-MOLECULAR REGIME

1. THE FREE-MOLECULAR DISTRIBUTION FUNCTION.

The theoretical treatment of gas flow problems for extremely high Knudsen numbers (the free-molecular regime) is very different from that typically encountered in conjunction with the hydrodynamics of media having the higher number densities most people are used to working with. The theoretical frameworks encountered at these extremes are connected via various fundamental relationships between the Kinetic Theory of Gases and the mechanics of media that exhibit continuum type behaviors (the continuum regime). The continuum equations may be obtained from the Boltzmann equation if one uses the summational invariants of encounters as molecular properties with which to construct a system of moment equations. This moment system, however, is indeterminate unless one knows the relationships between the basic hydrodynamic values and the stress tensor or the thermal flux vector. These relationships may be derived by using the Chapman-Enskog method of solution of the Boltzmann equation. It is very important in the continuum regime that such gas characteristics as the viscosity and thermal conductivity coefficients be determined independently of the boundary conditions.

The boundary problems for high Knudsen numbers are quite different. The kinetic analysis may only be used for the description of gas flows. The general solution of the Boltzmann equation requires usage of appropriate boundary conditions for the distribution function itself which ultimately results in a dependence of the hydrodynamic functions on the boundary parameters of the specific problem under consideration. Hence, in this regime there are no general relationships between the various macroscopic quantities.

The free-molecular regime will be examined in this chapter. A general characteristic of gas flows in this regime is that the presence of a body does not effectively disturb the distribution function of the impinging molecules since the collisions between molecules being reflected and those coming from infinity occur at distances on the order of λ from the surface of the body. Since $\lambda \gg R$, where R is the typical dimension of the body, the influence of these collisions can be neglected. Owing to this simplification, the theory describing free-molecular flows has been sufficiently developed that several useful solutions to applied transport problems have been obtained [1-6].

A rigorous theory of transport problems must start with the Boltzmann equation which, for stationary conditions, when $Kn \to \infty$, may be written as:

$$\mathbf{v} \cdot \frac{\partial f}{\partial \tilde{\mathbf{r}}} = 0 \ ,$$

or:

$$v_x \frac{\partial f}{\partial \tilde{x}} + v_y \frac{\partial f}{\partial \tilde{y}} + v_z \frac{\partial f}{\partial \tilde{z}} = 0 \ . \qquad (7\text{-}1)$$

Eq. (7-1) is a linear partial differential equation of the first-order. The auxiliary system of ordinary differential equations, usually called the characteristic or Lagrange system, is:

$$\frac{d\tilde{x}}{v_x} = \frac{d\tilde{y}}{v_y} = \frac{d\tilde{z}}{v_z} = \frac{df}{0} = dt \ , \qquad (7\text{-}2)$$

where t is an auxiliary parameter used to ease the solution of this system of equations. The characteristics of these equations are the straight lines specified by:

$$\tilde{\mathbf{r}} = \tilde{\mathbf{r}}_0 + \mathbf{v}\left(t - t_0\right) \ , \qquad (7\text{-}3)$$

where t denotes the value of the parameter on the characteristic (i.e. t is not time). These characteristics may be interpreted to be the molecular trajectories associated with this problem [7]. Eq. (7-2) means that:

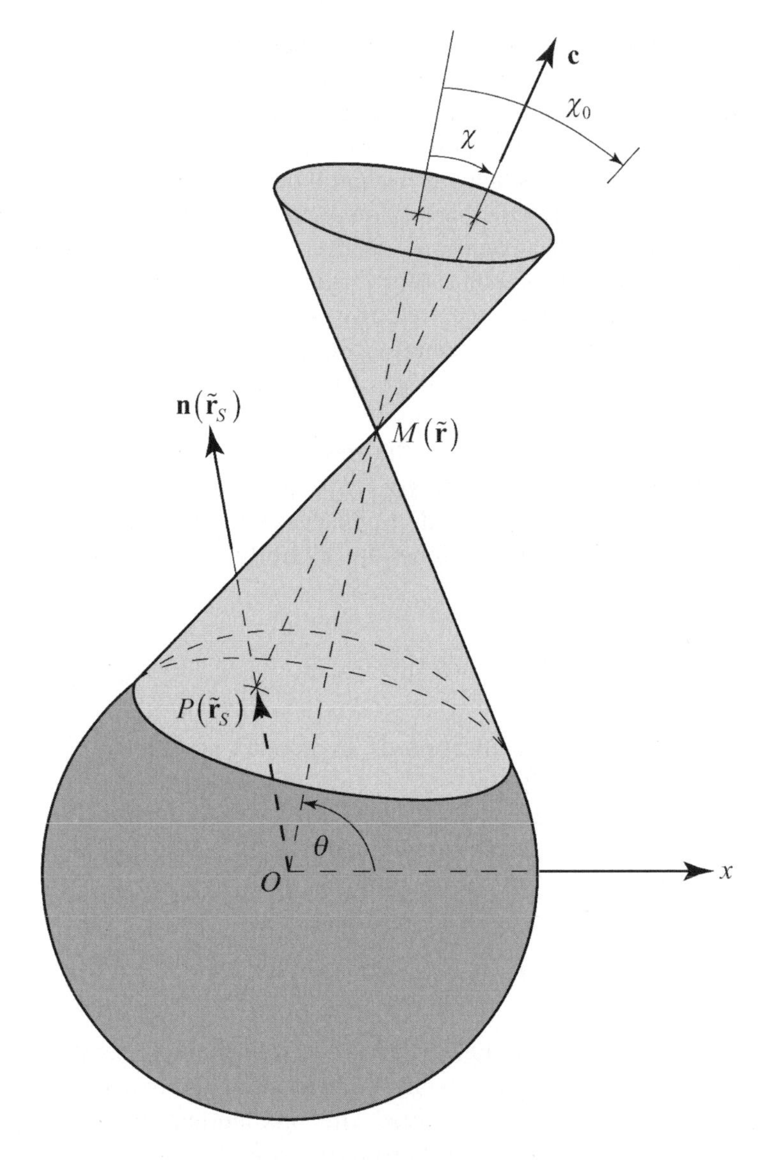

Figure 7-1. The cone of influence for a sphere.

$$\frac{df}{dt} = 0 \ . \tag{7-4}$$

which implies that f is constant along the molecular trajectories. For the molecular trajectories that start at infinity and at the surface of the body the distribution functions are, therefore, equal to the values of the distribution functions at infinity in the undisturbed flow and at the point, $P(\tilde{\mathbf{r}}_S)$, on the surface. These functions are not the same and, hence, the free-molecular distribution function is discontinuous in velocity space.

Consider the form of the distribution function at the point, $M(\tilde{\mathbf{r}})$, in various velocity domains when a spherical model of the particle is assumed. At each spatial point, $M(\tilde{\mathbf{r}})$, there exists a cone of influence, the surface of which is formed by the extensions of the tangents to the surface of the particle passing through the point $M(\tilde{\mathbf{r}})$ (see Fig. 7-1).

This conical surface in the velocity space is described by $\chi = \chi_0$, $\chi_0 = \arcsin(R/\tilde{r}) = \arcsin(1/r)$ where R is the radius of the particle and χ is the angle between vectors $\mathbf{c}$ and $\mathbf{r}$. In the velocity space, the surface of the cone of influence is the surface of discontinuity for the distribution function. This surface divides the molecules into two groups with distribution functions f_1 and f_2 (f_2 describes the molecules for which $v_r > v_r^*$ where $v_r^* = v\cos(\chi_0)$, and f_1 describes the remaining molecules). The free-molecular distribution function in the full velocity space may be expressed as [8]:

$$f = \tfrac{1}{2}(f_1 + f_2) + \tfrac{1}{2}(f_2 - f_1)\,\mathrm{sign}(v_r - v_r^*) \ . \tag{7-5}$$

This relationship is a generalization of expression previously proposed for the planar geometry [9,10].

On the particle surface, $v_r^* = 0$, and the distribution functions of the reflected and incident molecules may be written as:

$$f(\mathbf{v}, R\mathbf{n}) = \begin{cases} f^+(\mathbf{v}, R\mathbf{n}) = f_2 & ; \quad v_r > 0 \ , \\ f^-(\mathbf{v}, R\mathbf{n}) = f_1 & ; \quad v_r < 0 \ , \end{cases}$$

where $\mathbf{n}$ is a unit vector normal to the surface element and pointing into the gas. These distribution functions are connected by the boundary condition given in Eq. (6-4). Having derived the distribution function one may calculate the flux of any molecular property across the particle surface. For example, the force, $\mathbf{F}$, on the particle is given by integrating the net momentum transferred to the particle by the gas molecules over the particle surface, S, per unit time:

$$\mathbf{F} = -\int dS \sum_{\pm} \int_{\pm} m\mathbf{v}(\mathbf{v}\cdot\mathbf{n}) f^{\pm}(\mathbf{v}, \tilde{\mathbf{r}}_S)\, d\mathbf{v} \ , \tag{7-6}$$

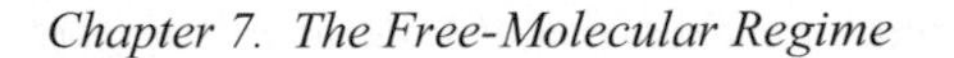

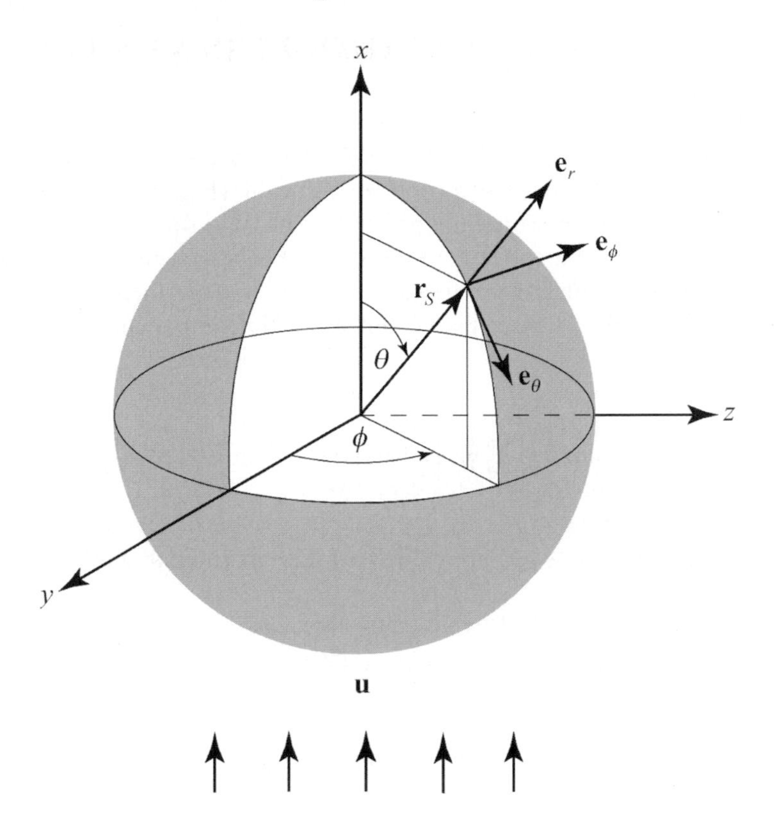

Figure 7-2. The geometry of the sphere drag problem.

where dS is the area of the surface element, f^- and f^+ indicate the distribution functions at the particle surface for both incident $(\mathbf{v}\cdot\mathbf{n}<0)$ and reflected $(\mathbf{v}\cdot\mathbf{n}>0)$ molecules, respectively, and $\tilde{\mathbf{r}}_S$ is the position vector of an arbitrary point on the surface.

For evaporation (condensation) problems one should know the flux of the number of molecules across the particle surface, S, per unit time. This flux may be given by:

$$N = \int dS \sum_{\pm} \int_{\pm} (\mathbf{v}\cdot\mathbf{n}) f^{\pm}(\mathbf{v},\tilde{\mathbf{r}}_S)\, d\mathbf{v} \ , \qquad (7\text{-}7)$$

where N is positive or negative depending on whether the process is connected with evaporating or condensing particles.

2. THE FORCE ON A PARTICLE IN A UNIFORM GAS FLOW.

Consider a small spherical particle $(R \ll \lambda)$ in a uniform gas flow for which the velocity at infinity is a constant value, $\mathbf{u}$ (Fig. 7-2). A coordinate system is introduced in which the polar axis has the same direction as the vector, $\mathbf{u}$. Then, $\mathbf{u} = u_r \mathbf{e}_r + u_\theta \mathbf{e}_\theta$ where $u_r = u\cos(\theta)$ and $u_\theta = -u\sin(\theta)$. From the symmetry of this problem, it is clear that the force has only one component in the direction of the vector $\mathbf{u}$. This component may be given by:

$$F_u = -2\pi R^2 m \int_0^\pi \sin(\theta) \left\{ \sum_\pm \int_\pm v_r \left(v_r \cos(\theta) - v_\theta \sin(\theta) \right) f^\pm d\mathbf{v} \right\} d\theta \ , \quad (7\text{-}8)$$

where the integration over all values of the azimuthal angle has been performed.

Suppose that $|\mathbf{u}| << (2kT/m)^{1/2}$. In this case of low velocity of the undisturbed flow, the following approximation for the distribution function of the incident molecules may be used:

$$f^- = f^{(0)}\left(1 + 2\mathbf{u}^* \cdot \mathbf{c}\right) = f^{(0)}\left\{1 + 2u^*\left[c_r \cos(\theta) - c_\theta \sin(\theta)\right]\right\} \ , \quad (7\text{-}9)$$

where:

$$\mathbf{u}^* = \left(\frac{m}{2kT}\right)^{1/2} \mathbf{u} \ , \ \mathbf{c} = \left(\frac{m}{2kT}\right)^{1/2} \mathbf{v} \ , \ f^{(0)} = n_0 \left(\frac{m}{2\pi kT}\right)^{3/2} \exp\left(-c^2\right) .$$

For the distribution function of the reflected molecules, one can use the Maxwellian model of the boundary conditions given by Eq. (6-4) in which the density of molecules reflected diffusely may be written as:

$$n_r(\mathbf{r}_S) = n_0\left(1 + \nu(\mathbf{r}_S)\right) \ ,$$

where $\nu(\mathbf{r}_S)$ is proportional to the small parameter u^* which is a linearization parameter. The linearized boundary condition may be expressed in the form:

$$f^{+}(\mathbf{v},\mathbf{r}_S)=(1-\alpha_\tau)f^{-}(-c_r,c_\theta,c_\phi,\mathbf{r}_S)$$
$$+\alpha_\tau n_0\left(\frac{m}{2\pi kT}\right)^{3/2}\exp(-c^2)(1+\nu(\mathbf{r}_S)) \ . \tag{7-10}$$

The unknown function $\nu(\mathbf{r}_S)$ may be found from the condition that there is no evaporation or condensation of molecules at the surface element surrounding the point, $\mathbf{r}_S$ (i.e. the particle surface is impenetrable for the gas molecules). This condition may be written as:

$$\mathbf{n}\cdot\int_{-}\mathbf{c}f^{-}d\mathbf{v}+\mathbf{n}\cdot\int_{+}\mathbf{c}f^{+}d\mathbf{v}=0 \ . \tag{7-11}$$

Having performed the integration in Eq. (7-11), one obtains:

$$\nu(\mathbf{r}_S)=-\sqrt{\pi}\left(\mathbf{u}^{*}\cdot\mathbf{n}\right) \ . \tag{7-12}$$

Substituting Eqs. (7-9), (7-10), and (7-12) into Eq. (7-8), the following expression may be written:

$$F_u=2\pi R^2\frac{\rho}{\pi^{3/2}}\left(\frac{2kT}{m}\right)u^{*}\int_0^{\pi}\sin(\theta)$$
$$\times\Big\{2(1-\alpha_\tau)\int_{+}c_r\left[c_r^2\cos^2(\theta)-c_\theta^2\sin^2(\theta)\right]\exp(-c^2)d\mathbf{c}$$
$$+\alpha_\tau\sqrt{\pi}\int_{+}c_r\left[c_r\cos^2(\theta)-c_\theta\sin(\theta)\cos(\theta)\right]\exp(-c^2)d\mathbf{c}$$
$$+2\int_{+}c_r\left[c_r\cos(\theta)+c_\theta\sin(\theta)\right]^2\exp(-c^2)d\mathbf{c}\Big\}d\theta \ .$$

After some straightforward integrations, one obtains:

$$\mathbf{F}=\tfrac{8}{3}R^2n_0(2\pi mkT)^{1/2}\left(1+\tfrac{1}{8}\alpha_\tau\pi\right)\mathbf{u} \ . \tag{7-13}$$

The same expression for the free-molecular force has been obtained by Waldmann [11].

3. CALCULATION OF MACROSCOPIC VALUES IN THE FREE-MOLECULAR REGIME.

Each macroscopic quantity may be expressed as the mean value of a molecular property, $\phi(\mathbf{v},\tilde{\mathbf{r}})$, by the relation:

$$n\bar{\phi} = \int \phi f d\mathbf{v} \ . \tag{7-14}$$

For the free-molecular regime, this integral contains the discontinuous distribution function given by Eq. (7-5). Substituting Eq. (7-5) into Eq. (7-14), one obtains:

$$n\bar{\phi} = \int\limits_{(1+2)} \phi f_1 d\mathbf{v} + \int\limits_{(2)} \phi\left(f_2 - f_1\right) d\mathbf{v} \ . \tag{7-15}$$

where, in the first integral, the integration extends over all values of the molecular velocity while, in the second integral, the integration extends only over that part of the velocity range for which $v_r > v_r^*$ (region 2 in the velocity space). For spherical geometries it is convenient to use spherical coordinates in the velocity space to perform the integration over region 2. These coordinates are given by:

$$c_r = c\cos(\chi) \ , \quad c_\theta = c\sin(\chi)\sin(\beta) \ , \quad c_\phi = c\sin(\chi)\cos(\beta) \ .$$

The integration over region 2 may be represented as:

$$\int\limits_{(2)} d\mathbf{c} = \int\limits_0^\infty c^2 dc \int\limits_0^{2\pi} d\beta \int\limits_0^w \sin(\chi) d\chi \ , \tag{7-16}$$

where $w = \arcsin(1/r)$ and $\mathbf{r} = \tilde{\mathbf{r}}/R$.

The main subtleties here are connected with the integration of the diffusely and specularly reflected parts of the distribution function. These parts always contain the scalar product of $\mathbf{n}_S = \mathbf{n}(\tilde{\mathbf{r}}_S)$ and the typical flow vector, $\mathbf{u}$. To complete the integration over region 2 one must find the connection between the unit vector $\mathbf{n}_S$ and the molecular velocity vector represented by the characteristic **PM** (see Fig. 7-1). The vector $\mathbf{n}_S$ can be expressed as:

$$\mathbf{n}_S = \frac{\tilde{\mathbf{r}}}{R} - \frac{\mathbf{c}}{c}\frac{|\mathbf{PM}|}{R} \ .$$

The value PM may be found from the triangle OPM (see Fig. 7-1) by solution of the quadratic equation:

$$PM^2 - \left(2\tilde{r}\cos(\chi)\right)PM + \left(\tilde{r}^2 - R^2\right) = 0 \ ,$$

that may be written as:

$$PM = \tilde{r}\cos(\chi) \pm R\sqrt{1 - \frac{\tilde{r}^2}{R^2}\sin^2(\chi)} \ .$$

In this formula, only the minus sign may be used in order that, as $\tilde{r} \to \infty$ and $\chi \to 0$, then $PM = \tilde{r} - R$. Then, the expression for $\mathbf{n}_S = \mathbf{n}_S(\tilde{\mathbf{r}}, \mathbf{c})$ may be written as:

$$\mathbf{n}_S(\tilde{\mathbf{r}}, \mathbf{c}) = \frac{\tilde{\mathbf{r}}}{R} - \frac{\mathbf{c}}{c}\left[\frac{\tilde{r}}{R}\cos(\chi) - \sqrt{1 - \frac{\tilde{r}^2}{R^2}\sin^2(\chi)}\right] \ . \tag{7-17}$$

Now, as an example, the integration required for the problem of free-molecular flow past a sphere that was discussed in Section 7.2 will be considered in order to calculate the number density of the flow. The distribution function may be written as:

$$f = \begin{cases} f_1 = f^{(0)}\left(1 + 2c_r u_r^* + 2c_\theta u_\theta^*\right) \ ; & c_r < c_r^* \ , \\ f_2 = \left(1 - \alpha_\tau\right) f^{(0)}\left(1 + 2\mathbf{c}' \cdot \mathbf{u}^*\right) & \\ \qquad + \alpha_\tau f^{(0)}\left[1 - \sqrt{\pi}\left(\mathbf{n}_S \cdot \mathbf{u}^*\right)\right] \ ; & c_r > c_r^* \ , \end{cases} \tag{7-18}$$

where $\mathbf{c}' = \mathbf{c} - 2\mathbf{n}_S(\mathbf{n}_S \cdot \mathbf{c})$ and for integration of the reflected distribution function one should use Eq. (7-17) for $\mathbf{n}_S$. To simplify what would otherwise be a very complicated set of integrations the conclusion of this calculation will be shown for the particular case when molecules are assumed to undergo only diffuse reflection at the surface ($\alpha_\tau = 1$). Then, the number density is expressed by:

$$n = \int_{(1+2)} f_1 d\mathbf{v} + \int_{(2)} (f_2 - f_1) d\mathbf{v} \ , \tag{7-19}$$

where the integrand of the second term has the form:

$$f_2 - f_1 = f^{(0)} \left[-\sqrt{\pi} \left(\mathbf{n}_S (\mathbf{r}, \mathbf{c}) \cdot \mathbf{u}^* \right) - 2c_r u_r^* - 2c_\theta u_\theta^* \right] .$$

After integration of the first term in Eq. (7-19) one obtains:

$$i_1 = \int_{(1+2)} f_1 d\mathbf{v} = n_0 \ .$$

The second integral may be written as:

$$i_2 = \frac{n_0}{\pi^{3/2}} \int_0^\infty c^2 \exp\left(-c^2\right) dc \int_0^{2\pi} d\beta \int_0^w \sin(\chi)$$
$$\left\{ -\sqrt{\pi} u_r^* \left[r \sin^2(\chi) + \cos(\chi) \sqrt{1 - r^2 \sin^2(\chi)} \right] \right. \tag{7-20}$$
$$\left. -2u_r^* c \cos(\chi) \right\} d\chi \ .$$

Basic integration gives:

$$i_2 = -n_0 u_r^* F(r) \ , \tag{7-21}$$

where:

$$F(r) = \tfrac{1}{2} \sqrt{\pi} \left\{ r(1-a) - \tfrac{1}{3} r \left(1 - a^3\right) + r^{-2} \left[\tfrac{1}{3} + 2\pi^{-1} \right] \right\} ,$$

and $a = \sqrt{1 - r^{-2}}$. The vanishing integrals have been eliminated from Eq. (7-20). Now, the number density may be expressed in the form:

$$n(r) = n_0 \left[1 - u^* F(r) \cos(\theta) \right] . \tag{7-22}$$

For the limiting case when $r \to \infty$ the number density $n(r) \to n_0$ which is equivalent to an undisturbed flow.

4. THERMOPHORESIS OF PARTICLES IN THE FREE-MOLECULAR REGIME.

It has been observed that a temperature gradient in a gas causes small particles suspended in the gas to migrate in direction of decreasing temperature. The force arising from the temperature gradient causing the migration of the particles is commonly termed the thermal force and the phenomenon is usually known as thermophoresis.

Previously, several theoretical descriptions of the thermophoresis of particles in the free-molecular regime, $Kn \to \infty$, have been given [11-16]. In these papers, calculations are made of the thermal force from the net momentum transferred by gas molecules striking and reflecting from the particle surface.

The physical model that is used is a single spherical particle of radius, R, in an infinite gas with a temperature gradient, ∇T, which is constant at large distances from the particle. It is assumed that the gas as a whole is in mechanical equilibrium so that at large distances from the particle there is no pressure gradient. Introduce a spherical coordinate system (r,θ,ϕ) with its origin at the center of the stationary spherical particle and a polar axis, $\mathbf{x}$, in the direction of the vector, $(\nabla T)_\infty$ (see Fig. 7-3).

Then, the temperature distribution at large distances from the particle has the form:

$$T = T(\tilde{\mathbf{r}}) = T_0 + (\nabla T)_x \tilde{r}\cos(\theta) = T_0\left(1 + \mathbf{q}\cdot\tilde{\mathbf{r}}\right) , \tag{7-23}$$

where $\mathbf{q} = (\nabla T)_\infty / T_0$.

The condition, $\lambda\left|\nabla \ln(T)\right| \ll 1$, is assumed to be satisfied at large distances from the particle. Then, the distribution function is the Chapman-Enskog distribution, that is given by:

$$f = f_{eq}\left\{1 - a_1^{(1)}(\mathbf{q}\cdot\mathbf{c})S_{3/2}^{(1)}\left(c^2\right)\right\} , \tag{7-24}$$

where $a_1^{(1)} = -\frac{15}{16}\sqrt{\pi}\lambda$ for rigid-sphere molecules and f_{eq} is the local Maxwellian distribution that may be written as:

$$f_{eq} = n(\tilde{\mathbf{r}})\left(\frac{m}{2\pi kT(\tilde{\mathbf{r}})}\right)^{3/2} \exp\left(-\frac{mv^2}{2kT(\tilde{\mathbf{r}})}\right) . \tag{7-25}$$

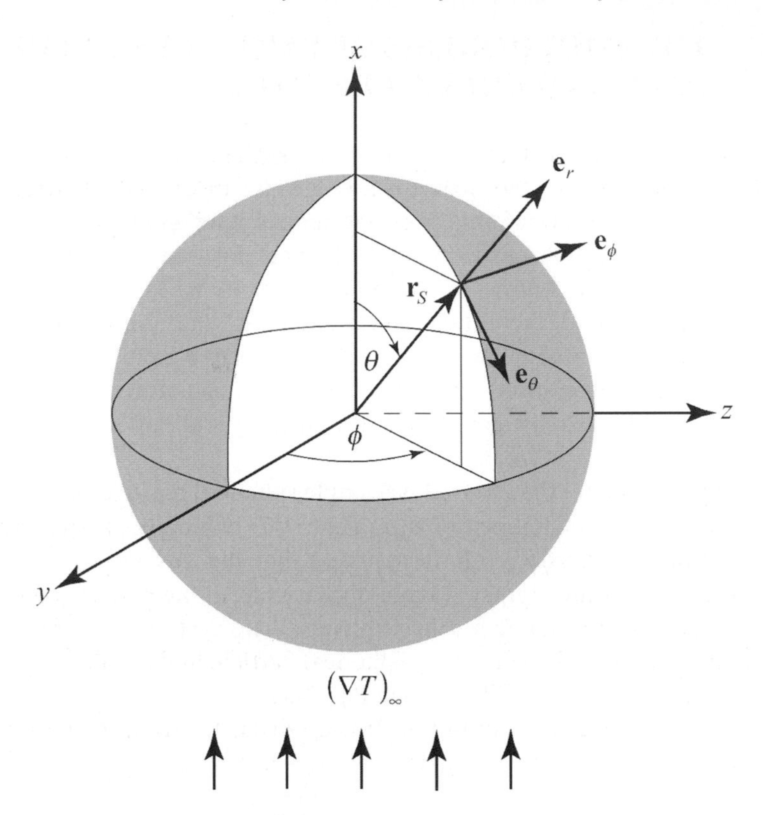

Figure 7-3. The geometry of the thermal force problem.

From the condition, $n_0 k T_0 = n(\tilde{\mathbf{r}}) k T(\tilde{\mathbf{r}})$, $n(\tilde{\mathbf{r}})$ in Eq. (7-25) may be represented as:

$$n(\tilde{\mathbf{r}}) = \frac{n_0 T_0}{T(\tilde{\mathbf{r}})} = \frac{n_0}{1 + (\mathbf{q} \cdot \tilde{\mathbf{r}})} .$$

For $\tilde{r} \sim \lambda$ the term $|\mathbf{q} \cdot \tilde{\mathbf{r}}| \ll 1$ and Eq. (7-25) may be represented as a power series in $(\mathbf{q} \cdot \tilde{\mathbf{r}})$. Considering only the linear term in this expansion, Eq. (7-25) may be written as:

$$f_{eq} = f^{(0)} \left[1 + \left(c^2 - \tfrac{5}{2} \right) (\mathbf{q} \cdot \tilde{\mathbf{r}}) \right] , \tag{7-26}$$

where:

$$f^{(0)} = n_0 \left(\frac{m}{2\pi k T_0} \right)^{3/2} \exp\left(-\frac{mv^2}{2kT_0} \right) .$$

From the symmetry of the problem under consideration, it is clear that the thermal force has only one component in the direction of the vector, $\mathbf{q}$. This component may be given by:

$$F_q = -2\pi R^2 m \int_0^{\pi} \sin(\theta) d\theta \times \left\{ \sum_{\pm} \int_{\pm} v_r \left(v_r \cos(\theta) - v_\theta \sin(\theta) \right) f^{\pm} d\mathbf{v} \right\} , \quad (7\text{-}27)$$

where $f^{\pm}$ are the distribution functions of the reflected and incident molecules, at the surface of the particle, respectively.

Using Eqs. (7-24) and (7-26), and the boundary condition given by Eq. (6-4), one can obtain the following expression for the distribution function at the surface of the particle:

$$f = \begin{cases} f^{-} = f^{(0)} \left[1 + \left(c^2 - \frac{5}{2} \right) (\mathbf{q} \cdot \mathbf{n}) R - a_1^{(1)} (\mathbf{q} \cdot \mathbf{c}) S_{3/2}^{(1)} \left(c^2 \right) \right] \\ f^{+} = (1 - \alpha_\tau) f^{(0)} \left[1 + \left(c^2 - \frac{5}{2} \right) (\mathbf{q} \cdot \mathbf{n}) R \right. \\ \qquad \left. - a_1^{(1)} g \left(-c_r \cos(\theta) - c_\theta \sin(\theta) \right) S_{3/2}^{(1)} \left(c^2 \right) \right] \\ \qquad + \alpha_\tau f^{(0)} \left[1 + v_r (\mathbf{r}_S) + \left(c^2 - \frac{3}{2} \right) \tau_r (\mathbf{r}_S) \right] . \end{cases} \quad (7\text{-}28)$$

The unknown functions, $v_r(\mathbf{r}_S)$ and $\tau_r(\mathbf{r}_S)$, are determined from the two boundary conditions given by:

$$\mathbf{n} \cdot \left\{ \int_{+} \mathbf{v} f^{+} d\mathbf{v} + \int_{-} \mathbf{v} f^{-} d\mathbf{v} \right\} = 0 , \quad (7\text{-}29)$$

and:

$$\mathbf{n} \cdot \left\{ \int_{+} \frac{1}{2} m v^2 \mathbf{v} f^{+} d\mathbf{v} + \int_{-} \frac{1}{2} m v^2 \mathbf{v} f^{-} d\mathbf{v} \right\} = -\kappa_p \left(\frac{\partial T_p}{\partial \tilde{r}} \right)_{\tilde{r}=R} , \quad (7\text{-}30)$$

where T_p is the temperature of the particle and κ_p is the thermal conductivity coefficient of the particle. The distribution of temperature inside the particle, in spherical coordinates, may be expressed in the form:

$$T_p = T_0 + \sum_{n=1}^{\infty} C_n \tilde{r}^n P_n\left(\cos(\theta)\right) . \tag{7-31}$$

Here, C_n are the coefficients of the Legendre polynomials, $P_n\left(\cos(\theta)\right)$. The left-hand-side of Eq. (7-30) contains the terms that are proportional only to $\cos(\theta)$, and therefore there is only one term in this sum and the temperature distribution inside the particle may be written as:

$$T_p = T_0\left(1+\tau_p(\tilde{\mathbf{r}})\right) = T_0\left(1+\frac{C_1 \tilde{r}\cos(\theta)}{T_0}\right) , \tag{7-32}$$

where C_1 is an unknown constant.

Before using the boundary conditions, one can transform the distribution function given by Eq. (7-28). Small terms of $O[R/\lambda]$ may be neglected in this regime. Eliminating these small terms, the following expression for the distribution function at the surface of the particle is obtained:

$$f = \begin{cases} f^- = f^{(0)}\left[1 - a_1^{(1)}(\mathbf{q}\cdot\mathbf{c}) S_{3/2}^{(1)}(c^2)\right] , \\ f^+ = (1-\alpha_\tau) f^{(0)} \\ \qquad \times\left[1 - a_1^{(1)} g\left(-c_r\cos(\theta) - c_\theta \sin(\theta)\right) S_{3/2}^{(1)}(c^2)\right] \\ \qquad +\alpha_\tau f^{(0)}\left[1 + \nu_r(\mathbf{r}_S) + \left(c^2 - \tfrac{3}{2}\right)\tau_r(\mathbf{r}_S)\right] . \end{cases} \tag{7-33}$$

The parameters of diffusely reflected molecules are assumed to deviate only slightly from the equilibrium state.

The boundary condition, Eq. (7-29), gives:

$$\nu_r + \tfrac{1}{2}\tau_r = 0 . \tag{7-34}$$

Eq. (7-29) is not altered in the case when the gas molecules are reflected from the surface of the particle under equilibrium conditions $\left(\tau_r = \tau_p\right)$. In this case one has:

$$\nu_p + \tfrac{1}{2}\tau_p = 0 \ . \tag{7-35}$$

After performing the integrations in Eq. (7-30), one obtains:

$$\nu_r + \tfrac{3}{2}\tau_r + \tfrac{5}{8}\sqrt{\pi}a_1^{(1)} q\cos(\theta) = -\frac{\kappa_p C_1}{\alpha_\tau \pi \alpha}\cos(\theta) \ , \tag{7-36}$$

where:

$$\alpha = \tfrac{1}{2}\pi^{-3/2} n_0 m \left(\frac{2kT_0}{m}\right)^{3/2} .$$

Using the definition of α_T in Eq. (7-30) yields:

$$\alpha_T\left(\nu_p + \tfrac{3}{2}\tau_p + \tfrac{5}{8}\sqrt{\pi}a_1^{(1)} q\cos(\theta)\right) = -\frac{\kappa_p C_1}{\alpha_\tau \pi \alpha}\cos(\theta) \ , \tag{7-37}$$

where:

$$\tau_p = \frac{C_1 R}{T_0}\cos(\theta) \ .$$

From Eqs. (7-34)-(7-37) it follows that:

$$\tau_r(\mathbf{r}_S) = -\tfrac{5}{8}\sqrt{\pi}a_1^{(1)}(\mathbf{q}\cdot\mathbf{n})\left[1 - \alpha_T \frac{1}{1+\varepsilon}\right] , \tag{7-38}$$

where:

$$\varepsilon = \frac{\alpha_\tau \alpha_T \kappa_g}{75\pi\kappa_p}\frac{R}{\lambda} ,$$

and κ_g is the thermal conductivity coefficient for the gas. Molecules are assumed to be rigid spheres. In the free-molecular regime the quantity ε may be neglected. Then the following expressions specify the boundary parameters, $\nu_r(\mathbf{r}_S)$ and $\tau_r(\mathbf{r}_S)$:

$$v_r(\mathbf{r}_S) = \tfrac{5}{16}\sqrt{\pi}(1-\alpha_T)a_1^{(1)}(\mathbf{q}\cdot\mathbf{n}) , \tag{7-39}$$

and:

$$\tau_r(\mathbf{r}_S) = -\tfrac{5}{8}\sqrt{\pi}(1-\alpha_T)a_1^{(1)}(\mathbf{q}\cdot\mathbf{n}) . \tag{7-40}$$

Using these relations, one may transform Eq. (7-27) into the form:

$$\begin{aligned} F_q = &-\frac{4n_0kT_0R^2a_1^{(1)}q}{\sqrt{\pi}}\int_0^{\pi}\sin(\theta)d\theta \\ &\times\left\{(1-\alpha_\tau)\int_0^{\infty}c_r\left(c_r^2\cos^2(\theta)-c_\theta^2\sin^2(\theta)\right)S_{3/2}^{(1)}(c^2)\exp(-c^2)d\mathbf{c}\right. \\ &+\tfrac{5}{16}\sqrt{\pi}\,\alpha_\tau(1-\alpha_T)\cos^2(\theta) \\ &\quad\times\left[\int_0^{\infty}c_r^2\exp(-c^2)d\mathbf{c}-2\int_0^{\infty}c_r^2\left(c^2-\tfrac{3}{2}\right)\exp(-c^2)d\mathbf{c}\right] \\ &\left.+\int_0^{\infty}c_r\left(c_r^2\cos^2(\theta)+c_\theta^2\sin^2(\theta)\right)S_{3/2}^{(1)}(c^2)\exp(-c^2)d\mathbf{c}\right\} . \end{aligned} \tag{7-41}$$

Having performed the necessary integrations, the expression for the thermal force may be written as:

$$\mathbf{F} = -\tfrac{5}{4}\pi R^2 n_0 k\lambda(\nabla T)_\infty\left[1+\alpha_\tau(1-\alpha_T)\tfrac{5}{32}\pi\right] . \tag{7-42}$$

This expression contains a new term, namely $\alpha_\tau(1-\alpha_T)\frac{5}{32}\pi$, as compared with the well known results [11-13]. This term for the particular case when $\alpha_\tau = 1$ has been obtained by Ivchenko and Yalamov [13] and later by Talbot et al. [14].

5. CONDENSATION ON A SPHERICAL DROPLET.

Consider the growth of a single droplet in an infinite expanse of gas that consists only of the gas phase of the liquid which forms the droplet. At large distances from the droplet the gas is assumed to have a constant temperature and to be supersaturated. The distribution function for these conditions has the form:

$$f = f^{(0)}(1+\nu_0) \ , \tag{7-43}$$

where:

$$f^{(0)} = n_0(T_0)\left(\frac{m}{2\pi k T_0}\right)^{3/2} \exp\left(-\frac{mv^2}{2kT_0}\right) ,$$

$n_0(T_0)$ is the number density of the saturated vapor at temperature, T_0, and ν_0 is the supersaturation. The distribution function on the surface of the droplet is assumed to be the Maxwellian function for the reflected molecules with the unknown number density and temperature. Let these parameters be deviated only slightly from the equilibrium state, and therefore one may use the linearized form of the distribution function. The distribution functions of incident and reflected molecules at the droplet may be expressed in the form:

$$f = \begin{cases} f^- = f^{(0)}(1+\nu_0) \ , \\ f^+ = f^{(0)}\left[1+\Phi^+(\mathbf{c},\tilde{\mathbf{r}}_S)\right] \\ \quad = \alpha_m f^{(0)}\left[1+\nu_S+\left(c^2-\frac{3}{2}\right)\tau_S\right] \\ \qquad +(1-\alpha_m) f^{(0)}\left[1+\nu_r+\left(c^2-\frac{3}{2}\right)\tau_r\right]. \end{cases} \tag{7-44}$$

where ν_S is the correction to the saturated vapor density at the temperature, $T_S = T_0(1+\tau_S)$. The unknown parameters, ν_r and τ_r, in the reflected distribution function may be found from the energy accommodation and impenetrability conditions which have sense only for those molecules reflected from the particle surface without a phase transition. The perfect representation of the reflected distribution function was obtained previously from Problem 6.6 and is expressible in the form:

$$\begin{aligned} \Phi^+(\mathbf{c},\tilde{\mathbf{r}}_S) &= \alpha_m\left[\nu_S+\left(c^2-\frac{3}{2}\right)\tau_S\right] \\ &+(1-\alpha_m)\left\{\left(2-c^2\right)\left[-\alpha_T\tau_S+(1-\alpha_T)\left(\frac{\delta_2}{\pi}-\frac{2\delta_1}{\pi}\right)\right]-\frac{2\delta_1}{\pi}\right\} , \end{aligned} \tag{7-45}$$

where:

$$\delta_1 = -\frac{1}{2}\pi\nu_0 \ \text{and} \ \delta_2 = -\pi\nu_0 \tag{7-46}$$

The correction to the equilibrium number density may be represented by:

$$\nu_S = \frac{2\gamma m}{\rho_1 k T_0 R} + a\tau_S \ , \tag{7-47}$$

where γ is the surface tension coefficient, ρ_1 is the density of the drop, R is the drop radius, $a = (\varsigma - 1)$, $\varsigma = h_{fg}/kT_0$, and h_{fg} is the latent heat of condensation per molecule. This equation is the sum of the linearized Kelvin and Clausius-Clapeyron terms [17,18]. The temperature, τ_S, may be found from the equation of energy conservation, which may be written as:

$$Q_r = -h_{fg} N_r \ , \tag{7-48}$$

or:

$$\alpha_T Q_{r,S} = -h_{fg} \alpha_m N_{r,S} \ . \tag{7-49}$$

Eq. (7-48) expresses the fact that, for stationary conditions, the heat of condensation must be transferred to the ambient gas by thermal conductivity. It also shows that the growth rate of the particle is controlled by the rate at which the liberated heat of condensation can be dissipated into the environment. Eq. (7-49) gives:

$$\tau_S = -\alpha_m \nu_0' \frac{\varsigma + 2}{2\alpha_m\left(a + \frac{3}{2}\right) + \alpha_m \varsigma \left(a + \frac{1}{2}\right) + 2\alpha_T\left(1 - \alpha_m\right)} \ , \tag{7-50}$$

where:

$$\nu_0' = \frac{2\gamma m}{\rho_1 k T_0 R} - \nu_0 \ .$$

The total flux of the number of molecules across the droplet surface may be written as:

$$\begin{aligned} N &= 4\pi R^2 N_r = 4\pi R^2 \alpha_m N_{r,S} \\ &= 4\pi R^2 \alpha_m \frac{n_0}{2\sqrt{\pi}} \left(\frac{2kT_0}{m}\right)^{1/2} \left[\nu_0' + \left(a + \tfrac{1}{2}\right)\tau_S\right] . \end{aligned} \tag{7-51}$$

Substituting τ_S in Eq. (7-51), one obtains:

$$N=\left(\frac{8\pi kT_0}{m}\right)^{1/2}R^2 n_0 v_0' \xi\left(\alpha_m,\alpha_T,\varsigma\right) , \tag{7-52}$$

where:

$$\xi\left(\alpha_m,\alpha_T,\varsigma\right)=4\alpha_m\frac{\alpha_m+\alpha_T\left(1-\alpha_m\right)}{\alpha_m\left(2\varsigma^2+3\varsigma+2\right)+4\alpha_T\left(1-\alpha_m\right)} .$$

The growth rate of the drop is determined by the equation:

$$\frac{dR}{dt}=-\frac{mN}{4\pi R^2\rho_1} . \tag{7-53}$$

Integration of this equation gives the dependence, $R=R(t)$, that has the form:

$$t=\frac{\rho_1}{n_0 v_0\xi}\left(\frac{2\pi}{mkT_0}\right)^{1/2}\left\{\left(R-R_0\right)+\frac{b}{v_0}\ln\left(\frac{v_0R-b}{v_0R_0-b}\right)\right\} , \tag{7-54}$$

where:

$$b=\frac{2\gamma m}{\rho_1 kT_0} .$$

6. NON-STATIONARY GAS FLOWS.

The properties of non-stationary flows of a rarefied gas may be described by solving the Cauchy problem for Boltzmann's equation that may be written in the form:

$$\frac{\partial f}{\partial t}+v_x\frac{\partial f}{\partial \tilde{x}}+v_y\frac{\partial f}{\partial \tilde{y}}+v_z\frac{\partial f}{\partial \tilde{z}}=0 . \tag{7-55}$$

The initial condition for this problem is given by:

$$f(\mathbf{v},\tilde{\mathbf{r}},0)=f_i(\mathbf{v},\tilde{\mathbf{r}}) \ . \tag{7-56}$$

The solution of this mathematical formulation begins with the construction of a general solution of Lagrange's auxiliary system of ordinary differential equations which may be expressed in the form:

$$\frac{dt}{1}=\frac{d\tilde{x}}{v_x}=\frac{d\tilde{y}}{v_y}=\frac{d\tilde{z}}{v_z}=\frac{df}{0} \ . \tag{7-57}$$

The general solution of Eq. (7-57) is:

$$f=\alpha=\text{const} \ , \tag{7-58}$$

$$\tilde{\mathbf{r}}-\mathbf{v}t=\tilde{\mathbf{r}}_0=\text{const} \ . \tag{7-59}$$

By virtue of Eqs. (7-58) and (7-59), the distribution function may be shown to be expressible in the following general form:

$$f(\mathbf{v},\tilde{\mathbf{r}},t)=\Psi(\tilde{\mathbf{r}}_0) \ ,$$

where Ψ is an arbitrary function. The initial condition then yields:

$$f(\mathbf{v},\tilde{\mathbf{r}},0)=\Psi(\tilde{\mathbf{r}}_0)=f_i(\mathbf{v},\tilde{\mathbf{r}}_0) \ . \tag{7-60}$$

Allowing for this relation, one can obtain the following solution of the Cauchy problem for the distribution function:

$$f(\mathbf{v},\tilde{\mathbf{r}},t)=f_i(\mathbf{v},\tilde{\mathbf{r}}-\mathbf{v}t) \ . \tag{7-61}$$

The knowledge of the distribution function allows one to calculate any macroscopic quantity characteristic of the gas flow. It is, however, very important to note that the calculation technique is very different from that described in Section 7.3. For a non-stationary flow, there is no cone of influence in the velocity space since the molecular trajectories passing through a fixed point, $M(\tilde{\mathbf{r}})$, at various moments of time, t, must start from different initial points, $\tilde{\mathbf{r}}_0$. The moments of the distribution function may be found by means of direct integration of Eq. (7-61) over all velocity space. The mean value of any molecular property, $\phi(\mathbf{v})$, is given by:

$$\bar{\phi} = n^{-1} \int \phi(\mathbf{v}) f_i(\mathbf{v}, \tilde{\mathbf{r}} - \mathbf{v}t) d\mathbf{v} \ . \tag{7-62}$$

To calculate this integral, it is convenient to change the variable of integration from $\mathbf{v}$ to a new variable, $\mathbf{r}'$, which is related to $\mathbf{v}$ by:

$$\mathbf{r}' = \tilde{\mathbf{r}} - \mathbf{v}t \ . \tag{7-63}$$

The volume elements of this transformation are connected through the Jacobian in the following manner:

$$d\mathbf{v} = \left| \frac{\partial(v_x, v_y, v_z)}{\partial(x', y', z')} \right| d\mathbf{r}' = t^{-3} d\mathbf{r}' \ . \tag{7-64}$$

Finally, the mean value of $\phi(\mathbf{v})$ may be expressed by [6]:

$$n\bar{\phi} = t^{-3} \int \phi\left(\frac{\tilde{\mathbf{r}} - \mathbf{r}'}{t}\right) f_i\left(\frac{\tilde{\mathbf{r}} - \mathbf{r}'}{t}, \mathbf{r}'\right) d\mathbf{r}' \ , \tag{7-65}$$

where integration is extended over the domain occupied by the gas at $t = 0$.

PROBLEMS

7.1. Determine the temperature of a sphere located in an infinite stream of a monatomic gas the velocity of which, at large distances from the body, is given by a vector, $\mathbf{u}$. Consider the case of arbitrary values of the ratio, $u/(2kT/m)^{1/2}$, if the particle is assumed to be a perfect heat conductor and molecules of a gas are considered to be reflected by the sphere surface with equilibrium conditions.

Solution: The distribution function at the surface of the particle can be expressed in the form:

$$f^- = n_0 \left(\frac{m}{2\pi k T_0}\right)^{3/2} \exp\left(-\frac{m(\mathbf{v} - \mathbf{u})^2}{2kT_0}\right),$$

$$f^{+}=\alpha_{\tau}n(\mathbf{r}_S)\left(\frac{m}{2\pi kT_S}\right)^{3/2}\exp\left(-\frac{mv^2}{2kT_S}\right)+(1-\alpha_{\tau})f^{-}(-c_r,c_\theta,c_\phi,\mathbf{r}_S)\ ,$$

where $n(\mathbf{r}_S)$ denotes the density of a fictitious gas at the point, $\mathbf{r}_S$, and T_S is the constant temperature of the particle. The unknown function, $n(\mathbf{r}_S)$, and the quantity, T_S, can be specified from two conditions expressing the impenetrability of the surface for gas molecules and the heat balance:

$$\mathbf{n}_S\cdot\int\mathbf{v}f d\mathbf{v}=0\ ,$$

$$\int_{(S)}d\mathbf{S}\cdot\int\tfrac{1}{2}mv^2\mathbf{v}f d\mathbf{v}=0\ .$$

From the first condition, one can obtain:

$$n(\mathbf{r}_S)=n_0\left(\frac{T_0}{T_S}\right)^{1/2}\left\{\exp\left(-u_r^{*2}\right)-\sqrt{\pi}u_r^*\left(1-\operatorname{erf}\left(u_r^*\right)\right)\right\}\ ,$$

where $u_r^*=\mathbf{u}^*\cdot\mathbf{n}_S=u^*\cos(\theta)$. The direction of the vector, $\mathbf{u}$, is assumed to be along the polar axis. Taking into account both conditions, one can obtain:

$$T_S=T_0\frac{\left(1+\frac{1}{2}u^{*2}\right)+\frac{1}{2}\sqrt{\pi}\left(\frac{5}{2}u^*+u^{*3}\right)\nu(u^*)}{1+\sqrt{\pi}u^*\nu(u^*)}\ ,$$

where:

$$\nu(u^*)=\frac{\int_0^1 t\operatorname{erf}(u^*t)\,dt}{\int_0^1\exp\left(-u^{*2}t^2\right)dt}\ ,\ \text{and }\mathbf{u}^*=\mathbf{u}\left(\frac{2kT_0}{m}\right)^{-1/2}\ .$$

A more detailed analysis is contained in [3,19].

7.2. Using the same conditions as in Problem 7.1, determine the drag force on the sphere.

Solution: Substituting the distribution function described in Problem 7.1 in Eq. (7-8), one obtains:

$$\mathbf{F}_D = \tfrac{8}{3} R^2 n_0 \left(2\pi mkT_0\right)^{1/2} \mathbf{u} \left\{ \nu_1\left(u^*\right) + \alpha_\tau \left[\tfrac{1}{8}\pi \left(\frac{T_S}{T_0}\right)^{1/2} + \nu_2\left(u^*\right) \right] \right\} ,$$

where $\mathbf{u}^*$ is the dimensionless velocity of the gas, α_τ is the tangential momentum accommodation coefficient and the following notations have been introduced:

$$\nu_1\left(u^*\right) = \tfrac{3}{2} \int_0^1 \left[t^2 \exp\left(-u^{*2}t^2\right) + \tfrac{1}{2}\sqrt{\pi}\left(\frac{t}{u^*} + 2u^*t^3\right) \operatorname{erf}\left(u^*t\right) \right] dt ,$$

$$\nu_2\left(u^*\right) = -\tfrac{3}{2} \int_0^1 \left[\left(t^2 - \tfrac{1}{2}\right) \exp\left(-u^{*2}t^2\right) \right.$$
$$\left. + \tfrac{1}{2}\sqrt{\pi}\left(\tfrac{1}{2}\frac{t}{u^*} - u^*t + 2u^*t^3\right) \operatorname{erf}\left(u^*t\right) \right] dt .$$

The temperature of the sphere, T_S, was determined in Problem 7.1.

7.3. Determine the mean velocity distribution (both components) for a linearized uniform flow of a gas past a sphere if the mean gas velocity far from the sphere is $\mathbf{u}$. Consider the specific case when $\alpha_\tau = 1$ and $|\mathbf{u}| \ll \left(2kT/m\right)^{1/2}$.

Solution: For this problem the distribution function given in Eq. (7-18) is employed. The component of the mean velocity vector may be calculated from Eq. (7-15). For example, the radial component may be determined in the following manner:

$$u_r^*\left(r,\theta\right) = \frac{1}{n_0} \int_{(1+2)} c_r f_1 d\mathbf{v} + u_r^* \pi^{-3/2} \int_0^{2\pi} d\beta$$
$$\times \int_0^\infty c^3 \exp\left(-c^2\right) \int_0^w \cos(\chi)\sin(\chi)$$
$$\times \left\{ -\sqrt{\pi}\left(r\sin^2(\chi) + \cos(\chi)\left[1 - r^2\sin^2(\chi)\right]^{1/2} \right) - 2c\cos(\chi) \right\} dc\, d\chi ,$$

where $w = \arcsin(1/r)$ and $u_r^* = u^* \cos(\theta)$. An analogous relationship exists for the other component, $u_\theta^*(r,\theta)$. Having performed the indicated integrations, one obtains:

$$u_r(r,\theta) = u(1 - g(r))\cos(\theta), \text{ and } u_\theta(r,\theta) = -u(1 + h(r))\sin(\theta) .$$

Here, the functions, $g(r)$ and $h(r)$, may be written as:

$$g(r) = \tfrac{1}{4}r^{-3} + r\left[\tfrac{1}{4}r^{-3} + \tfrac{1}{8}r^{-1}a^2 - \tfrac{1}{8}a^4 \ln\left(\frac{1+r^{-1}}{a}\right)\right] + \tfrac{1}{2}\left(1 - a^3\right) , \text{ and:}$$

$$h(r) = \tfrac{1}{2}\left\{g(r) - r\left[\tfrac{1}{2}r^{-1} - \tfrac{1}{2}a^2 \ln\left(\frac{1+r^{-1}}{a}\right)\right] - \tfrac{3}{2}(1-a)\right\} ,$$

where $a = \left(1 - r^{-2}\right)^{1/2}$. More detailed analyses may be found in [7,20].

7.4. Prove that the velocity field, as defined in Problem 7.3, is not the potential field.

Solution: For the potential flow field, the following equation must be satisfied: $\nabla \times \mathbf{u}(\mathbf{r}) = 0$. This quantity, for the free-molecular velocity field described in Problem 7.3, may be expressed in the form:

$$\nabla \times \mathbf{u}(r,\theta) = \begin{vmatrix} \dfrac{\mathbf{e}_r}{r^2 \sin(\theta)} & \dfrac{\mathbf{e}_\theta}{r \sin(\theta)} & \dfrac{\mathbf{e}_\phi}{r} \\ \dfrac{\partial}{\partial r} & \dfrac{\partial}{\partial \theta} & \dfrac{\partial}{\partial \phi} \\ u_r(r,\theta) & r u_\theta(r,\theta) & 0 \end{vmatrix}$$

$$= -\mathbf{e}_\phi u \sin(\theta)\left[\frac{h(r) + g(r)}{r} + \frac{dh}{dr}\right] \neq 0$$

From this one can see that the velocity field has no potential.

7.5. In Problem 7.3, determine the pressure and temperature distribution. Check the validity of the relationship: $p = nkT$.

Solution: The same distribution function and calculational techniques as in Problem 7.3 yield:

$$p = p_0 \left\{ 1 - \tfrac{1}{2}\sqrt{\pi}u_r^* \left(\psi_1(r) + \psi_2(r) \right) \right\} ,$$

$$\psi_1(r) = r\left[(1-a) - \tfrac{1}{3}\left(1-a^3\right) \right] ,$$

$$\psi_2(r) = \tfrac{1}{3} r^{-2} \left(1 + 8\pi^{-1} \right) , \text{ and:}$$

$$T(r,\theta) = T_0 \left(1 - \tfrac{1}{3}\pi^{-1/2} u_r^* r^{-2} \right) ,$$

where $a = \left(1 - r^{-2}\right)^{1/2}$ and $u_r^* = u^* \cos(\theta)$. It is easy to see that the equation of state is not altered for this free-molecular flow.

7.6. A sphere suspended in a gas has radius, R, and temperature, $T_0 + \Delta T$ $(\Delta T \ll T_0)$. Determine the heat flux and the temperature distribution of a gas if its number density and temperature far from the sphere are n_0 and T_0, respectively. Solve this problem for the free-molecular regime.

Solution: A distribution function should be sought in the form:

$$f_1 = f^{(0)} ,$$

$$f_2 = \left(1 - \alpha_\tau\right) f^{(0)} + \alpha_\tau f^{(0)} \left\{ 1 + \nu_r + \left(c^2 - \tfrac{3}{2}\right)\tau_r \right\} ,$$

where:

$$f^{(0)} = n_0 \left(\frac{m}{2\pi k T_0} \right)^{3/2} \exp\left(\frac{mv^2}{2kT_0} \right) ,$$

is the distribution function at large distances from the heated sphere. The quantities, ν_r and τ_r, may be determined from the boundary relationships

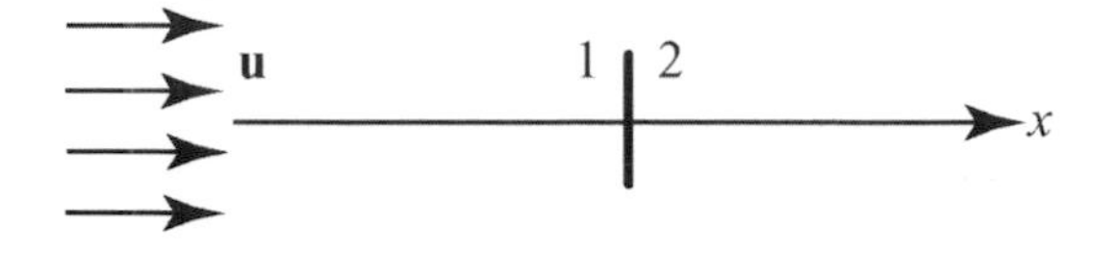

Figure 7-4. The geometry to be used for Problem 7.7 in determining the force on a round disk perpendicular to a uniform gas flow.in determining the force on a round disk perpendicular to a uniform gas flow

$N_r = 0$, $N_{r\,eq} = 0$, and $Q_r = \alpha_T Q_{r\,eq}$. Performing some simple integrations and algebraic transformations, one can obtain:

$$Q = 4\pi R^2 n_0 k T_0 \pi^{-1/2} \left(\frac{2kT_0}{m} \right)^{1/2} \alpha_\tau \alpha_T \frac{\Delta T}{T_0} ,$$

where Q is the heat flux through the sphere surface. The temperature distribution can be expressed in the form:

$$T(r) = T_0 \left\{ 1 + \tfrac{3}{4} \alpha_\tau \alpha_T \left[1 - \left(1 - r^{-2} \right)^{1/2} \right] \frac{\Delta T}{T_0} \right\} .$$

7.7. Calculate the force on a round disk the surface of which is perpendicular to the velocity vector of a uniform gas flow at infinity. The radius of the disk, R, is assumed to be much less than the mean free path and $|\mathbf{u}| \ll (2kT/m)^{1/2}$ where $\mathbf{u}$ and T are the mean gas velocity and temperature, respectively. Use the geometry in Fig. 7-4.

Solution: The force on the left and right sides of the disk may be expressed as:

$$\mathbf{F}_{1,2} = -\pi R^2 \sum_{\mp} \int_{\mp} m\mathbf{v} \left(\mathbf{v}\mathbf{n}_{1,2} \right) f_{1,2}^{\mp}(\mathbf{v}) d\mathbf{v} , \tag{P-1}$$

where $\mathbf{n}_i$ is the external (pointing into the gas) normal vector to the disk surface. The reflected distribution functions are given by:

$$f_1^-(\mathbf{v}) = \alpha_\tau n_1 \left(\frac{m}{2\pi kT} \right)^{3/2} \exp\left(-c^2\right) + \left(1 - \alpha_\tau\right) f_1^+\left(-c_x, c_y, c_z\right) ,$$

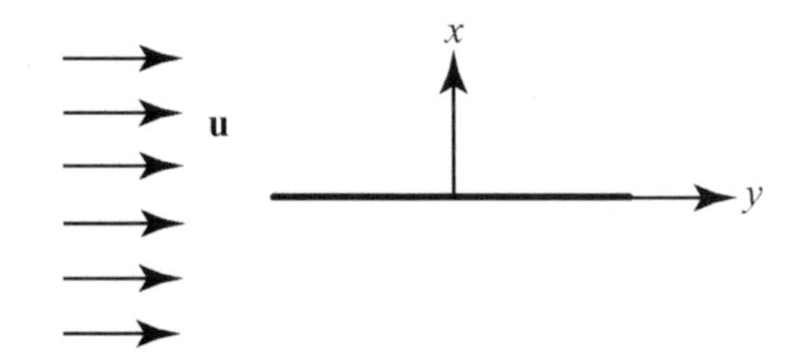

Figure 7-5. The geometry to be used for Problem 7.8 in determining the drag on a plate parallel to a steady, free-molecular gas flow.

$$f_2^+(\mathbf{v}) = \alpha_\tau n_2 \left(\frac{m}{2\pi kT}\right)^{3/2} \exp\left(-c^2\right) + \left(1-\alpha_\tau\right) f_2^-\left(-c_x, c_y, c_z\right) ,$$

where the notation (±) refers to those regions in the velocity space in which $c_x > 0$ and $c_x < 0$, respectively. The incident distribution function may be written as:

$$f_1^+(\mathbf{v}) = f_2^-(\mathbf{v}) = n\left(\frac{m}{2\pi kT}\right)^{3/2} \exp\left(-c^2\right)\left(1+2c_x u^*\right) .$$

The number density, $n_i = n\left(1+\nu_i\right)$, may be found from the impenetrability condition of the disk surface which yields: $\nu_{1,2} = \pm\sqrt{\pi}u^*$. Then, after some simple integration, the expression for the force becomes:

$$\mathbf{F} = 4R^2 n\left(2\pi kTm\right)^{1/2}\left[1+\alpha_\tau\left(\tfrac{1}{8}\pi - \tfrac{1}{2}\right)\right]\mathbf{u} .$$

7.8. A plate of area, S, is situated in a steady gas flow and is oriented parallel to the velocity vector, $\mathbf{u}$, as shown in Fig. 7-5. In the free-molecular regime, determine the drag on the plate for arbitrary values of the speed ratio, $|\mathbf{u}|/\left(2kT/m\right)^{1/2}$, and of the tangential momentum accommodation coefficient.

Solution: In this problem, the drag is given by:

$$\mathbf{F} = -2S\sum_{\pm}\int_{\pm} m\mathbf{v}v_x f^{\pm}(\mathbf{v})d\mathbf{v} .$$

The incident and reflected distribution functions associated with the upper side of the plate are specified by:

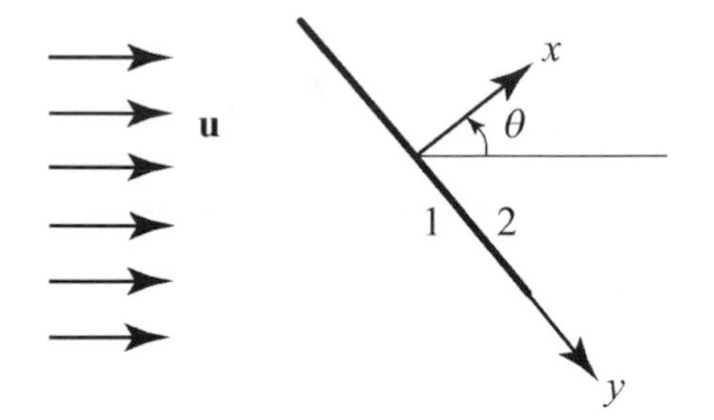

Figure 7-6. The geometry to be used for Problem 7.9 in determining the drag on a plate oriented at an angle to a steady, free-molecular gas flow.

$$f^{-}(\mathbf{v}) = n\left(\frac{m}{2\pi kT}\right)^{3/2} \exp\left(-\frac{m(\mathbf{v}-\mathbf{u})^2}{2kT}\right), \text{ and:}$$

$$f^{+}(\mathbf{v}) = \alpha_\tau n\left(\frac{m}{2\pi kT}\right)^{3/2} \exp\left(-\frac{mv^2}{2kT}\right) + \\ (1-\alpha_\tau) n\left(\frac{m}{2\pi kT}\right)^{3/2} \exp\left[-\frac{m}{2kT}\left(v_x^2 + (v_y - u)^2 + v_z^2\right)\right].$$

Using these, the drag is found to be:

$$\mathbf{F} = \alpha_\tau pS\left(\frac{2m}{\pi kT}\right)^{1/2} \mathbf{u}.$$

7.9. A plate of area, S, is situated in a steady gas flow and is oriented such that the gas stream is incident on the plate at an angle θ as shown in Fig. 7-6. In the free-molecular regime, determine the drag on the plate for arbitrary values of the speed ratio, $|\mathbf{u}|/(2kT/m)^{1/2}$, and of the tangential momentum accommodation coefficient.

Solution: The net drag component in direction of the flow vector, $\mathbf{u}$, is given by:

$$F_u = S\sum_{\pm}\int_{\pm} m\left(v_x \cos(\theta) + v_y \sin(\theta)\right) v_x \left(f_1^{\pm} - f_2^{\pm}\right) d\mathbf{v},$$

where the indices 1 and 2 refer to the upstream and downstream sides of the plate, respectively. The distribution functions of incident and reflected molecules may be expressed in the form:

$$f_1^+(\mathbf{v}) = f_2^-(\mathbf{v}) = n\left(\frac{m}{2\pi kT}\right)^{3/2} \exp\left(-\frac{m}{2kT}(\mathbf{v}-\mathbf{u})^2\right),$$

$$f_1^-(\mathbf{v}) = \alpha_\tau n_1\left(\frac{m}{2\pi kT}\right)^{3/2} \exp\left(-\frac{m}{2kT}\mathbf{v}^2\right) + (1-\alpha_\tau)n\left(\frac{m}{2\pi kT}\right)^{3/2}$$
$$\times\exp\left\{-\frac{m}{2kT}\left[\left(v_x + u\cos(\theta)\right)^2 + \left(v_y - u\sin(\theta)\right)^2 + v_z^2\right]\right\},$$

and:

$$f_2^+(\mathbf{v}) = \alpha_\tau n_2\left(\frac{m}{2\pi kT}\right)^{3/2} \exp\left(-\frac{m}{2kT}\mathbf{v}^2\right) + (1-\alpha_\tau)n\left(\frac{m}{2\pi kT}\right)^{3/2}$$
$$\times\exp\left\{-\frac{m}{2kT}\left[\left(v_x + u\cos(\theta)\right)^2 + \left(v_y - u\sin(\theta)\right)^2 + v_z^2\right]\right\}.$$

The values, n_1 and n_2, are specified by the impenetrability condition. The drag is found to be:

$$\mathbf{F} = pS\left(\frac{2m}{\pi kT}\right)^{1/2} \mathbf{u}\Big\{(2-\alpha_\tau)\cos(\theta)\Big[\cos(\theta)\exp\left(-u^{*2}\cos^2(\theta)\right) +$$
$$\frac{\sqrt{\pi}}{u^*}\left(\tfrac{1}{2} + u^{*2}\cos^2(\theta)\right)\operatorname{erf}\left(u^*\cos(\theta)\right)\Big] + \alpha_\tau\Big[\tfrac{1}{2}\pi\cos^2(\theta) +$$
$$\sin^2(\theta)\left(\exp\left(-u^{*2}\cos^2(\theta)\right) + \sqrt{\pi}u^*\cos(\theta)\operatorname{erf}\left(u^*\cos(\theta)\right)\right)\Big]\Big\},$$

where $u^* = (m/2kT)^{1/2}|\mathbf{u}|$.

7.10. Determine the force on a round disk located in a rarefied gas if the surfaces of the disk have different temperatures, T_1 and T_2. The disk has a radius of R which is assumed to be much less than the mean free path and $T_1 - T_2 \ll T_0$ where T_0 is the gas temperature. Use the geometry in Fig. 7-7.

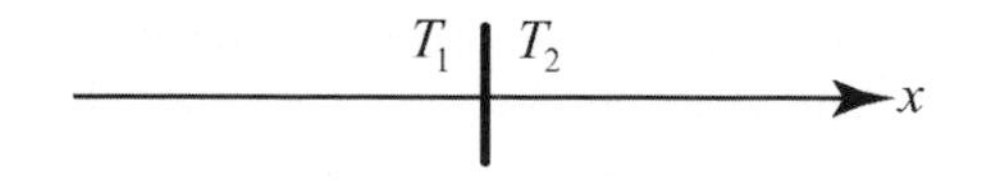

Figure 7-7. The geometry to be used for Problem 7.10 in determining the force on a round disk in a rarefied gas when the two sides of the disk have different temperatures.

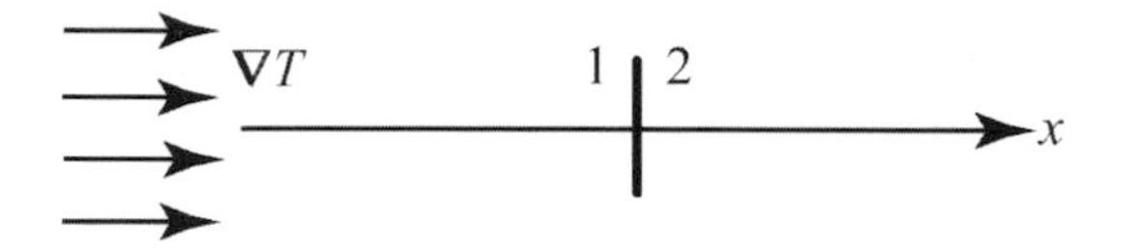

Figure 7-8. The geometry to be used for Problem 7.11 in determining the thermophoretic force on a thin disk located in a rarefied gas having a constant temperature gradient.

Solution: Using the same technique as in Problem 6.5, one can obtain:

$$\Phi_1^- = -\alpha_\tau \alpha_T \left(2 - c^2\right)\tau_{1w} \quad ; \quad \tau_{1w} = \frac{T_1 - T_0}{T_0} \ ,$$

$$\Phi_2^+ = -\alpha_\tau \alpha_T \left(2 - c^2\right)\tau_{2w} \quad ; \quad \tau_{2w} = \frac{T_2 - T_0}{T_0} \ ,$$

where Φ_1^- and Φ_2^+ are the corrections to the reflected distribution functions associated with the left- and right-hand surfaces of the disk, respectively, and $\Phi_1^+ = \Phi_2^- = 0$. The force on the disk may be calculated from the equation given in the solution to Problem 7.7 [Eq. (P-1)]. This gives:

$$F_x = \tfrac{1}{4}\pi R^2 p_0 \alpha_\tau \alpha_T \frac{T_1 - T_2}{T_0} \ .$$

7.11. Determine the thermophoretic force on a thin absolutely thermal conductive disk located in a rarefied gas with a constant temperature gradient, ∇T. The radius of the disk is R and is assumed to be much less than the mean free path. Use the geometry in Fig. 7-8.

Solution: The corrections to the reflected distribution functions may be written as (Problem 6.5):

$$\Phi_1^-(\mathbf{c},0)=(1-\alpha_\tau)\Phi_1^+(-c_x,c_y,c_z,0)$$
$$+\alpha_\tau\left\{(2-c^2)\left[(1-\alpha_T)\left(\frac{\delta_2^{(1)}}{\pi}-\frac{2\delta_1^{(1)}}{\pi}\right)\right]-\frac{2\delta_1^{(1)}}{\pi}\right\},$$

and:

$$\Phi_2^+(\mathbf{c},0)=(1-\alpha_\tau)\Phi_2^-(-c_x,c_y,c_z,0)$$
$$+\alpha_\tau\left\{(2-c^2)\left[(1-\alpha_T)\left(\frac{\delta_2^{(2)}}{\pi}-\frac{2\delta_1^{(2)}}{\pi}\right)\right]-\frac{2\delta_1^{(2)}}{\pi}\right\}.$$

Here, $\Phi_1^+(\mathbf{c},0)$, $\Phi_2^-(\mathbf{c},0)$, $\delta_1^{(i)}$, and $\delta_2^{(i)}$, are defined by:

$$\Phi_1^+(\mathbf{c},0)=-a_1^{(1)}qc_xS_{3/2}^{(1)}(c^2)\ ,\ \delta_1^{(1)}=0\ ,\ \delta_2^{(1)}=\tfrac{5}{8}\pi^{3/2}a_1^{(1)}q\ ,\text{ and:}$$

$$\Phi_2^-(\mathbf{c},0)=-a_1^{(1)}qc_xS_{3/2}^{(1)}(c^2)\ ,\ \delta_1^{(2)}=0\ ,\ \delta_2^{(2)}=\tfrac{5}{8}\pi^{3/2}a_1^{(1)}q\ ,$$

with:

$$q=T_0^{-1}\frac{\partial T}{\partial x},$$

where the indices 1 and 2 are associated with the left- and right-hand surfaces of the disk, respectively. Substituting these into the force equation from Problem 7.7 [Eq. (P-1)], one can obtain:

$$\mathbf{F}=-\tfrac{15}{8}\pi R^2 n_0 k\lambda(\nabla T)_\infty\left[1-\tfrac{1}{2}\alpha_\tau+\alpha_\tau(1-\alpha_T)\tfrac{5}{32}\pi\right].$$

7.12. Determine the number density of a gas at a point, A, located behind a round disk of radius, R, which is much less than the mean free path, i.e. $R\ll\lambda$. The mean gas velocity, $\mathbf{u}$, is assumed to be much less than the thermal molecular velocity. Use the geometry in Fig. 7-9.

Solution: Since $\mathbf{n}_S=\text{const}$ for all molecular trajectories beginning at the disk surface, $\mathbf{v}'=(-v_x,v_y,v_z)$. Therefore, the distribution function has a very simple form for arbitrary values of the tangential accommodation

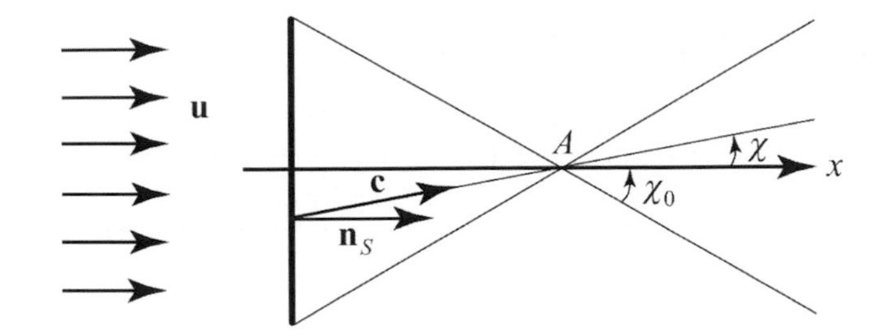

Figure 7-9. The geometry to be used for Problem 7.12 in determining the number density behind a round disk perpendicular to a free-molecular gas flow in which the mean gas velocity is much less than the thermal molecular velocity.

coefficient. The distribution function outside and inside of the cone of influence is expressed as:

$$f_1 = f^{(0)}\left(1 + 2c_x u^*\right),$$

$$f_2 = \alpha_\tau f^{(0)}\left(1 - \sqrt{\pi}u^*\right) + \left(1 - \alpha_\tau\right) f^{(0)}\left(1 - 2c_x u^*\right).$$

The number density is then given by:

$$n(x) = n_0\left\{1 + \frac{2}{\sqrt{\pi}}\int_0^\infty c^2 \exp\left(-c^2\right)dc \int_0^{\chi_0}\sin(\chi)\,d\chi \right.$$

$$\left.\left[-\alpha_\tau\sqrt{\pi}u^* - 2\left(2 - \alpha_\tau\right)u^* c\cos(\chi)\right]\right\}$$

$$= n_0\left\{1 - u\left(\frac{m}{2kT_0}\right)^{1/2}\left[\alpha_\tau\frac{\sqrt{\pi}}{2}\left(1 - \frac{x}{\sqrt{R^2 + x^2}}\right)\right.\right.$$

$$\left.\left. + \left(2 - \alpha_\tau\right)\frac{1}{\sqrt{\pi}}\frac{R^2}{R^2 + x^2}\right]\right\}.$$

7.13. A gas with a number density, n_0, is flowing into vacuum across a round hole of radius, $R \ll \lambda$. Determine the number density at a point, A. Use the geometry in Fig. 7-10.

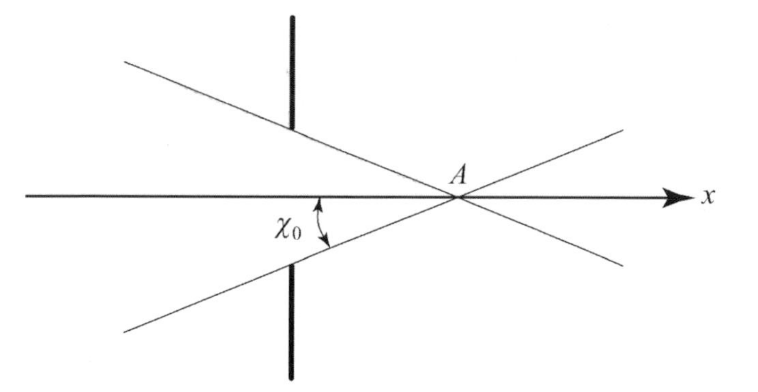

Figure 7-10. The geometry to be used for Problem 7.13 in determining the number density of a free-molecular flow exiting through a round hole.

Solution: The distribution function outside and inside of the cone of influence may be expressed in the form:

$$f_1 = 0 \ ,$$

$$f_2 = n_0 \left(\frac{m}{2\pi kT} \right)^{3/2} \exp\left(-c^2\right) .$$

From this, the number density at A is determined to be:

$$n(x) = \tfrac{1}{2} n_0 \left[1 - \frac{x}{\sqrt{R^2 + x^2}} \right] .$$

7.14. A gas with number density, n_0, passes outward through a small slot by means of effusion (pressure driven motion of a gas from a vessel to a vacuum through a small hole). Determine the gas number density at a large distance, $\tilde{r}$, from the slot ($\tilde{r} \gg \sqrt{S_0}$, where S_0 is the slot area). Use the geometry in Fig. 7-11.

Solution: The distribution functions outside (1) and inside (2) of the cone of influence are given by:

$$f_1 = 0 \text{ and } f_2 = n_0 \left(\frac{m}{2\pi kT} \right)^{3/2} \exp\left(-c^2\right), \text{ respectively.}$$

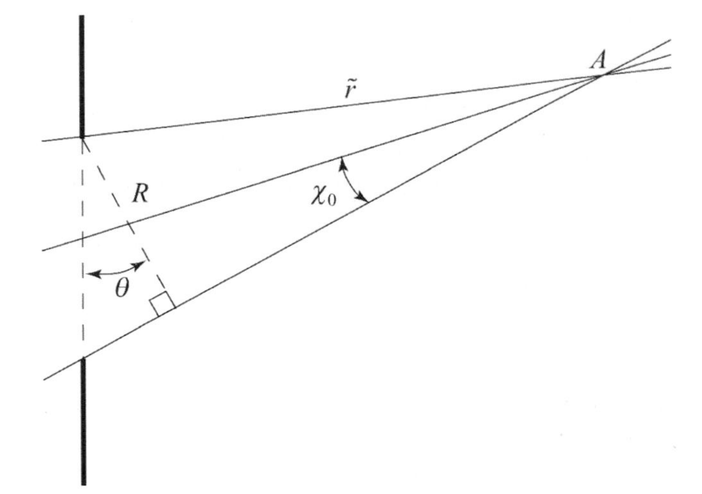

Figure 7-11. The geometry to be used for Problem 7.14 in determining the number density of an effusing gas at large distances from a small slot.

Taking into account the condition, $\tilde{r} \gg R$, after integration one can obtain:

$$n(\tilde{r}) = \frac{n_0}{4}\frac{R^2}{\tilde{r}^2} = \frac{n_0 S}{4\pi\tilde{r}^2} = \frac{n_0 S_0 \cos(\theta)}{4\pi\tilde{r}^2}.$$

7.15. Determine the number density of a gas at a point, A, behind a round disk of radius, $R \ll \lambda$. The mean gas velocity, $|\mathbf{u}|$, is assumed to be arbitrary. The flow geometry is given in Problem 7.12 (see Fig. 7-9). Consider the particular case for which:

$$u \gg (2kT_0/m)^{1/2} \text{ and } \cos(\chi_0) \gg (2kT_0/m)^{1/2} u^{-1}.$$

Solution: The distribution functions outside and inside of the cone of influence are given by:

$$f_1 = n_0 \left(\frac{m}{2\pi k T_0}\right)^{3/2} \exp\left(-\left(\mathbf{c} - \mathbf{u}^*\right)^2\right),$$

$$f_2 = \left(\frac{m}{2\pi kT_0}\right)^{3/2}$$
$$\times\left\{\alpha_\tau n_r \exp\left(-c^2\right) + \left(1-\alpha_\tau\right) n_0 \exp\left(-\left(c_x + u^*\right)^2 - c_y^2 - c_z^2\right)\right\} .$$

The number density, n_r, of reflected molecules may be specified from the impenetrability condition which yields:

$$n_r = n_0\left[\exp\left(-u^{*2}\right) - \sqrt{\pi}u^*\left(1 - \text{erf}\left(u^*\right)\right)\right] .$$

Having employed standard integration techniques (see Section 7.3), one can obtain:

$$\begin{aligned} n(x) = n_0 \Big\{ & \tfrac{1}{2}(2-\alpha_\tau) + \tfrac{1}{2}\alpha_\tau\left(1-\cos(\chi_0)\right) \\ & \times\left[\exp\left(-u^{*2}\right) - \sqrt{\pi}u^*\left(1-\text{erf}\left(u^*\right)\right)\right] \\ & + \tfrac{1}{2}\alpha_\tau \cos(\chi_0)\exp\left(-u^{*2}\sin^2(\chi_0)\right) \\ & \tfrac{1}{2}(2-\alpha_\tau)\left[\cos(\chi_0)\,\text{erf}\left(u^*\cos(\chi_0)\right)\right. \\ & \left.\times\exp\left(-u^{*2}\sin^2(\chi_0)\right) - \text{erf}\left(u^*\right)\right]\Big\} , \end{aligned}$$

where:

$$\cos(\chi_0) = \frac{x}{\sqrt{R^2 + x^2}} .$$

In the particular case when $u^* \cos(\chi_0) \gg 1$, this expression reduces to:

$$n(x) = n_0 \cos(\chi_0)\exp\left(-u^{*2}\sin^2(\chi_0)\right) .$$

7.16. Determine the frictional force on the lower plate in a system where the upper of two parallel plates is moving horizontally. Use the geometry in Fig. 7-12 and assume that $|\mathbf{u}| \ll (2kT/m)^{1/2}$.

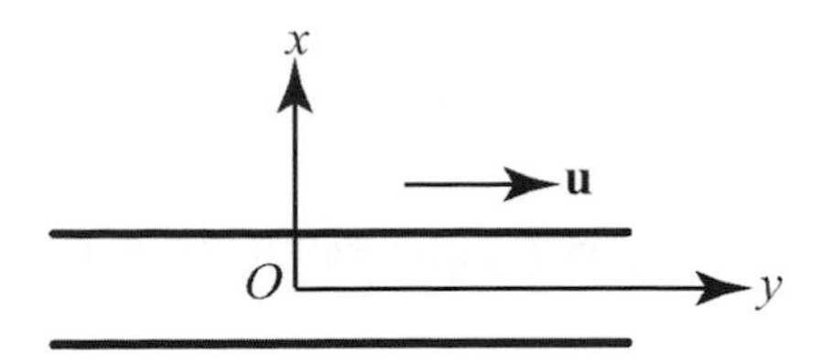

Figure 7-12. The geometry to be used for Problem 7.16 in determining the frictional force on a plate due to the horizontal motion of a second, parallel plate.

Solution: The distribution function for $v_x > 0$ and $v_x < 0$ may be expressed as $f^{\pm} = f^{(0)}\left(1 + a^{\pm} c_y\right)$ where $a^{\pm} = \text{const}$. Using the Maxwellian boundary condition for each plate, one can obtain:

$$a^{+} = \frac{2(1-\alpha_{\tau})}{2-\alpha_{\tau}} u^{*} \quad , \quad a^{-} = -\frac{2}{2-\alpha_{\tau}} u^{*} \quad , \quad u^{*} = \left(\frac{m}{2kT}\right)^{1/2} u \; .$$

The force of friction per unit surface area is given by:

$$F_y = -\sum_{\pm} \int_{\pm} m v_x v_y f^{\pm} d\mathbf{v} = \frac{\alpha_{\tau}}{2-\alpha_{\tau}} \left(\frac{m}{2\pi kT}\right)^{1/2} pu \; ,$$

where p is the hydrostatic pressure.

7.17. Assume that a gas is flowing in the y-direction between two infinite planes separated in the x-direction by a distance, d, where $d \ll \lambda$ and λ is the mean free path of the gas molecules. Assume that the number density of the gas is maintained at different steady values at two different locations along the y-axis and that the temperature of this system is uniform. Determine the number of molecules per unit area per second flowing in y-direction.

Solution: The distribution function may be chosen in the form:

$$f = f^{(0)}\left(1 + \tilde{y} g_n + \Phi(\mathbf{c}, \tilde{x})\right), \text{ where } g_n = (1/n_0)(dn/d\tilde{y})$$

The correction to the distribution function, $\Phi(\mathbf{c}, \tilde{x})$, which describes the gas-surface interactions may be specified from the Boltzmann equation such that:

$$c_x \frac{\partial \Phi}{\partial \tilde{x}} + c_y g_n = 0 .$$

The correction, $\Phi(\mathbf{c}, \tilde{x})$, may be expressed in the form:

$$\Phi(\mathbf{c}, \tilde{x}) = \tfrac{1}{2}\left(a^+(\tilde{x}) + a^-(\tilde{x})\right)c_y + \tfrac{1}{2}\left(a^+(\tilde{x}) - a^-(\tilde{x})\right)c_y \operatorname{sign}(c_x)$$

The half-range moment equations are obtained by multiplying the Boltzmann equation by $c_y\left(1 \pm \operatorname{sign}(c_x)\right)\exp\left(-c^2\right)d\mathbf{c}$ and integrating over all velocities. These equations are found to be:

$$\pm \frac{da^{\pm}(\tilde{x})}{d\tilde{x}} = -\sqrt{\pi}\, g_n .$$

The boundary conditions are given by $a^{\pm}(\mp d/2) = (1-\alpha_\tau)a^{\mp}(\mp d/2)$. After simple calculations one finds:

$$J_y = n_0 u_y = \tfrac{1}{2} n_0 \left(\frac{2kT}{m}\right)^{1/2}\left(a^+ + a^-\right) = -\frac{2-\alpha_\tau}{\alpha_\tau} \tfrac{1}{8}\pi d \left(\frac{8kT}{\pi m}\right)^{1/2} \frac{\partial n}{\partial \tilde{y}} .$$

7.18. Two vessels having volumes, V' and V'', are joined by a long capillary where R and L are the capillary radius and length, respectively, $L \gg R$, and $R \ll \lambda$ where λ is the mean free path of the gas molecules. The two vessels are maintained at different temperatures, T' and T'' with $T' > T''$. At the beginning $(t = 0)$ the vessels are filled with different gases to the same pressure. When the capillary is opened a pressure difference, $\delta p = p'' - p'$, will arise. Determine the time dependence of this pressure difference.

Solution: To describe this transport problem one should start with the basic equations of balance for the number of molecules of each of the two gases which are given by:

$$-V' \frac{dn_i'}{dt} = J_{zi} , \qquad \text{(P-2)}$$

$$V'' \frac{dn_i''}{dt} = J_{zi} , \qquad \text{(P-3)}$$

where J_{zi} is the i-th constituent molecular number flux across the tube cross section per second. For the free-molecular regime, J_{zi}, may be expressed in the form [21-23]:

$$J_{zi} = -\frac{2-\alpha_{i\tau}}{\alpha_{i\tau}} \tfrac{2}{3}\pi R^3 \sqrt{\frac{8k}{\pi m_i}} \frac{d}{d\tilde{z}} n_i \sqrt{T} ,$$

where $\alpha_{i\tau}$ is the tangential momentum accommodation coefficient of the i-th constituent. This flux may be approximately represented by:

$$J_{zi} = -a_i \left(n_i'' \sqrt{T''} - n_i' \sqrt{T'} \right), \text{ where:}$$

$$a_i = \frac{2-\alpha_{i\tau}}{\alpha_{i\tau}} \frac{2\pi R^3}{3L} \sqrt{\frac{8k}{\pi m_i}} .$$

Then, the basic equations describing kinetics of this system take the form:

$$\frac{dn_i'}{dt} = -\frac{a_i \sqrt{T'}}{V'} n_i' + \frac{a_i \sqrt{T''}}{V'} n_i'' , \tag{P-4}$$

$$\frac{dn_i''}{dt} = \frac{a_i \sqrt{T'}}{V''} n_i' - \frac{a_i \sqrt{T''}}{V''} n_i'' . \tag{P-5}$$

The necessary initial conditions are given by:

$$n_1'(0) = 0,\ n_1'' = \frac{p(0)}{kT''},\ n_2'(0) = \frac{p(0)}{kT'}, \text{ and } n_2'' = 0 .$$

This formulation then yields:

$$\delta p(t)=k\left[T''(n_1''+n_2'')-T'(n_1'+n_2')\right]=$$

$$\frac{p(0)}{\delta_0}\left\{\left(1+\frac{V'}{V''}\frac{T''}{T'}\right)\left(\sqrt{\frac{T'}{T''}}-\frac{T'}{T''}\right)\right.$$

$$\left.-\left(\frac{V'}{V''}+\frac{T'}{T''}\right)\left(\sqrt{\frac{T''}{T'}}\exp(-\omega_2 t)-\exp(-\omega_1 t)\right)\right\},$$

where:

$$\delta_0=\sqrt{\frac{T'}{T''}}+\frac{V'}{V''} \text{ and } \omega_i=a_i\left[\frac{\sqrt{T'}}{V'}+\frac{\sqrt{T''}}{V''}\right].$$

7.19. Two vessels having volumes, V' and V'', are joined by a long capillary where R and L are the capillary radius and length, respectively, $L \gg R$, and $R \ll \lambda$ where λ is the mean free path of the gas molecules. The two vessels are maintained at different temperatures, T' and T'' with $T' > T''$. At the beginning $(t=0)$ both vessels are filled with a simple gas to the same pressure. When the capillary is opened a pressure difference, $\delta p(t)=p''-p'$, will arise. Determine the time dependence of this pressure difference and the steady-state pressure difference that develops.

Solution: The pressure difference may be obtained from solution of Problem 7.18. in which one should substitute $\omega_1=\omega_2=\omega$. This yields:

$$\delta p(t)=-\frac{p(0)}{\delta_0}\left(1-\sqrt{\frac{T''}{T'}}\right)\left(\frac{V'}{V''}+\frac{T'}{T''}\right)\left[1-\exp(-\omega t)\right], \text{ where:}$$

$$\omega=\frac{2-\alpha_\tau}{\alpha_\tau}\frac{2\pi R^3}{3L}\left(\frac{8kT''}{\pi m}\right)^{1/2}\left(\frac{1}{V''}+\frac{1}{V'}\sqrt{\frac{T'}{T''}}\right).$$

The steady-state (or stationary) pressure difference is then given by:

$$\Delta p_{stat}=\left|\delta p(\infty)\right|=\frac{p(0)}{\delta_0}\left(1-\sqrt{\frac{T''}{T'}}\right)\left(\frac{V'}{V''}+\frac{T'}{T''}\right), \text{ where:}$$

$$\delta_0=\sqrt{\frac{T'}{T''}}+\frac{V'}{V''}.$$

7.20. Two vessels having volumes, V' and V'', are joined by a long capillary where R and L are the capillary radius and length, respectively, $L \gg R$, and $R \ll \lambda$ where λ is the mean free path of the gas molecules. The two vessels are maintained at the same constant temperature, $T' = T'' = T$ and are filled at the beginning $(t=0)$ with a simple gas to different initial pressures, $p'(0)=p$ and $p''(0)=p+\Delta p$. When the capillary is opened the pressure difference, $\delta p(t)= p'' - p'$, will change as a function of time. Determine the time dependence of this pressure difference. Derive a formula for the tangential momentum accommodation coefficient if the pressure difference, $\delta p(t^*)$, is known at a particular time, t^*.

Solution: Using the following initial conditions:

$$n'(0)=\frac{p}{kT} \text{ and } n''(0)=\frac{p+\Delta p}{kT}$$

in Eqs. (P-4) and (P-5) (Problem 7.18), one can obtain:

$$\delta p(t)=\Delta p \exp(-\omega t) \text{ where:}$$

$$\omega=\frac{2-\alpha_\tau}{\alpha_\tau}\omega_d\,,\ \omega_d=\frac{2\pi R^3}{3LV}\left(\frac{8kT}{\pi m}\right)^{1/2} \text{ and } V=\frac{V'V''}{V'+V''}.$$

The expression for Δp yields:

$$\alpha_\tau=2\left\{1+\frac{1}{\omega_d t^*}\ln\left(\frac{\Delta p}{\delta p(t^*)}\right)\right\}^{-1}.$$

7.21. Two vessels having volumes, V' and V'', are joined by a long capillary where R and L are the capillary radius and length, respectively, $L \gg R$, and $R \ll \lambda$ where λ is the mean free path of the gas molecules. The two vessels are maintained at the same constant temperature, $T' = T'' = T$ and are filled at the beginning $(t=0)$ with different gases to the same initial pressure. When the capillary is opened a pressure difference, $\delta p = p'' - p'$, will develop as a function of time. Determine the time dependence of this pressure difference and the maximum value of the pressure difference that develops.

Solution: The time dependence of the number densities in the two vessels is described by Eqs. (P-4) and (P-5) (Problem 7.18). The initial conditions are given by:

$$n_1'(0)=0,\ n_1''(0)=\frac{p(0)}{kT},\ n_2'(0)=\frac{p(0)}{kT},\ \text{and}\ n_2''(0)=0$$

This formulation yields a pressure difference of:

$$\delta p(t)=p(0)\left[\exp(-\omega_1 t)-\exp(-\omega_2 t)\right]\ \text{where:}$$

$$\omega_i=\frac{2-\alpha_{i\tau}}{\alpha_{i\tau}}\frac{2\pi R^3}{3LV}\left(\frac{8kT}{\pi m_i}\right)^{1/2}\ \text{and}\ V=\frac{V'V''}{V'+V''}.$$

The subtraction of the two decreasing exponential terms results in the existence of a pressure difference maximum, $\Delta p=|\delta p|_{\max}$. This phenomenon is usually called the diffusion-pressure effect. The value, Δp, is found to be:

$$\frac{\Delta p}{p(0)}=f(\omega^*)=\left(1-\frac{1}{\omega^*}\right)\exp(-u).$$

The time at which the maximum pressure difference, Δp, occurs is given by $t^*=u/\omega_2$. Here, the following notations have been introduced:

$$\omega^*=\frac{\alpha_{2\tau}(2-\alpha_{1\tau})}{\alpha_{1\tau}(2-\alpha_{2\tau})}\sqrt{\frac{m_2}{m_1}}\ \text{and}\ u=\frac{\ln(\omega^*)}{\omega^*-1}.$$

7.22. For the free-molecular regime, determine the gas temperature between two parallel plates having temperatures T_1 and T_2. Molecules are assumed to be reflected diffusely with arbitrary accommodation of energy at each surface. Use the geometry in Fig. 7-13:

Solution: The distribution functions for the sets of molecules having positive ($v_x>0$) and negative ($v_x<0$) motions are expressible as:

$$f^+=n_1\left(\frac{m}{2\pi kT_{1r}}\right)^{3/2}\exp\left(-\frac{mv^2}{2kT_{1r}}\right)\ \text{and:}$$

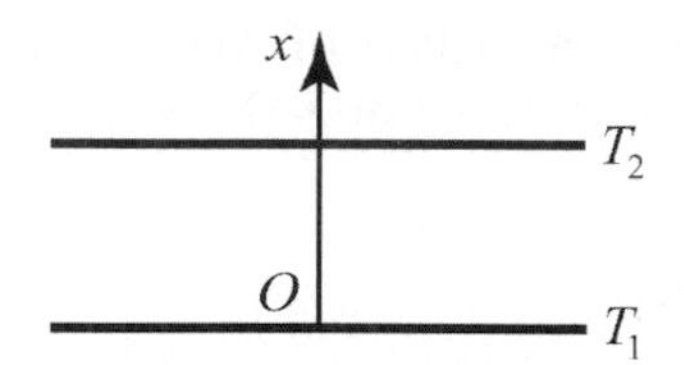

Figure 7-13. The geometry to be used in Problem 7.22 in determining the gas temperature between two parallel plates in the free-molecular regime.

$$f^- = n_2 \left(\frac{m}{2\pi k T_{2r}} \right)^{3/2} \exp\left(-\frac{mv^2}{2kT_{2r}} \right).$$

The number densities, n_1, and n_2 , may be found from the relations:

$$N_x = \int_+ v_x f^+ d\mathbf{v} + \int_- v_x f^- d\mathbf{v} = 0 \text{ and } n = n_1 + n_2,$$

where n is the gas number density. These yield:

$$n_1 = n\frac{a}{1+a} \text{ and } n_2 = n\frac{1}{1+a} \text{ where } a = \left(\frac{T_{2r}}{T_{1r}} \right)^{1/2}.$$

The quantities, T_{1r} , and T_{2r} , are specified by the boundary conditions:

$$\alpha_T = \frac{Q_x}{[Q_{1x}]_{eq}} = \frac{Q_x}{[Q_{2x}]_{eq}} \text{ where:}$$

$$[Q_{1x}]_{eq} = \tfrac{1}{2} m \left[\int_+ v^2 v_x f_{eq}^+ d\mathbf{v} + \int_- v^2 v_x f^- d\mathbf{v} \right] \text{ and:}$$

$$[Q_{2x}]_{eq} = \tfrac{1}{2} m \left[\int_+ v^2 v_x f^+ d\mathbf{v} + \int_- v^2 v_x f_{eq}^- d\mathbf{v} \right].$$

The additional unknown parameters, $[n_1]_{eq}$ and $[n_2]_{eq}$, which are included in f_{eq}^+ and f_{eq}^- can be expressed in terms of T_{1r}, and T_{2r} using the impenetrability conditions $[N_{1x}]_{eq}=0$ and $[N_{2x}]_{eq}=0$. The gas temperature is found to be:

$$T=\frac{m}{3nk}\int v^2 f d\mathbf{v}=(T_{1r}T_{2r})^{1/2}=$$

$$(2-\alpha_T)^{-1}\left[(T_1+(1-\alpha_T)T_2)(T_2+(1-\alpha_T)T_1)\right]^{1/2}.$$

7.23. Determine the free-molecular heat flux in a monatomic gas between two coaxial cylinders the radii and temperatures of which are $R_1, T_0+\Delta T$, and R_2, T_0 respectively, and where $R_1<R_2$. Assume that $\Delta T \ll T_0$ and that the number density of the gas is n_0 at $\tilde{r}=R_1$. The tangential momentum and energy accommodation coefficients are assumed to be different for the inner and outer cylinders.

Solution: The zone of influence around a cylindrical body is analogous to the cone of influence found around a spherical body or associated with a circular orifice. It is formed by the intersection of two planes that lie tangent to the cylinder in question. The distribution functions outside and inside the zone of influence of the inner cylinder may be chosen in the form:

$$f_1=f^{(0)}\left(1+\nu_1+c^2-\tfrac{3}{2}\tau_1\right) \text{ and } f_2=f^{(0)}\left(1+\nu_2+c^2-\tfrac{3}{2}\tau_2\right).$$

The boundary conditions are given by:

$$\alpha_{1T}=\frac{Q_r(1)}{Q_r^-(1)+\left[Q_r^+(1)\right]_{eq}},\quad \alpha_{2T}=\frac{Q_r(z)}{Q_r^+(z)+\left[Q_r^-(z)\right]_{eq}}$$

$$N_r(1)=0,\quad n(1)=n_0,\quad N_r(z)=0,$$

$$N_r^-(1)+\left[N_r^+(1)\right]_{eq}=0, \text{ and } N_r^+(z)+\left[N_r^-(z)\right]_{eq}=0,$$

where $z=R_2/R_1$. In calculating the fluxes using this formulation, one has to allow for the discontinuity of the distribution functions. For example, the heat flux, if the reflected molecules have the same temperature as the surface of the outer cylinder, may be calculated in the following way:

$$Q_r^+(z)+\left[Q_r^-(z)\right]_{eq}=\frac{p_0}{\pi^{3/2}}\left(\frac{2kT_0}{m}\right)^{1/2}\alpha_{2\tau}$$

$$\times\left\{-\pi\nu_{eq}^- + \int_0^\infty c_\rho^2\exp\left(-c_\rho^2\right)dc_\rho \int_{-\infty}^{\infty}\left(c_\rho^2+c_z^2\right)\exp\left(-c_z^2\right)dc_z\right.$$

$$\times\left[\int_0^\alpha \Phi_1^+(\mathbf{c},z)\sin(\theta)d\theta+\int_\alpha^{\pi-\alpha}\Phi_2^+(\mathbf{c},z)\sin(\theta)d\theta+\right.$$

$$\left.\left.+\int_{\pi-\alpha}^{\pi}\Phi_1^+(\mathbf{c},z)\sin(\theta)d\theta\right]\right\},$$

where $\Phi_i^+(\mathbf{c},z)=\nu_i+\left[\left(c_\rho^2+c_z^2\right)-\frac{3}{2}\right]\tau_i$ and $\alpha=\arccos\left(z^{-1}\right)$. The heat flux per unit length of the cylinders is found to be:

$$Q_{FM}=\left[\frac{1}{2\pi R_1}\frac{1}{\alpha_{1\tau}\alpha_{1T}}+\frac{1}{2\pi R_2}\left(\frac{1}{\alpha_{2\tau}\alpha_{2T}}-1\right)\right]^{-1}\left(\frac{2k}{\pi m T_0}\right)^{1/2}p_0\Delta T\,.$$

7.24. Determine the free-molecular heat flux in a monatomic gas between two concentric spheres the radii and temperatures of which are $R_1, T_0+\Delta T$, and R_2, T_0, respectively, and where $R_1 < R_2$. Assume $\Delta T \ll T_0$ and that the number density of the gas is n_0 at $\tilde{r}=R_1$. The tangential momentum and energy accommodation coefficients are assumed to be different for the inner and outer spheres.

Solution: The distribution functions outside and inside the cone of influence of the inner sphere may be chosen in the form:

$$f_1=f^{(0)}\left(1+\nu_1+\left(c^2-\tfrac{3}{2}\right)\tau_1\right) \text{ and } f_2=f^{(0)}\left(1+\nu_2+\left(c^2-\tfrac{3}{2}\right)\tau_2\right).$$

The same boundary conditions are used as were used in Problem 7.23, but the integration technique is necessarily different from that used with the cylindrical geometry. An example of this integration is given by:

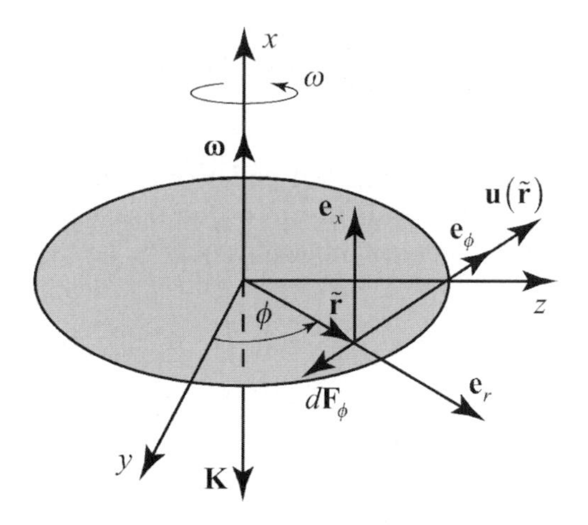

Figure 7-14. The geometry to be used for Problem 7.25 in determining the torque on a flat, rotating disk in the free-molecular regime.

$$Q_r^+(z)+\left[Q_r^-(z)\right]_{eq}=\frac{p_0}{\pi^{3/2}}\left(\frac{2kT_0}{m}\right)^{1/2}\alpha_{2\tau}$$

$$\times\left\{\left[-\pi\nu_{eq}^- + 2\pi\int_0^\infty c^5\exp\left(-c^2\right)dc\int_0^w \Phi_2^+(\mathbf{c},z)\sin(\chi)\cos(\chi)d\chi\right.\right.$$

$$\left.\left.+\int_w^{\pi/2}\Phi_1^+(\mathbf{c},z)\sin(\chi)\cos(\chi)d\chi\right]\right\},$$

where $\Phi_i^+(\mathbf{c},z)=\nu_i+\left(c^2-\frac{3}{2}\right)\tau_i$ and $w=\arcsin\left(z^{-2}\right)$. The heat flux is then found to be:

$$Q_{FM}=\left[\frac{1}{4\pi R_1^2}\frac{1}{\alpha_{1\tau}\alpha_{1T}}+\frac{1}{4\pi R_2^2}\left(\frac{1}{\alpha_{2\tau}\alpha_{2T}}-1\right)\right]^{-1}\left(\frac{2k}{\pi m T_0}\right)^{1/2}p_0\Delta T\ .$$

7.25. A flat disk of radius, $R \ll \lambda$, revolves in a rarefied gas with a constant angular speed, ω, where $\omega R \ll (2kT/m)^{1/2}$. Determine the torque on the disk, $\mathbf{K}$. Use the geometry in Fig. 7-14.

Solution: The torque, $\mathbf{K}$, in this geometry has only one component, K_x, which may be expressed as:

$$K_x = 2\int_0^R r dF_\phi = -4\pi \int_0^R r^2 dr \sum_{\pm} \int_{\pm} m v_\phi v_x f^{\pm}(\mathbf{v}) d\mathbf{v} \ ,$$

where the (±) notation refers to $v_x > 0$ and $v_x < 0$, respectively, and dF_ϕ is the force component acting on one side of the disk only. The distribution function of reflected and incident molecules may be written as:

$$f^+(\mathbf{c}) = \alpha_\tau f^{(0)} \left(1 + (2m/kT)^{1/2} c_\phi \omega r\right) + (1-\alpha_\tau) f^{(0)} (c_r, c_\phi, -c_x) \ ,$$

$$f^-(\mathbf{c}) = f^{(0)} \ .$$

Substituting these functions into the above expression, one can obtain:

$$K_x = -\tfrac{1}{2}\alpha_\tau \omega R^4 n (2\pi mkT)^{1/2} \ .$$

7.26. An infinite cylinder of radius, $R \ll \lambda$, revolves in a rarefied gas with a constant angular speed, ω, where $\omega R \ll (2kT/m)^{1/2}$. Determine the torque per unit length of the cylinder, $\mathbf{K}$.

Solution: The x-component of the torque is specified by:

$$K_x = -2\pi R^2 \sum_{\pm} \int_{\pm} m v_r v_\phi f^{\pm}(\mathbf{v}) d\mathbf{v} \ ,$$

where:

$$f^+ = \alpha_\tau f^{(0)} \left(1 + (2m/kT)^{1/2} c_\phi \omega R\right) + (1-\alpha_\tau) f^{(0)} (-c_r, c_\phi, c_x) \ ,$$

$$f^- = f^{(0)} \ .$$

and the direction of the x-axis is the same as the direction of $\boldsymbol{\omega}$. Simple integrations then yield:

$$K_x = -\alpha_\tau \omega R^3 n (2\pi mkT)^{1/2} \ .$$

7.27. A sphere of radius, $R \ll \lambda$, revolves in a rarefied gas with a constant angular speed, ω, where $\omega R \ll (2kT/m)^{1/2}$. Determine the torque, $\mathbf{K}$, on the sphere if the polar axis, x, is in the same direction as $\boldsymbol{\omega}$.

Solution: The vector, $\mathbf{K}$, has one component, K_x, which may be found from:

$$K_x = -2\pi R^3 \int_0^{\pi} \sin^2(\theta) d\theta \sum_{\pm} \int_{\pm} m v_r v_\phi f^{\pm}(\mathbf{v}) d\mathbf{v} \ , \text{ where:}$$

$$f^+ = \alpha_\tau f^{(0)} \left(1 + (2m/kT)^{1/2} c_\phi \omega R \sin(\theta)\right) + (1-\alpha_\tau) f^{(0)} \left(-c_r, c_\theta, c_\phi\right) ,$$

$$f^- = f^{(0)} \ .$$

After integration, one obtains:

$$K_x = -\tfrac{4}{3} \alpha_\tau \omega R^4 n (2\pi mkT)^{1/2} \ .$$

7.28. An equilibrium gas occupies a semi-infinite expanse, $x < 0$ and is contained by an infinite planar surface at $x = 0$. Determine the number density of the gas as a function of time in the neighborhood of the surface ($x \ll \lambda$) if the gas is expanding into a vacuum (the region of $x > 0$) after the sudden removal of the surface (which disappears at $t = 0$).

Solution: If $x \ll \lambda$, molecular encounters may be neglected. Therefore, the Boltzmann equation is given by:

$$\frac{\partial f}{\partial t} + v_x \frac{\partial f}{\partial x} = 0 \ .$$

The Lagrange auxiliary system is then:

$$\frac{dt}{1} = \frac{dx}{v_x} = \frac{df}{0} \ .$$

The two independent solutions of this system may be written as $x - v_x t = x_0$, and $f = \alpha$, where x_0 and α are arbitrary constants. The general solution of the Boltzmann equation is then expressed as $\Phi(x_0, \alpha) = 0$ where $\Phi(x_0, \alpha)$ is an arbitrary function. This yields:

$$f = \Psi(x_0) = \Psi(x - v_x t) .$$

Taking into account the initial condition, one can represent the distribution function in the form:

$$f = \begin{cases} f^{(0)} & ; \quad v_x > x/t \quad (x_0 < 0) , \\ 0 & ; \quad v_x < x/t \quad (x_0 > 0) , \end{cases}$$

where:

$$f^{(0)}(v) = n_0 \left(\frac{m}{2\pi k T_0} \right)^{3/2} \exp\left(-\frac{mv^2}{2kT_0} \right) .$$

Now, the number density may be calculated according to the following:

$$n(x,t) = n_0 \left(\frac{m}{2\pi k T_0} \right)^{3/2} \int_{-\infty}^{\infty} \int_{-\infty}^{\infty} \int_{x/t}^{\infty} \exp(-c^2) dv_y dv_z dv_x$$

$$= \frac{n_0}{\sqrt{\pi}} \int_{\frac{x}{t}\sqrt{\frac{m}{2kT_0}}}^{\infty} \exp(-c_x^2) dc_x = \frac{n_0}{2} \left(1 - \text{erf}\left(\frac{x}{t} \sqrt{\frac{m}{2kT_0}} \right) \right) .$$

7.29. A sphere of radius, $R \ll \lambda$ contains a gas with a constant density, n_0, and constant temperature, T_0. At some initial time, $t = 0$, the sphere vanishes releasing the gas into an infinite vacuum. Determine the number density, $n(r,t)$.

Solution: Given Eq. (7-65) in the text, the number density may be expressed as:

$$n(\tilde{\mathbf{r}},t) = t^{-3} \int f_i \left(\frac{\tilde{\mathbf{r}} - \mathbf{r}'}{t}, \mathbf{r}' \right) d\mathbf{r}' ,$$

where:

$$f_i = n_0(\mathbf{r}') \left(\frac{m}{2\pi k T_0} \right)^{3/2} \exp\left(-\frac{m}{2kT_0} \left(\frac{\tilde{\mathbf{r}} - \mathbf{r}'}{t} \right)^2 \right) ,$$

$$n_0(\mathbf{r}') = \begin{cases} n_0 & ; \quad r' \le R \ , \\ 0 & ; \quad r' > R \ . \end{cases}$$

This integration may be performed easily in spherical coordinates when the polar angle is taken as the angle between $\tilde{\mathbf{r}}$ and $\mathbf{r}'$. Having integrated over all of the angular variables, one obtains:

$$n(\tilde{r},t) = \frac{n_0}{\tilde{r}} \left(\frac{m}{2\pi k T_0 t^2} \right)^{1/2} \int_0^R r' \times \left[\exp\left(-\frac{m}{2kT_0 t^2} (\tilde{r} - r')^2 \right) - \exp\left(-\frac{m}{2kT_0 t^2} (\tilde{r} + r')^2 \right) \right] dr' \ .$$

Further integration with respect to r' yields:

$$n(\tilde{r},t) = \tfrac{1}{2} n_0 \left\{ \operatorname{erf}\left(\alpha (\tilde{r} + R) \right) - \operatorname{erf}\left(\alpha (\tilde{r} - R) \right) + \frac{1}{\alpha \sqrt{\pi} \tilde{r}} \left[\exp\left(-\alpha^2 (\tilde{r} + R)^2 \right) - \exp\left(-\alpha^2 (\tilde{r} - R)^2 \right) \right] \right\} ,$$

where $\alpha = \left(m / 2kT_0 t^2 \right)^{1/2}$.

REFERENCES

1. Kennard, E.H., *Kinetic Theory of Gases* (McGraw-Hill, New York, 1938).
2. Heineman, M., "Theory of Drag in Highly Rarefied Gases," *Comm. Pure Appl. Math.* **1(3)**, 259-273 (1948).
3. Patterson, G.N., *Molecular Flow of Gases* (Wiley and Sons, New York, 1956).
4. Schaaf, S.A. and Chambré, P.L., *Flow of Rarefied Gases* (Princeton University Press, Princeton, New Jersey, 1961).
5. Kogan, M.N., *Rarefied Gas Dynamics* (Plenum Press, NY, 1969).
6. Bird, G.A., *Molecular Gas Dynamics* (Clarendon Press, Oxford, 1976).
7. Szymanski, Zd., "Some Flow Problems of Rarefied Gases," *Arch. Mech. Stos. (Warsaw)* **8**, 449-470 (1956).
8. Ivchenko, I.N., "Generalization of the Lees Method in Boundary Problems of Transfer," *J. Coll. Interface Sci.* **135(1)**,16-19 (1990).
9. Wang-Chang, C.S. and Uhlenbeck, G.E., "Transport Phenomena in Very Dilute Gases," Report CM-579, UMH-3-F, University of Michigan, 1949.
10. Ziering, S., *On Transport Theory of Rarefied Gases* (Ph.D. Dissertation, Syracuse University, 1958).

11. Waldmann, L., "Uber die Kraft Eines Inhomogenen Gases auf Kleine Suspendierte Kugeln," *Z. fur Naturforsch.* **14a**, 589-599 (1959).
12. Brock, J.R., "The Thermal Force in the Transition Region," *J. Coll. Interface Sci.* **23(3)**, 448-452 (1967).
13. Ivchenko, I.N. and Yalamov, Yu.I., "On the Thermophoresis of Aerosol Particles in the Almost-Free-molecular Regime," *Izv. AN SSSR, M. Zh. G.* **(3)**, 3-7 (1970).
14. Talbot, L., Cheng, R.K., Schefer, R.W., and Willis, D.R., "Thermophoresis of Particles in a Heated Boundary Layer," *J. Fluid Mech.* **101(4)**, 737-758 (1980).
15. Williams, M.M.R., "On the Motion of Small Spheres in Gases, II: Thermo-phoresis, Diffusio-phoresis and Related Phenomena," *Z. fur Naturforsch.* **27a(12)**, 1804-1811 (1972).
16. Mason, E.A. and Malinauskas, A.P., *Gas Transport in Porous Media: The Dusty-Gas Model* (Elsevier, Amsterdam, Oxford, N.Y., 1983).
17. Fuchs, N.A., *Evaporation and Droplet Growth in Gaseous Media* (Pergamon Press, Oxford, 1959).
18. Mason, B.J., *Clouds, Rain and Rainmaking* (Cambridge University Press, 1962).
19. Williams, M.M.R., "On the Motion of Small Spheres in Gases III: Drag and Heat Transfer," *J. Phys. D: Appl. Phys.* **6**, 744-758 (1973).
20. Wang-Chang, C.S., "Transport Phenomena in Very Dilute Gases, II," Report CM-654, NORD 7924, UMH-3-F, University of Michigan, 1950.
21. Smoluchowski, M., "Zur Kinetischen Theorie der Transpiration und Diffusion Verdünnter Gase," *Annalen der Physik* **33(16)**, 1559-1570 (1910).
22. Waldmann, L. and Schmitt, K.H., "Über das bei der Gasdiffusion auftretende Druckgefälle," *Z. fur Naturforsch.* **16a**, 1343-1354 (1961).

Chapter 8

METHODS OF SOLUTION OF PLANAR PROBLEMS

1. MAXWELL'S METHOD.

For the last thirty years, the most progress in investigating transport problems has been achieved for one-dimensional, stationary transport problems. In this context, several analytical methods have been developed which have been applied to the classical slip problems. This has resulted in several analytical expressions for the slip coefficients. The recent development of direct numerical solutions for these same problems allows one to estimate the accuracies of the various analytical methods.

For simplicity, only planar transport problems are described in this chapter and all molecules are assumed to act as rigid spheres. Although the methods that are discussed in this chapter may be generalized for arbitrary intermolecular potential models, this chapter does not include the specific details necessary for such generalizations. However, some generalization details for certain selected methods are included in Chapter 9. The reason that the details of the generalization process have been limited in this book to selected methods is that the details necessary for the generalization of each method will typically be specific to that method and some of these generalizations can be very complex and difficult to achieve. The half-range moment method is a case in point and no attempt has been made in this book to generalize it although the basic method is described in Section 8.3.

In the case of planar transport problems there exist some simple approximate methods based on the use of conservation laws as exact moment solutions of the Boltzmann equation. First, consider the approximate method proposed by Maxwell to solve some boundary transport

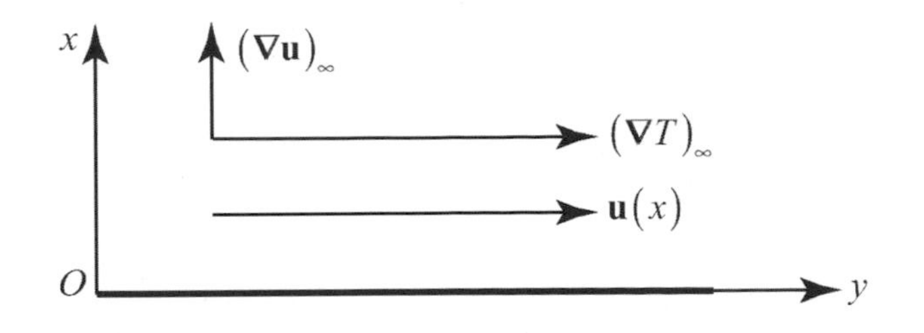

Figure 8-1. The geometry of slip problems.

problems [1]. The main features of this method will be analyzed by the solving of the slip boundary problems. First of all, the velocity-slip and thermal-creep problems will be examined.

One considers a semi-infinite expanse of a gas bounded by a flat plate located at $\tilde{x}=0$, and lying in the $y-z$ plane. Far from the plate the gas is maintained at a constant velocity gradient $\left(\partial u/\partial \tilde{x}\right)_\infty$, normal to the plate and at a constant temperature gradient $\left(\partial T/\partial \tilde{y}\right)_\infty$, tangential to the wall. The geometry of these slip-flow problems is illustrated in Fig. 8-1. These gradients are assumed to be small such that the relations $\left|\lambda \nabla u/u\right| \ll 1$ and $\left|\lambda \nabla \ln\left(T\right)\right| \ll 1$ are satisfied. Then, at large distances from the wall, the gas is described by the Chapman-Enskog distribution function.

The distribution function near the wall is a discontinuous function in the velocity space. Maxwell was the first to notice this feature of the distribution function and to show that by a simple consideration of the constancy of the stress and heat flux, some meaningful results for the slip terms could be obtained.

The influence of the wall near its surface may be taken into account by the introduction of the discontinuous term, $\Phi^{\pm}\left(\mathbf{v},\tilde{\mathbf{r}}\right)$, in the distribution function. This distribution function, for all planar and linearized transport problems, may be written as:

$$f^{\pm} = f_{eq}\left\{1+\Psi\left(\mathbf{v},\tilde{\mathbf{r}}\right)+\Phi^{\pm}\left(\mathbf{v},\tilde{\mathbf{r}}\right)\right\} , \tag{8-1}$$

where f_{eq} is the local Maxwellian distribution function given by Eq. (5-8) and $\Psi\left(\mathbf{v},\tilde{\mathbf{r}}\right)$ is the Chapman-Enskog correction to the distribution function.

For the slip-flow and thermal-creep problems the distribution functions f_{eq} and $\Psi\left(\mathbf{v},\tilde{\mathbf{r}}\right)$ may be written as:

$$f_{eq} = f^{(0)}\left[1+\left(c^2-\tfrac{5}{2}\right)\tilde{y}q_y+2c_y u^*\left(\tilde{x}\right)\right], \tag{8-2}$$

where:

$$f^{(0)} = n_0 \left(\frac{m}{2\pi k T_0} \right)^{3/2} \exp\left(-c^2\right)$$

and:

$$\Psi\left(\mathbf{c}, \tilde{\mathbf{r}}\right) = \Psi_1\left(\mathbf{c}, \tilde{\mathbf{r}}\right) + \Psi_2\left(\mathbf{c}, \tilde{\mathbf{r}}\right) = c_y \phi_T\left(c\right) q_y + c_x c_y \phi_u\left(c\right) h \ , \tag{8-3}$$

where $q_y = T_0^{-1}\left(\partial T / \partial \tilde{y}\right)_\infty$, T_0 is the temperature in the plane $\tilde{y} = 0$, and $h = \left(\partial u^* / \partial \tilde{x}\right)_\infty$. The remaining quantities are given by:

$$u^*\left(\tilde{x}\right) = \left(\frac{m}{2kT_0} \right)^{1/2} u\left(\tilde{x}\right) \ ,$$

$$\phi_T\left(c\right) = -A\left(c\right) \ ,$$

and:

$$\phi_u\left(c\right) = -2\left(\frac{2kT}{m} \right)^{1/2} B\left(c\right) \ .$$

The first approximations to the functions, $A\left(c\right)$ and $B\left(c\right)$, have been found in Chapter 5 to be:

$$A\left(c\right) = -\tfrac{15}{16}\sqrt{\pi}\lambda S_{3/2}^{(1)}\left(c^2\right) \ ,$$

and:

$$B\left(c\right) = \tfrac{5}{8}\sqrt{\pi}\lambda\left(\frac{m}{2kT} \right)^{1/2} \ .$$

where molecules are assumed to be rigid spheres.

The Navier-Stokes equations are used in the Maxwellian analysis to obtain the dependence of the gas velocity on the normal coordinate. These equations are the moment equations which for the current problem may be written as:

$$\frac{\partial}{\partial \tilde{x}} \int v_x f d\mathbf{v} = 0 \ , \tag{8-4a}$$

$$\frac{\partial}{\partial \tilde{x}} \int v_x^2 f d\mathbf{v} = 0 \ , \tag{8-4b}$$

$$\frac{\partial}{\partial \tilde{x}} \int v_x v_y f d\mathbf{v} = 0 \ . \tag{8-4c}$$

If the influence of the wall could be neglected ($\Phi^{\pm} = 0$ in Eq. (8-1)), then from Eq. (8-4c) one would have a linear dependence of the mean velocity of the gas on the x-coordinate. Then, allowing for the velocity of the wall equal to zero, one would obtain:

$$P_{xy} = \text{const} \ , \quad \left(\partial u^* / \partial \tilde{x}\right)_{\infty} = \text{const} \ , \quad u^*\left(\tilde{x}\right) = \tilde{x}\left(\partial u^* / \partial \tilde{x}\right)_{\infty} \ . \tag{8-5}$$

Now, the equation for the correction to the distribution function, $\Phi^{\pm}\left(\mathbf{c}, \tilde{x}\right)$, may be derived. Substituting Eq. (8-1) into the Boltzmann equation, one obtains:

$$c_y\left(c^2 - \tfrac{5}{2}\right) q_y + 2 c_x c_y h + c_x \frac{\partial \Phi^{\pm}}{\partial \tilde{x}} = \mathrm{J}\left(\Psi_1 + \Psi_2 + \Phi^{\pm}\right) \ . \tag{8-6}$$

Moreover, one has the following equations for the functions, $\phi_T\left(c\right)$ and $\phi_u\left(c\right)$:

$$c_y\left(c^2 - \tfrac{5}{2}\right) = \mathrm{J}\left(c_y \phi_T\left(c\right)\right) \ , \tag{8-7}$$

$$2 c_x c_y = \mathrm{J}\left(c_x c_y \phi_u\left(c\right)\right) \ , \tag{8-8}$$

where:

$$\mathrm{J}\left(\varphi\right) = -\frac{n_0^2}{f^{(0)}}\left(\frac{m}{2kT}\right)^{1/2} I\left(\varphi\right) \ ,$$

and $I(\varphi)$ is the standard collision operator given by Eq. (5-13).

Allowing for these equations, one may derive the basic equation for the correction, $\Phi(\mathbf{c},\tilde{x})$:

$$c_x \frac{\partial \Phi}{\partial \tilde{x}} = \mathrm{J}(\Phi) \ . \tag{8-9}$$

Now, consider the boundary condition for this equation. From the general form of the boundary conditions given by Eq. (6-4) one obtains for these slip problems:

$$\Phi^+(\mathbf{c},0) = -\alpha_\tau \Psi_1 - (2-\alpha_\tau)\Psi_2 + (1-\alpha_\tau)\Phi^-(\mathbf{c}',0) \ , \tag{8-10}$$

where $\mathbf{c}' = \{-c_x, c_y, c_z\}$. It is useful here to introduce a Hilbert space defined by the scalar product:

$$(\rho_1(x,\mathbf{c}), \rho_2(x,\mathbf{c})) = \int \rho_1(x,\mathbf{c}) \exp\{-c^2\} \rho_2(x,\mathbf{c}) d\mathbf{c} \ . \tag{8-11}$$

The Maxwellian analysis takes into account the influence of the wall by the choice of a definite form for the distribution function. Maxwell used the following approach for the distribution function of incident molecules:

$$\Phi^-(\mathbf{c},0) = a_0 c_y \ ,$$

$$\Phi^-(\mathbf{c},\infty) = \Phi^+(\mathbf{c},\infty) = a_0 c_y \ ,$$

where $a_0 = \mathrm{const}$. This says that the incident distribution function at the wall is identical to the distribution function far from the wall which implies that, in the Knudsen layer, collisions between the incident and reflected molecules have been neglected.

Now, to find an approximate solution for a_0, one proceeds in the following way. Taking the scalar product of Eq. (8-9) with c_y (in the manner of Eq. (8-11)) gives:

$$\frac{\partial}{\partial \tilde{x}} \left(c_x c_y \, , \Phi(\mathbf{c},\tilde{x}) \right) = 0 \ ,$$

i.e.:

$$\left(c_x c_y, \Phi(\mathbf{c}, \tilde{x})\right) = \text{const} \ . \tag{8-12}$$

This constant can be evaluated by the use of the asymptotic values, $a_0 c_y$, for $\Phi^{\pm}(\mathbf{c}, \infty)$ and thus, the relationship expressing conservation of tangential momentum may be written as:

$$\left(c_x c_y, \Phi^{\pm}(\mathbf{c}, 0)\right) = \left(c_x c_y, \Phi^{\pm}(\mathbf{c}, \infty)\right) = 0 \ . \tag{8-13}$$

The constant a_0 may be found by the use of the boundary condition, Eq. (8-10), in the scalar product, Eq. (8-13). It is the main supposition of the Maxwellian analysis that the discontinuity of the distribution function is allowed for only by using the boundary distribution function for reflected molecules. The Maxwellian distribution function does not, itself, satisfy the boundary condition exactly, but rather, the boundary condition is satisfied only in an integral sense.

Having performed the integration in Eq. (8-13), one obtains:

$$a_0 = \frac{2-\alpha_\tau}{\alpha_\tau} \tfrac{5}{8} \pi \lambda \left(\frac{\partial u^*}{\partial \tilde{x}}\right)_\infty + \tfrac{15}{32} \sqrt{\pi} \lambda T_0^{-1} \left(\frac{\partial T}{\partial \tilde{y}}\right)_\infty \ . \tag{8-14}$$

This expression results in relations for the slip-flow and thermal-creep coefficients. In accordance with the definitions of these coefficients, the mean velocity of the gas at large distances from the wall may be presented in the form:

$$u(\infty) = \left(c_m \lambda_\mu + \tilde{x}\right) \left(\frac{\partial u}{\partial \tilde{x}}\right)_\infty + c_{Tsl} \nu^* T_0^{-1} \left(\frac{\partial T}{\partial \tilde{y}}\right)_\infty \ , \tag{8-15}$$

where $\nu^* = \nu \lambda_\kappa / \lambda_\mu$ and ν is the kinematic viscosity. Here, two other representations of the mean free path have been introduced; specifically λ_μ and λ_κ. Later, it will be useful to have the following expressions for these:

$$\lambda_\mu = \tfrac{8}{5} \frac{\mu}{\sqrt{\pi} p} \left(\frac{2kT}{m}\right)^{1/2} = \lambda \mu / \mu^{(1)} \ , \tag{8-16a}$$

and:

$$\lambda_\kappa = \tfrac{64}{75}\frac{\kappa}{\sqrt{\pi}\,p}\left(\frac{mT}{2k}\right)^{1/2} = \lambda\kappa/\kappa^{(1)} \ , \tag{8-16b}$$

where $\mu^{(1)}$ and $\kappa^{(1)}$ are the viscosity and thermal conductivity coefficients obtained from the first-order approximation in the Chapman-Enskog theory. This choice for the Maxwellian analysis provides independence of the slip-flow coefficients on the order of the Chapman-Enskog solution. Since only the first-order Chapman-Enskog solution is employed in this section ($\lambda_\mu = \lambda_\kappa = \lambda$), the slip factors in Eq. (8-15) are given by:

$$c_m = \frac{2-\alpha_\tau}{\alpha_\tau}\tfrac{5}{16}\pi \ , \tag{8-17a}$$

and:

$$c_{Tsl} = \tfrac{3}{4} \ , \tag{8-17b}$$

Next, consider the temperature-jump problem. One may use the Maxwell method to solve this problem. Consider a semi-infinite expanse of gas bounded by a plate oriented in the $y-z$ plane and located at $\tilde{x}=0$. Assume that, far from the plate, the temperature of the gas depends only on the x-coordinate and that the temperature gradient is small such that the relationship $|\lambda\nabla\ln(T)| \ll 1$ is satisfied. If this condition holds one may use the Chapman-Enskog distribution function at large distances from the wall and $\partial\ln(T)/\partial\tilde{x}$ may be expressed as $q_x = T_0^{-1}\,\partial T/\partial\tilde{x}$, where T_0 is the equilibrium temperature. The temperature is assumed to deviate only slightly from the equilibrium value.

To write the correct form of the distribution function far from the wall one should take into account the conservation law obtained from the Boltzmann equation which, for this problem, is given by:

$$\frac{\partial}{\partial\tilde{x}}\int v_x v^2 f d\mathbf{v} = 0 \ , \tag{8-18}$$

This implies that:

$$\int v_x v^2 f d\mathbf{v} = \text{const} \ . \tag{8-19}$$

From this equation, if $\tilde{x} \to \infty$ such that $f \to f_{eq}\left(1 + c_x \phi_T(\mathbf{c}) q_x\right)$, then $(\partial T / \partial \tilde{x})_\infty = \text{const}$. The distribution function for this problem may be written as:

$$f^{\pm} = f^{(0)} \left\{ 1 + \left(c^2 - \tfrac{5}{2}\right) \tilde{x} q_x + c_x \phi_T(\mathbf{c}) q_x + \left(c^2 - \tfrac{5}{2}\right) a_1^{\pm}(\tilde{x}) \right\} , \tag{8-20}$$

where $a_1^{\pm}(\tilde{x})$ is a correction to the distribution function which takes into account the influence of the wall.

The following approach has been used in the Maxwellian analysis:

$$a_1^{\pm}(\infty) = a_1 \ , \ a_1^{-}(0) = a_1 \ . \tag{8-21}$$

The distribution function of the molecules incident on the wall is assumed to be the same as that far from the wall. The distribution function of the reflected molecules on the wall may be found by the use of a boundary condition, which may be written as:

$$\begin{aligned} f^{+}(\mathbf{v},0) &= (1 - \alpha_\tau) f^{-}(-v_x, v_y, v_z, 0) \\ &\quad + \alpha_\tau f^{(0)} \left(1 + v_r + \left(c^2 - \tfrac{3}{2}\right) \tau_r\right) , \end{aligned} \tag{8-22}$$

where v_r and τ_r are unknown functions which may be determined by using the impenetrability condition and introducing the thermal accommodation coefficient. For the equilibrium conditions at the wall, $\tau_w = 0$, but $v_w \neq 0$. This correction to the equilibrium distribution function may be found from the condition:

$$\int v_x f(\mathbf{v},0) d\mathbf{v} = \alpha_\tau \left\{ \int_{-} v_x f^{-}(\mathbf{v},0) d\mathbf{v} + \int_{+} v_x f^{(0)} (1 + v_w) d\mathbf{v} \right\} = 0 \ . \tag{8-23}$$

From Eq. (8-23) one may obtain:

$$v_w = -\tfrac{1}{2} a_1 \ . \tag{8-24}$$

The conservation law given by Eq. (8-19) may be written as:

$$\int v_x v^2 f(\mathbf{v},0) d\mathbf{v} = \int v_x v^2 f(\mathbf{v},\infty) d\mathbf{v} \ ,$$

and, using the boundary conditions, one can transform this relation to the form:

$$\alpha_\tau \alpha_T \left\{ \int_- v_x v^2 f^- (\mathbf{v},0) d\mathbf{v} + \int_+ v_x v^2 f^{(0)} (1+\nu_w) d\mathbf{v} \right\} = \int v_x v^2 f(\mathbf{v},\infty) d\mathbf{v} \ . \tag{8-25}$$

Performing the integrations in Eq. (8-25), one can obtain:

$$\Delta T = a_1 T_0 = \frac{2-\alpha_\tau \alpha_T}{\alpha_\tau \alpha_T} \tfrac{75}{128} \pi \lambda \left(\frac{\partial T}{\partial \tilde{x}} \right)_\infty \ . \tag{8-26}$$

Thus, the Maxwellian method may be used successfully to obtain expressions for the slip coefficients. Within the framework of the moment theory, this method may be classified as a one-moment approach because only one moment has been used. The distribution function for the incident molecules at the wall is the same as that in the ambient gas at large distances from the wall and, therefore, the distribution function is independent of the normal coordinate. This does not permit one to obtain spatial distributions of macroscopic values for the gas in the Knudsen layer near the wall; a serious situation which comprises the main deficiency of the Maxwellian method.

2. LOYALKA'S METHOD.

The Maxwellian method for determining the distribution function of incident molecules at the wall may be generalized by a simple modification suggested by Loyalka [2]. Consider the details of this generalization in the solutions of the slip-flow and temperature-jump problems. To obtain simple solutions for the slip coefficients, one considers a gas flow bounded by a flat wall for the conditions described in Section 8.1.

The distribution function for these slip problems has the form:

$$f = f^{(0)} \left\{ 1 + \left(c^2 - \tfrac{5}{2} \right) \tilde{y} q_y + 2 c_y \tilde{x} h + c_y \phi_T(c) q_y + c_x c_y \phi_u(c) h + \Phi^{\pm}(\mathbf{c}, \tilde{x}) \right\} , \tag{8-27}$$

where $q_y = T_0^{-1}\left(\partial T/\partial \tilde{y}\right)_\infty$, $h = \left(\partial u^*/\partial \tilde{x}\right)_\infty$, and $c_y\phi_T(c)q_y$ and $c_xc_y\phi_u(c)h$ are the Chapman-Enskog corrections to the distribution function for thermal conductivity and viscosity, respectively. The correction that allows for the influence of the wall may be found from the equation:

$$c_x \frac{\partial \Phi^\pm}{\partial \tilde{x}} = \mathrm{J}\left(\Phi^\pm\right). \tag{8-28}$$

For $\tilde{x} \to \infty$, the functions $\Phi^\pm(\mathbf{c}, \tilde{x})$ have the form:

$$\lim_{\tilde{x}\to\infty} \Phi^\pm(\mathbf{c}, \tilde{x}) = a_0 c_y \ ,$$

where a_0 is a constant. If the two-moment approach is used, then the correction, $\Phi^\pm(\mathbf{c}, \tilde{x})$, may be expressed as:

$$\Phi^\pm(\mathbf{c}, \tilde{x}) = a_0^\pm(\tilde{x}) c_y \ . \tag{8-29}$$

The Maxwellian analysis is constructed on the assumption that $a_0^\pm(\infty) = a_0^-(0) = a_0$. Loyalka has proposed a simple generalization of Maxwell's method which assumes that the distribution function of incident molecules has the following form at the wall:

$$\Phi^-(\mathbf{c}, 0) = a_0^-(0) c_y = a_w c_y \ , \tag{8-30}$$

where a_w is an unknown constant. From a physical viewpoint, the relation $a_w \neq a_0$ implies that the incident stream has a mass velocity different from that of the hydrodynamic solution at the wall.

The unknown constants, a_0 and a_w, may be found by the consideration of some scalar products of Eq. (8-28). First, the scalar product given by Eq. (8-13) is used. Next, the scalar product of Eq. (8-28) with $c_xc_y\phi_u(c)$ gives:

$$\frac{\partial}{\partial \tilde{x}}\left(c_x^2 c_y \phi_u(c), \Phi^\pm(\mathbf{c}, \tilde{x})\right) = \left(c_x c_y \phi_u(c), \mathrm{J}\left[\Phi^\pm(\mathbf{c}, \tilde{x})\right]\right) =$$
$$\left(\mathrm{J}\left[c_x c_y \phi_u(c)\right], \Phi^\pm(\mathbf{c}, \tilde{x})\right) = 2\left(c_x c_y \, , \Phi^\pm(\mathbf{c}, \tilde{x})\right) = 0 \ ,$$

and, therefore:

$$\left(c_x^2 c_y \phi_u(c), \Phi^\pm(\mathbf{c}, \tilde{x})\right) = \text{const} \ . \tag{8-31}$$

To obtain this expression, Eq. (8-8) and the commutative property of the standard bracket integrals (which is due to the self-adjointness of the collision operator) have been used. Now there are two equations that may be represented as:

$$\left(c_x c_y , \Phi^{\pm}(\mathbf{c},0)\right) = 0 \ , \tag{8-32}$$

and:

$$\left(c_x^2 c_y \phi_u(c), \Phi^{\pm}(\mathbf{c},0)\right) = \left(c_x^2 c_y \phi_u(c), \Phi^{\pm}(\mathbf{c},\infty)\right) . \tag{8-33}$$

Using the boundary condition given by Eq. (8-10) with Eqs. (8-32) and (8-33), one can obtain:

$$a_w = \tfrac{15}{32}\sqrt{\pi}\lambda q_y + \frac{2-\alpha_\tau}{\alpha_\tau}\tfrac{5}{8}\pi\lambda h \ , \tag{8-34}$$

and:

$$a_0 = \left(1 + \tfrac{1}{2}\alpha_\tau\right)\tfrac{15}{32}\sqrt{\pi}\lambda q_y + \frac{2-\alpha_\tau}{\alpha_\tau}\left(1 + \frac{4-\pi}{2\pi}\alpha_\tau\right)\tfrac{5}{8}\pi\lambda h \ . \tag{8-35}$$

Taking into account Eq. (8-15), the following expressions for the slip coefficients may be written:

$$c_m = \frac{2-\alpha_\tau}{\alpha_\tau}\left(1 + \frac{4-\pi}{2\pi}\alpha_\tau\right)\tfrac{5}{16}\pi \ , \tag{8-36}$$

$$c_{Tsl} = \tfrac{3}{4}\left(1 + \tfrac{1}{2}\alpha_\tau\right) . \tag{8-37}$$

In particular, for the very important case when $\alpha_\tau = 1$, the slip velocities are given by:

$$\Delta u = (1.116)\lambda\left(\frac{\partial u}{\partial \tilde{x}}\right)_\infty \ , \tag{8-38}$$

$$u_{sl} = \tfrac{9}{8}\nu T_0^{-1}\left(\frac{\partial T}{\partial \tilde{y}}\right)_{\infty} . \tag{8-39}$$

The analytical expressions for the slip coefficients are not altered for arbitrary models of the intermolecular potential. This assumption is correct within the framework of this method if one uses the first-order Chapman-Enskog solutions for thermal conductivity and viscosity to derive these formulas.

Next, consider the solution of the temperature-jump problem by Loyalka's method. The same conditions are used for the gas at large distances from the wall as were used in the Maxwellian analysis. Therefore, the distribution function may be written as:

$$f^{\pm} = f^{(0)}\left\{1 + \left(c^2 - \tfrac{5}{2}\right)\tilde{x}q_x + c_x\phi_T(c)q_x + \Phi^{\pm}(\mathbf{c},\tilde{x})\right\} , \tag{8-40}$$

where $q_x = T_0^{-1}\left(\partial T/\partial \tilde{x}\right)_{\infty}$ and $\Phi^{\pm}(\mathbf{c},\tilde{x})$ is the correction that allows for the influence of the wall. This correction is assumed to have the form:

$$\Phi^{\pm}(\mathbf{c},\tilde{x}) = a_1^{\pm}(\tilde{x})\left(c^2 - \tfrac{5}{2}\right) , \tag{8-41}$$

where:

$$\lim_{\tilde{x}\to\infty} \Phi^{\pm}(\mathbf{c},\tilde{x}) = \text{const}\times\left(c^2 - \tfrac{5}{2}\right) .$$

The following approach has been used in the Loyalka analysis:

$$a_1^{\pm}(\infty) = a_1 \ , \quad a_1^{-}(0) = a_{1w} \ , \quad a_1 \neq a_{1w} \ ,$$

such that there are now two independent constants, a_1 and a_{1w}, instead of the single constant, a_1, in the Maxwellian analysis. This generalization gives a more correct form for the incident molecular distribution function at the wall than that used by Maxwell.

The unknown constants, a_1 and a_{1w}, may be obtained by the use of two scalar products of the Boltzmann equation, Eq. (8-28), and the boundary conditions for the correction to the distribution function. First, the scalar product given by Eq. (8-18) results in one equation for the two unknown constants:

$$\left(c_x c^2, \Phi^{\pm}(\mathbf{c},0)\right)=\left(c_x c^2, \Phi^{\pm}(\mathbf{c},\infty)\right)=0 \ . \tag{8-42}$$

Next, the scalar product of Eq. (8-28) with $c_x \phi_T(c)$ yields:

$$\begin{aligned}\frac{\partial}{\partial \tilde{x}}\left(c_x^2 \phi_T(c), \Phi^{\pm}(\mathbf{c},\tilde{x})\right) &= \left(c_x \phi_T(c), \mathrm{J}\left[\Phi^{\pm}(\mathbf{c},\tilde{x})\right]\right) \\ &= \left(\Phi^{\pm}(\mathbf{c},\tilde{x}), \mathrm{J}\left[c_x \phi_T(c)\right]\right) \\ &= \left(\Phi^{\pm}(\mathbf{c},\tilde{x}), c_x\left(c^2-\tfrac{5}{2}\right)\right) \\ &= \left(c_x c^2, \Phi^{\pm}(\mathbf{c},\tilde{x})\right)=0 \ .\end{aligned}$$

This implies that:

$$\left(c_x^2 \phi_T(c), \Phi^{\pm}(\mathbf{c},\tilde{x})\right)=\text{const} \ . \tag{8-43}$$

Thus, the second equation for the unknown functions, a_1 and a_{1w}, may be written as:

$$\left(c_x^2 \phi_T(c), \Phi^{\pm}(\mathbf{c},0)\right)=\left(c_x^2 \phi_T(c), \Phi^{\pm}(\mathbf{c},\infty)\right) \ . \tag{8-44}$$

The corrections to the distribution function of the reflected molecules at the wall may be found by using the Maxwellian model of the boundary conditions which, for this problem, may be expressed in the form:

$$\begin{aligned}\Phi^{+}(\mathbf{c},0) = &-(2-\alpha_\tau)c_x \phi_T(c) q_x + (1-\alpha_\tau)\Phi^{-}(\mathbf{c}',0) \\ &+\alpha_\tau\left[\nu_r + \left(c^2-\tfrac{3}{2}\right)\tau_r\right] ,\end{aligned} \tag{8-45}$$

where $\mathbf{c}'=\left\{-c_x, c_y, c_z\right\}$, and ν_r and τ_r are corrections to the number density and temperature for the diffusely reflected molecules. The corrections, ν_r and τ_r, may be obtained from the three additional conditions:

$$\int_{-} v_x f^{-}(\mathbf{v},0)d\mathbf{v} + \int_{+} v_x f^{+}(\mathbf{v},0)d\mathbf{v}=0 \ , \tag{8-46}$$

$$\int_{-} v_x f^{-}(\mathbf{v},0)\,d\mathbf{v} + \int_{+} v_x f_w^{+}(\mathbf{v},0)\,d\mathbf{v} = 0 \ , \tag{8-47}$$

$$\alpha_T = \frac{Q_x}{Q_{xw}} \ . \tag{8-48}$$

In these expressions, one must use the general form of the boundary condition (Eq. (8-22)). Then, using Eq. (8-41), after integration, one obtains:

$$v_r = -\tfrac{75}{256}\pi\left(1-\alpha_T\right)\lambda q_x - \tfrac{1}{2}\left(2-\alpha_T\right)a_{1w} \ , \tag{8-49}$$

$$\tau_r = \tfrac{75}{128}\pi\left(1-\alpha_T\right)\lambda q_x + \left(1-\alpha_T\right)a_{1w} \ . \tag{8-50}$$

Now, Eqs. (8-42) and (8-44) may be employed to derive the relations for a_1 and a_{1w}. Substituting Eqs. (8-49) and (8-50) into Eq. (8-45), and performing the integrations indicated by Eqs. (8-42) and (8-44), one can obtain:

$$a_{1w} = \tfrac{75}{128}\pi\frac{2-\alpha_\tau\alpha_T}{\alpha_\tau\alpha_T}\lambda q_x \ , \tag{8-51}$$

$$a_1 = a_{1w}\left[1+\frac{\alpha_\tau\alpha_T\left(2-\alpha_\tau\right)}{\left(2-\alpha_\tau\alpha_T\right)}\left(\tfrac{52}{25}\pi^{-1}-\tfrac{1}{2}\right)\right] . \tag{8-52}$$

Taking into account these expressions, one can obtain the following formula for the temperature-jump:

$$\Delta T = c_T\lambda\left(\frac{\partial T}{\partial\tilde{x}}\right)_\infty \ . \tag{8-53}$$

Here, c_T is the temperature-jump coefficient given by:

$$c_T = \tfrac{75}{128}\pi\frac{2-\alpha_\tau\alpha_T}{\alpha_\tau\alpha_T}\left[1+\frac{\alpha_\tau\alpha_T\left(2-\alpha_\tau\right)}{\left(2-\alpha_\tau\alpha_T\right)}\left(\tfrac{52}{25}\pi^{-1}-\tfrac{1}{2}\right)\right] . \tag{8-54}$$

This expression for the temperature-jump coefficient given in Eq. (8-54) is more general than that reported in [2]. For the specific case when $\alpha_T = 1$, however, both expressions are identical. Eq. (8-54) is applicable for arbitrary models of the intermolecular potential if one uses the first-order Chapman-Enskog solution. A more complete description of the temperature-jump phenomenon may be obtained by solving Problem 8.6.

The preceding analysis of the slip-flow transport problems shows that Loyalka's method may be classified as a two-moment approximation. Although this method eliminates the main deficiency of the Maxwellian analysis, it does not permit one to obtain the dependence of macroscopic values on the normal coordinate in the Knudsen layer. Nevertheless, it is very important to note that the conservation laws used in Loyalka's analysis are the exact moment solutions of the Boltzmann equation. Each exact solution for the correction to the distribution function must satisfy these moment relationships which are the most general properties of the Boltzmann equation. The use of exact moment solutions allows one to anticipate highly accurate results based on what is a reasonably simple analytical analysis. Moreover, for this method there would appear to be no difficulties in generalizing the analysis for arbitrary models of the intermolecular potential. To this end, it is necessary only that Chapman-Enskog solutions of sufficiently high-order be employed.

3. THE HALF-RANGE MOMENT METHOD.

Another approximate method was first proposed by Wang-Chang and Uhlenbeck [3] and was then further developed by Gross and Ziering, among others, to solve planar boundary transport problems [4-13]. Known initially as the half-range moment method it is also now commonly referred to as the Gross-Ziering method after those principally responsible for its further development. To demonstrate the various features of this method as described in [13], the velocity-slip problem is examined. The conditions for the gas are assumed to be the same as described in Section 8.1. For the linearized problem the distribution function may be given by:

$$f^{\pm} = f^{(0)}\left\{1 + 2c_y\tilde{x}\left(\frac{\partial u^*}{\partial \tilde{x}}\right)_\infty + c_x c_y \phi_u(c)\left(\frac{\partial u^*}{\partial \tilde{x}}\right)_\infty + \Phi^{\pm}(\mathbf{c},\tilde{x})\right\}, \qquad (8\text{-}55)$$

where $\Phi^{\pm}(\mathbf{c},\tilde{x})$ is the correction to the distribution function which accounts for the influence of the wall and $\phi_u(c)$ is specified by Eq. (8-3). The correction, $\Phi^{\pm}(\mathbf{c},\tilde{x})$, may be found from the linearized Boltzmann equation:

$$c_x \frac{\partial \Phi}{\partial \tilde{x}} = \mathrm{J}(\Phi) \;, \tag{8-56}$$

where:

$$\mathrm{J}(\Phi) = \int b db d\varepsilon \int |\mathbf{c}_1 - \mathbf{c}| f_1^{(0)} (\Phi_1' + \Phi' - \Phi_1 - \Phi) d\mathbf{v}_1 \;.$$

In the half-range moment method, the distribution function is approximated by half-range polynomials in velocity space and the space-dependent coefficients are determined by solving half-range moment equations. It might appear that the most natural systematic scheme for construction of half-range moment equations is to introduce a complete orthonormal set of polynomials, defined over the half-range in velocity space, such that the distribution function may then be expanded in terms of these polynomials and characterized by the space-time dependent expansion coefficients.

Unfortunately, this scheme requires the use of a large number of expansion terms which results in great mathematical difficulty. The complexity of the calculations increases rapidly with the number of terms and, therefore, this calculation scheme is not always practical for use in transport problems. However, the distribution function can sometimes satisfy the boundary conditions exactly such that very accurate results can be obtained by low-order approximations employing only a few terms in the expansion. Such approximations can give the exponential variation of the macroscopic values in the Knudsen layer near the wall. When the boundary conditions are not satisfied exactly, it is necessary to use the integral forms of the boundary conditions.

For the current slip-flow problem, assume that the correction, $\Phi^{\pm}(\mathbf{c}, \tilde{x})$, may be represented in the half-range velocity space as:

$$\Phi^{\pm}(\mathbf{c}, \tilde{x}) = a_0^{\pm}(\tilde{x}) c_y + a_1^{\pm}(\tilde{x}) c_x c_y \;. \tag{8-57}$$

If the following function is introduced:

$$\operatorname{sign}(c_x) = \begin{cases} 1 \;; & c_x > 0 \;, \\ -1 \;; & c_x < 0 \;, \end{cases}$$

the correction to the distribution function, $\Phi(\mathbf{c}, \tilde{x})$, is conveniently expressed for the full velocity space by:

$$\Phi^{\pm}(\mathbf{c},\tilde{x}) = \Phi^{+}\frac{1+\operatorname{sign}(c_x)}{2} + \Phi^{-}\frac{1-\operatorname{sign}(c_x)}{2}$$

$$= \frac{a_0^+ + a_0^-}{2}c_y + \frac{a_0^+ - a_0^-}{2}c_y \operatorname{sign}(c_x) + \frac{a_1^+ + a_1^-}{2}c_x c_y + \frac{a_1^+ - a_1^-}{2}c_x c_y \operatorname{sign}(c_x) \ . \tag{8-58}$$

Substituting this correction to the distribution function into Eq. (8-56), multiplying both sides alternately by:

$$c_y\left(1 \pm \operatorname{sign}(c_x)\right)\exp\left(-c^2\right)d\mathbf{c} \ ,$$

and:

$$c_x c_y\left(1 \pm \operatorname{sign}(c_x)\right)\exp\left(-c^2\right)d\mathbf{c} \ ,$$

and, finally, integrating over all velocities, one obtains the following system of half-range moment equations:

$$\frac{da_0^+}{d\tilde{x}} = a_0^+ A_1 - a_0^- A_1 + a_1^+ A_2 + a_1^- A_3 \ , \tag{8-59a}$$

$$\frac{da_0^-}{d\tilde{x}} = a_0^+ A_1 - a_0^- A_1 + a_1^+ A_3 + a_1^- A_2 \ , \tag{8-59b}$$

$$\frac{da_1^+}{d\tilde{x}} = a_0^+ B_1 - a_0^- B_1 + a_1^+ B_2 + a_1^- B_3 \ , \tag{8-59c}$$

$$\frac{da_1^-}{d\tilde{x}} = -a_0^+ B_1 + a_0^- B_1 - a_1^+ B_3 - a_1^- B_2 \ , \tag{8-59d}$$

where the coefficients, A_i and B_i, may be expressed as linear combinations of some of the moments of the collision integral:

$$A_1 = b\left[I_1 - \tfrac{1}{2}\sqrt{\pi}I_2\right],$$

$$A_2 = b\left[I_2 - \tfrac{1}{2}\sqrt{\pi}\left(I_3 + I_4\right)\right],$$

$$A_3 = b\left[I_2 - \tfrac{1}{2}\sqrt{\pi}\left(I_3 - I_4\right)\right],$$

$$B_1 = b\left[I_2 - \tfrac{1}{2}\sqrt{\pi}I_1\right],$$

$$B_2 = b\left[\left(I_3 + I_4\right) - \tfrac{1}{2}\sqrt{\pi}I_2\right],$$

$$B_3 = b\left[\left(I_3 - I_4\right) - \tfrac{1}{2}\sqrt{\pi}I_2\right],$$

where:

$$b = \frac{4}{\pi\left(4-\pi\right)} .$$

Introducing the following notation:

$$\left[\Phi(\mathbf{v}), \Psi(\mathbf{v})\right]^* = \int \Phi(\mathbf{v})\mathrm{J}\left(\Psi(\mathbf{v})\right)\exp\left(-c^2\right)d\mathbf{c} .$$

the bracket integrals, I_j, may be written as:

$$I_1 = \left[c_y \operatorname{sign}(c_x), c_y \operatorname{sign}(c_x)\right]^* ,$$

$$I_2 = \left[c_y \operatorname{sign}(c_x), c_x c_y\right]^* ,$$

$$I_3 = \left[c_x c_y \, , c_x c_y \right]^* \, ,$$

$$I_4 = \left[c_x c_y \, \mathrm{sign}(c_x), c_x c_y \, \mathrm{sign}(c_x) \right]^* \, .$$

The following additional bracket integrals that one might expect to encounter are identically zero and do not appear in this moment system:

$$\left[c_x c_y \, , c_x c_y \, \mathrm{sign}(c_x) \right]^* = \left[c_y \, \mathrm{sign}(c_x), c_x c_y \, \mathrm{sign}(c_x) \right]^* = 0 \, .$$

The bracket integrals, I_j, correspond to the original notation used by Gross and Ziering [5] and are different from the Chapman-Enskog bracket integrals usually encountered in kinetic theory. The relationship between these two kinds of bracket integrals is:

$$[\Phi \, , \Psi]^* = -n\pi^{3/2} \left(\frac{m}{2kT} \right)^{1/2} [\Phi \, , \Psi] \, , \tag{8-60}$$

where $[\Phi \, , \Psi]$ is the standard bracket integral.

The values of the integrals, I_j, are dependent on the intermolecular interaction potentials. A method of calculation of these integrals, for rigid-sphere molecules, is given in Appendix A and is substantially simpler than previous methods [14, 15]. For rigid-sphere molecules, this method results in the following bracket integrals:

$$I_1 = \frac{1 - 8\sqrt{2}}{12} \frac{\pi}{\lambda} \, , \quad I_2 = -\frac{3\pi + 8}{32} \sqrt{\frac{2}{\pi}} \frac{\pi}{\lambda} \, ,$$

$$I_3 = -\tfrac{2}{5} \frac{\pi}{\lambda} \, , \quad I_4 = \frac{-46 + 17\sqrt{2}}{120} \frac{\pi}{\lambda} \, .$$

The value for the integral, I_4, originally reported by Gross and Ziering [5] is in error. The absolute value of the Gross-Ziering integral is about 9.3 times greater than that reported here. The same analytical expressions being used in this analysis were previously reported by Porodnov and Suetin [9].

Now, the solution of the moment system, Eqs. (8-59a)-(8-59d), takes the form:

$$a_0^+(\tilde{x}) = A + \alpha_0^+ B \exp(-\alpha\tilde{x}) \ , \tag{8-61a}$$

$$a_0^-(\tilde{x}) = A + B \exp(-\alpha\tilde{x}) \ , \tag{8-61b}$$

$$a_1^+(\tilde{x}) = \alpha_1^+ B \exp(-\alpha\tilde{x}) \ , \tag{8-61c}$$

$$a_1^-(\tilde{x}) = \alpha_1^- B \exp(-\alpha\tilde{x}) \ , \tag{8-61d}$$

where:

$$\alpha_0^+ = (4.19294) \ , \quad \alpha_1^+ = (-4.12697) \ ,$$
$$\alpha_1^- = (0.52413) \ , \quad \alpha = (2.20153)\lambda^{-1} \ .$$

The constants of integration, A and B, may be found from the boundary conditions (Eq. (8-10)) which may be written as:

$$\begin{aligned} a_0^+(0) &= (1-\alpha_\tau) a_0^-(0) \ , \\ a_1^+(0) &= -(2-\alpha_\tau)\phi_u(c)\left(\frac{\partial u^*}{\partial \tilde{x}}\right)_\infty - (1-\alpha_\tau) a_1^-(0) \ . \end{aligned} \tag{8-62}$$

From these boundary conditions one then obtains:

$$A = \tfrac{5}{4}\sqrt{\pi}\lambda\left(\frac{\partial u^*}{\partial \tilde{x}}\right)_\infty \frac{2-\alpha_\tau}{\alpha_\tau} \frac{(1-\alpha_\tau)-\alpha_0^+}{\alpha_1^+ + (1-\alpha_\tau)\alpha_1^-} \ ,$$

$$B = \tfrac{5}{4}\sqrt{\pi}\lambda\left(\frac{\partial u^*}{\partial \tilde{x}}\right)_\infty \frac{2-\alpha_\tau}{\alpha_1^+ + (1-\alpha_\tau)\alpha_1^-} \ ,$$

The solution of the moment system, Eqs. (8-61a)-(8-61d), can now be employed in the following expression for the mean velocity of the gas:

$$u(\tilde{x}) = n^{-1} \int v_y f(\mathbf{v}, \tilde{x}) d\mathbf{v}$$

$$= \left(\frac{2kT}{m}\right)^{1/2} \left\{ \tilde{x}\left(\frac{\partial u^*}{\partial \tilde{x}}\right)_\infty + \tfrac{1}{4}\left[a_0^+ + a_0^- + \frac{1}{\sqrt{\pi}}\left(a_1^+ - a_1^-\right)\right]\right\} . \tag{8-63}$$

which, at large distances from the wall, has the linear profile:

$$u(\tilde{x})\Big|_{\tilde{x}\to\infty} = \tilde{x}\left(\frac{\partial u}{\partial \tilde{x}}\right)_\infty + \tfrac{1}{4}\left(\frac{2kT}{m}\right)^{1/2}\left(a_0^+(\infty) + a_0^-(\infty)\right) . \tag{8-64}$$

The solution of the Navier-Stokes equations, if one uses the slip boundary conditions, has the following form:

$$u(\tilde{x}) = (c_m \lambda + \tilde{x})\left(\frac{\partial u}{\partial \tilde{x}}\right)_\infty . \tag{8-65}$$

A comparison of Eqs. (8-64) and (8-65) indicates that:

$$c_m = \frac{2-\alpha_\tau}{\alpha_\tau}\tfrac{5}{8}\sqrt{\pi}\,\frac{(3.193)+\alpha_\tau}{(3.603)+(0.524)\alpha_\tau} . \tag{8-66}$$

For the particular case when $\alpha_\tau = 1$, one has $c_m = (1.125)$. Additionally, for this case, the mean velocity of the gas may be expressed as:

$$u(\tilde{x}) = \left[\tilde{x} + (1.125)\lambda - (0.345)\lambda \exp\left(-\frac{(2.202)\tilde{x}}{\lambda}\right)\right]\left(\frac{\partial u}{\partial \tilde{x}}\right)_\infty . \tag{8-67}$$

In summary, the four-moment approximation to the half-range moment method described in this section satisfies the boundary conditions exactly and results in an exponential profile for the mean velocity of the gas in the Knudsen layer (near the wall). This equates to substantially more information about the state of the gas near the wall than is obtained from the methods of Maxwell and Loyalka described in Sections 8.1 and 8.2.

4. FEATURES OF THE BOUNDARY CONDITIONS FOR THE MOMENT EQUATIONS.

In Sec. 8.3, the boundary slip-flow problem was solved by using the Maxwellian boundary conditions for the distribution function. Unfortunately, these boundary conditions have an essential deficiency connected with the execution of the conservation laws for the Boltzmann equation. It was stated in Section 8.2 that a correction to the distribution function is needed to satisfy the two conservation laws expressed by Eqs. (8-32) and (8-33). The satisfaction of these conditions is a general requirement of the Boltzmann equation in planar boundary transport problems. If the exact form of the boundary conditions is used, the conservation law expressed in Eq. (8-33) is not satisfied. Therefore, one must use another form of the boundary conditions. To explore this problem in more detail, consider a microscopic boundary condition with the following form:

$$f^{+} = \mathrm{A} f^{-} \ . \tag{8-68}$$

The Boltzmann equation near the flat wall may be represented by [11]:

$$v_x \frac{\partial f}{\partial \tilde{x}} = \frac{\delta f}{\delta t} + \delta(\tilde{x})\left[\eta(-v_x) v_x f^{-} + \eta(v_x) v_x \mathrm{A} f^{-}\right] , \tag{8-69}$$

where $\delta(\tilde{x})$ is the Dirac delta function and:

$$\eta(v_x) = \begin{cases} 1 \ ; & v_x > 0 \ , \\ 0 \ ; & v_x < 0 \ . \end{cases}$$

Introducing the boundary condition into the right-hand-side of Eq. (8-69) and performing the integration of this equation over x from 0 to $\varepsilon > 0$ gives:

$$\lim_{\varepsilon \to 0} \int_0^{\varepsilon} v_x \frac{\partial f}{\partial \tilde{x}} d\tilde{x} = v_x f = \begin{cases} v_x \mathrm{A} f^{-} \ ; & v_x > 0 \ , \\ v_x f^{-} \ ; & v_x < 0 \ . \end{cases} \tag{8-70}$$

From this relation one may conclude that to obtain the boundary conditions to the moment equations one should use the following form:

$$\int_{+} v_x \phi_i(\mathbf{v}) f^+ d\mathbf{v} = \int_{+} v_x \phi_i(\mathbf{v}) \mathrm{A} f^- d\mathbf{v} \ , \tag{8-71}$$

where ϕ_i are molecular properties which are used to construct the moment system.

It would appear that there is now a complete set of boundary conditions for the moment equations. Unfortunately, these boundary conditions do not simultaneously satisfy both conservation laws which are expressed as Eqs. (8-32), (8-33) and (8-42), (8-44), for the slip-flow and thermal transport problems, respectively. These conservation equations are general properties of the Boltzmann equation and, therefore, to obtain the correct results, the two conservation laws must be taken into account. One possible solution to this problem is to use these conservation laws as integral expressions of the boundary conditions.

To examine this idea, consider the slip-flow problem for which there exists a solution by the half-range moment method. Instead of the boundary conditions expressed by Eq. (8-62) there are two integral relations that may be written as:

$$\left(c_x c_y \, , \Phi^{\pm}(\mathbf{c},0)\right) = 0 \ , \tag{8-72}$$

$$\left(c_x^2 c_y \phi_u(c) \, , \Phi^{\pm}(\mathbf{c},0)\right) = \left(c_x^2 c_y \phi_u(c) \, , \Phi^{\pm}(\mathbf{c},\infty)\right) . \tag{8-73}$$

The correction to the reflected distribution function at the wall surface has the following form:

$$\begin{aligned} \Phi^+(\mathbf{c},0) = &-(2-\alpha_\tau) c_x c_y \phi_u(c) \left(\frac{\partial u^*}{\partial \tilde{x}} \right)_\infty \\ &+ (1-\alpha_\tau)\left[a_0^-(0) c_y - a_1^-(0) c_x c_y \right] , \end{aligned} \tag{8-74}$$

where $\phi_u(c) = -2(2kT/m)^{1/2} B(c)$ and the function, $B(c)$, corresponds to the second-order Chapman-Enskog solution. The second-order terms in this quantity improve the accuracy of the analysis. This function, for rigid-sphere molecules, may be written as [17]:

$$\phi_u(c) = -\tfrac{5}{4}\sqrt{\pi}\lambda_\mu \left[1 + b_2^{(2)*} S_{5/2}^{(1)}(c^2) \right] , \tag{8-75}$$

where $b_2^{(2)*} = (0.05854)$, and $\lambda_\mu = \frac{205}{202}\lambda$.

The correction to the distribution function has been obtained in Section 8.4 and is given by Eqs. (8-61a)-(8-61d). The use of the new boundary conditions, however, results in new values for the constants, A and B. By using the new boundary model of Eqs. (8-72)-(8-73), one obtains:

$$A = \frac{2-\alpha_\tau}{\alpha_\tau}\frac{\Delta_1}{\Delta}\tfrac{5}{4}\sqrt{\pi}\lambda_\mu h \ , \quad B = \frac{\Delta_2}{\Delta}\tfrac{5}{4}\sqrt{\pi}\lambda_\mu h \ , \tag{8-76}$$

where Δ_1, Δ_2, and Δ may be represented for rigid-sphere molecules by the following expressions:

$$\Delta_1 = -\left[(0.6690) + \alpha_\tau (0.1775)\right] \ , \tag{8-77a}$$

$$\Delta_2 = (0.3413) - \alpha_\tau (0.1706) \ , \tag{8-77b}$$

$$\Delta = -\left[(0.7549) + \alpha_\tau (0.09714)\right] \ . \tag{8-77c}$$

The mean velocity of the gas is expressed by Eq. (8-63). Using Eqs. (8-77), the relation for the mean velocity may be given by:

$$u(\tilde{x}) = \left[c_m \lambda_\mu + \tilde{x} - \lambda_\mu u_d^*(\tilde{x})\right]\left(\frac{\partial u}{\partial \tilde{x}}\right)_\infty \ , \tag{8-78}$$

where c_m is the isothermal-creep coefficient and $u_d^*(\tilde{x})$ is the dimensionless velocity defect that describes the deflection of the mean velocity from the linear profile in the Knudsen layer. The expressions for these quantities, with an arbitrary tangential momentum accommodation coefficient, may be written as:

$$c_m = \frac{2-\alpha_\tau}{\alpha_\tau}\tfrac{5}{8}\sqrt{\pi}\frac{\Delta_1}{\Delta} \ , \tag{8-79}$$

$$u_d^*(\tilde{x}) = -(1.4229)\frac{\Delta_2}{\Delta}\exp\left(-\frac{(2.2015)\tilde{x}}{\lambda}\right) \ . \tag{8-80}$$

Table 8-1. Values of the slip-flow coefficient, c_m .

α_τ	c_m Eq. (8-79)	c_m Eq. (8-81)	δ † (%)	c_m Eq. (8-66)	δ † (%)
0.2	9.0710	9.0277	0.48	9.1235	0.80
0.4	4.1310	4.0976	0.81	4.1759	1.51
0.6	2.4650	2.4400	1.01	2.5027	2.14
0.8	1.6185	1.6006	1.11	1.6496	2.71
1.0	1.1006	1.0884	1.11	1.1254	3.22

† $\delta\% = \left(\left|a-a^*\right|/a^*\right)\times 100$; $a^* = c_m$ (from Eq. (8-79).

Table 8-1 contains the comparison of results which are obtained by the use of the various methods for several values of the coefficient, α_τ. Results obtained by Loyalka's analysis take into account the correction of the second-order to the Chapman-Enskog distribution function, and therefore the expression for the slip coefficient differs from that represented by Eq. (8-35). For this case, Loyalka's formula has the form:

$$c_m = \frac{2-\alpha_\tau}{\alpha_\tau}\left(1+\frac{4\omega_1-\pi}{2\pi}\alpha_\tau\right)\tfrac{5}{16}\pi \;;\; \omega_1 = 1-b_2^{(2)*}+\tfrac{17}{4}\left(b_2^{(2)*}\right)^2 . \qquad (8\text{-}81)$$

The Gross-Ziering analysis does not permit the use of the second-order Chapman-Enskog solution owing to the form of the boundary conditions.

The comparison of these results shows a good agreement between the values of c_m calculated by the use of Eq. (8-79) and by Loyalka's formula. Equation (8-66), which was obtained using the common forms of the boundary conditions, yields results that do not agree so well with Loyalka's formula. However, this comparison does not allow one to draw any rigorous conclusions regarding the accuracy of any these models because the exact analytical solutions are not known. It should be possible, however, to draw conclusions regarding the accuracy of the different methods by comparing their results to reasonably exact direct numerical solutions of this slip-flow problem. This will be considered in Section 9.3, where a complete analysis of the accuracy of the various methods and boundary models will be made.

5. SOLUTION OF THE THERMAL-CREEP PROBLEM BY THE HALF-RANGE MOMENT METHOD.

The flow of a gas in the y-direction, over a plane surface lying at $\tilde{x}=0$, is considered for the conditions described in Section 8.1. Far from the plate the gas is maintained at a constant temperature gradient, $(\partial T/\partial \tilde{y})_{\infty}$. For these conditions, the thermal-creep problem is described by the following distribution function:

$$f = f^{(0)}\left[1+\left(c^2-\tfrac{5}{2}\right)\tilde{y}q_y - c_y A(c) q_y + \Phi^{\pm}(\mathbf{c},\tilde{x})\right] , \tag{8-82}$$

where $A(c)$ may be written as [17]:

$$A(c) = a_1^{(2)}\left[S_{3/2}^{(1)}\left(c^2\right) + a_2^{(2)*} S_{3/2}^{(2)}\left(c^2\right)\right] ,$$

where $a_2^{(2)*} = (0.08889)$, $a_1^{(2)} = -\frac{15}{16}\sqrt{\pi}\lambda_\kappa$, and $\lambda_\kappa = \frac{45}{44}\lambda$. In this formula molecules are assumed to be rigid spheres.

The integral form of the boundary conditions allows one to employ the same correction to the distribution function for the thermal-creep problem as in the slip-flow problem. This form of the distribution function takes into account the viscosity effects in a sufficiently correct form. Let the correction to the distribution function be assigned by the relation:

$$\Phi^{\pm}(\mathbf{c},\tilde{x}) = a_0^{\pm}(\tilde{x}) c_y + a_1^{\pm}(\tilde{x}) c_x c_y , \tag{8-83}$$

where the functions, $a_i^{\pm}$ may be written as:

$$a_0^{+}(\tilde{x}) = C_1 + \alpha_0^{+} C_2 \exp(-\alpha\tilde{x}) , \tag{8-84a}$$

$$a_0^{-}(\tilde{x}) = C_1 + C_2 \exp(-\alpha\tilde{x}) , \tag{8-84b}$$

$$a_1^{+}(\tilde{x}) = \alpha_1^{+} C_2 \exp(-\alpha\tilde{x}) , \tag{8-84c}$$

$$a_1^{-}(\tilde{x}) = \alpha_1^{-} C_2 \exp(-\alpha\tilde{x}) . \tag{8-84d}$$

The values, $\alpha_i^{\pm}, \alpha$, for rigid spheres are given by:

$$\alpha_0^{+} = (4.19294) , \quad \alpha_1^{+} = -(4.12697) ,$$
$$\alpha_1^{-} = (0.52413) , \quad \alpha = (2.20153)\lambda^{-1} .$$

The constants C_1 and C_2 may be obtained from boundary conditions which have the following form:

$$\left(c_x c_y\,,\varPhi^{\pm}\left(\mathbf{c},0\right)\right)=0\ ,\tag{8-85a}$$

$$\left(c_x^2 c_y \phi_u\left(c\right),\varPhi^{\pm}\left(\mathbf{c},0\right)\right)=\left(c_x^2 c_y \phi_u\left(c\right),\varPhi^{\pm}\left(\mathbf{c},\infty\right)\right)\ .\tag{8-85b}$$

where $\varPhi^{+}\left(\mathbf{c},0\right)$ is represented by:

$$\begin{aligned}\varPhi^{+}\left(\mathbf{c},0\right)=\alpha_\tau a_1^{(2)} q_y c_y\left[S_{3/2}^{(1)}\left(c^2\right)+a_2^{(2)*}S_{3/2}^{(2)}\left(c^2\right)\right]\\ +\left(1-\alpha_\tau\right)\left[a_0^{-}\left(0\right)c_y-a_1^{-}\left(0\right)c_x c_y\right]\ ,\end{aligned}\tag{8-86}$$

and the standard notation for the scalar product from Eq. (8-11) has been used. In the case of $\phi_u\left(c\right)$, Eq. (8-75) is used. This integral form of the boundary conditions makes allowance for the features of the Boltzmann equation near a flat wall.

Having performed integration in Eqs. (8-85) one obtains:

$$C_1=\frac{\Delta_1}{\Delta}\ ,\quad C_2=\frac{\Delta_2}{\Delta}\ ,$$

where Δ_1, Δ_2, and Δ may be written as:

$$\Delta_1=-a_1^{(2)} q_y\left[\left(0.4354\right)+\left(0.2179\right)\alpha_\tau\right]\ ,$$

$$\Delta_2=\left(0.3022\right)\alpha_\tau a_1^{(2)} q_y\ ,$$

$$\Delta=\left(0.8518\right)+\left(0.1096\right)\alpha_\tau\ .$$

For this problem it is convenient to represent the mean velocity of the gas in the form:

$$u(\tilde{x}) = c_{Tsl} \nu^* q_y \times \left\{ 1 + \tfrac{1}{2} \left[\left(\alpha_0^+ + 1 \right) + \frac{1}{\sqrt{\pi}} \left(\alpha_1^+ - \alpha_1^- \right) \right] \frac{\Delta_2}{\Delta_1} \exp(-\alpha \tilde{x}) \right\}, \tag{8-87}$$

where c_{Tsl} is given by:

$$c_{Tsl} = \tfrac{3}{2} \frac{(0.4354) + (0.2179)\alpha_\tau}{(0.8518) + (0.1096)\alpha_\tau} . \tag{8-88}$$

The dimensionless velocity defect may be defined as:

$$u_d^*(\tilde{x}) = \frac{u(\infty) - u(\tilde{x})}{\nu^* q_y} . \tag{8-89}$$

This quantity may then be expressed by:

$$u_d^*(\tilde{x}) = \alpha_\tau \frac{(0.5822)}{(0.8518) + (0.1096)\alpha_\tau} \exp\left(-\frac{(2.2015)\tilde{x}}{\lambda} \right) . \tag{8-90}$$

For the particular case when $\alpha_\tau = 1$, one obtains $c_{Tsl} = (1.0193)$. The full analysis of results obtained here will be described later. It is of considerable interest to investigate the influence of the various forms of the boundary conditions on the thermal-creep coefficient and the velocity defect. This analysis will be described in the next section.

6. INFLUENCE OF THE BOUNDARY MODELS ON THE THERMAL-CREEP COEFFICIENT.

Here, expressions are derived for the thermal-creep coefficient and the velocity defect for the case when one uses the ordinary moment form of the boundary conditions that may be represented by:

$$\int_+ c_x c_y \Phi^+(\mathbf{c}, 0) \exp\left(-c^2\right) d\mathbf{c} = \int_+ c_x c_y \mathrm{A} \Phi^-(\mathbf{c}, 0) \exp\left(-c^2\right) d\mathbf{c} , \tag{8-91a}$$

$$\int_{+} c_x^2 c_y \Phi^+(\mathbf{c},0)\exp\left(-c^2\right)d\mathbf{c} = \int_{+} c_x^2 c_y \mathrm{A}\Phi^-(\mathbf{c},0)\exp\left(-c^2\right)d\mathbf{c} \ , \tag{8-91b}$$

where:

$$\begin{aligned}\mathrm{A}\Phi^-(\mathbf{c},0) = {} & \alpha_\tau a_1^{(2)} q_y \left[S_{3/2}^{(1)}\left(c^2\right) + a_2^{(2)*} S_{3/2}^{(2)}\left(c^2\right) \right] \\ & + \left(1-\alpha_\tau\right)\left[a_0^-(0) c_y - a_1^-(0) c_x c_y \right] .\end{aligned} \tag{8-92}$$

Substituting the functions given by Eqs. (8-84a)-(8-84d) into Eqs. (8-91a) and (8-91b), one obtains the constants C_1 and C_2 which determine the correction to the distribution function. This analysis yields:

$$c_{Tsl} = \tfrac{3}{2} \frac{(1.5464) + (1.1741)\alpha_\tau}{(3.0927) + (0.4499)\alpha_\tau} \ , \tag{8-93}$$

$$u_d^*(\tilde{x}) = \alpha_\tau \frac{(3.4149)}{(3.0927) + (0.4499)\alpha_\tau} \exp\left(-\frac{(2.2015)\tilde{x}}{\lambda} \right) . \tag{8-94}$$

For example, if $\alpha_\tau = 1$, then $c_{Tsl} = (1.1519)$. This value is about 15% greater than that given by Eq. (8-96). But a reliable analysis of the accuracy of the model considered here, as was indicated previously, may only be performed by making a comparison with the numerical solution for this boundary transport problem. The accuracy of the various methods and boundary models for the thermal-creep problem will be investigated in Section 9.5.

PROBLEMS

8.1. Determine the mean velocity profile, $u(\tilde{x})$, and the pressure tensor component, P_{xy}, for Couette flow by means of the Maxwell method. Derive an expression for the apparent viscosity coefficient, μ', that is determined by:

$$\mu' \frac{u_w}{d} = \mu \frac{\partial u}{\partial \tilde{x}} \ .$$

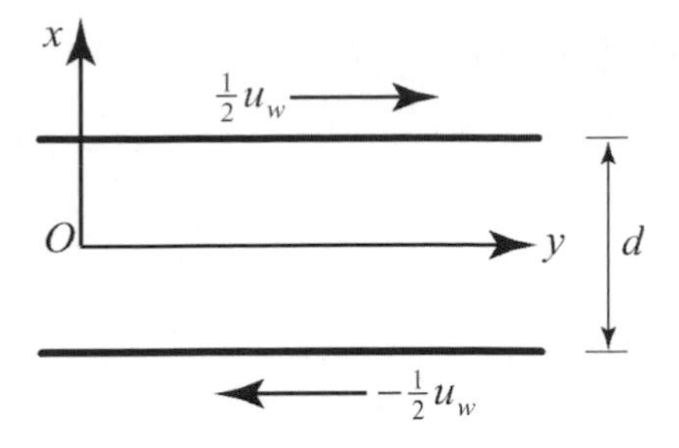

Figure 8-2. The geometry to be used in Problem 8.1 in determining the mean velocity profile, the pressure tensor component, and the apparent viscosity for Couette flow.

Use the flow geometry in Fig. 8-2.

Solution: For this analysis the distribution function should be chosen in the form:

$$f = f^{(0)}\left\{1 + c_x c_y \phi_u(c) A + 2A\tilde{x}c_y\right\} ,$$

which provides a linear profile for the mean velocity of the gas. Conservation of tangential momentum at the wall may be written as:

$$\left(c_x c_y , \Phi^{\pm}\left(\mathbf{c}, -\tfrac{1}{2}d\right)\right) = 0 .$$

Using the Maxwellian boundary condition for the rebounded distribution function, one can obtain:

$$u(\tilde{x}) = \frac{u_w}{d}\tilde{x}\zeta(Kn) ,$$

$$\zeta(Kn) = \left(1 + \frac{2-\alpha_\tau}{\alpha_\tau}\tfrac{5}{8}\pi\frac{\lambda}{d}\right)^{-1} ,$$

$$P_{xy} = -\mu'\frac{u_w}{d} ,$$

and $\mu' = \mu\zeta(Kn)$.

8.2. Determine the mean velocity profile, $u(\tilde{x})$, and the pressure tensor component, P_{xy}, for Couette flow by means of Loyalka's method. Derive an expression for the apparent viscosity coefficient, μ', that is determined by:

$$\mu' \frac{u_w}{d} = \mu \frac{\partial u}{\partial \tilde{x}} .$$

Use the same flow geometry as in Problem 8.1 (see Fig. 8-2).

Solution: One should use a distribution function of the form:

$$f = f^{(0)} \left\{ 1 + \Phi^{\pm}(\mathbf{c}, \tilde{x}) \right\} ,$$

where $\Phi^{\pm}(\mathbf{c},\tilde{x}) = \frac{1}{2}\left(a_0^{+}(\tilde{x}) + a_0^{-}(\tilde{x})\right)c_y + \frac{1}{2}\left(a_0^{+}(\tilde{x}) - a_0^{-}(\tilde{x})\right)c_y \,\mathrm{sign}(c_x)$. As a result of this choice, the boundary conditions are satisfied exactly. The conservation laws may then be expressed in the form $\left(c_x c_y, \Phi^{\pm}(\mathbf{c},\tilde{x})\right) = A_1$ and $\left(c_x^2 c_y \phi_u(c), \Phi^{\pm}(\mathbf{c},\tilde{x})\right) = 2A_1\tilde{x} + A_2$ where A_1 and A_2 are constants which may be found from the two boundary conditions. From the conservation laws, one obtains:

$$a_0^{+}(\tilde{x}) = -u_w^{*} \frac{\frac{5}{8}\pi \frac{\lambda}{d} - 2\frac{\tilde{x}}{d}}{1 + \frac{2-\alpha_\tau}{\alpha_\tau} \frac{5}{8}\pi \frac{\lambda}{d}} ,$$

and:

$$a_0^{-}(\tilde{x}) = u_w^{*} \frac{\frac{5}{8}\pi \frac{\lambda}{d} + 2\frac{\tilde{x}}{d}}{1 + \frac{2-\alpha_\tau}{\alpha_\tau} \frac{5}{8}\pi \frac{\lambda}{d}} .$$

Using these expressions, one can obtain the same relations for $u(\tilde{x})$ and P_{xy} as in Problem 8.1. However, the distribution function obtained here describes the Knudsen limit regime better than that obtained in Problem 8.1.

8.3. Determine the mean velocity profile, $u(\tilde{x})$, and the pressure tensor component, P_{xy}, for Couette flow by means of the Gross-Ziering method for the two-moment approach. Take the correction to the distribution function in the form $\Phi^{\pm}(\mathbf{c},\tilde{x}) = a_0^{\pm}(\tilde{x}) c_y$. Derive an expression for the apparent viscosity coefficient, μ', that is determined by:

$$\mu' \frac{u_w}{d} = P_{xy} \ .$$

Use the same flow geometry as in Problem 8.1 (see Fig. 8-2).

Solution: For this two-moment approach, the half-range moment system is obtained by multiplying the Boltzmann equation, Eq. (8-56), by $c_y \left[1 \pm \text{sign}(c_x)\right] \exp\left(-c^2\right) d\mathbf{c}$ and integrating over all velocities. The solution of this moment system yields:

$$u(\tilde{x}) = \frac{u_w}{d} \tilde{x} \zeta(Kn) \ ,$$

$$\zeta(Kn) = \left(1 - \frac{2-\alpha_\tau}{\alpha_\tau} \frac{\pi}{I_1 d}\right)^{-1} \ ,$$

$$P_{xy} = \frac{p\sqrt{\pi}}{I_1} \left(\frac{m}{2kT}\right)^{1/2} \frac{u_w}{d} \zeta(Kn) \ ,$$

$$I_1 = \frac{1-8\sqrt{2}}{12} \frac{\pi}{\lambda} \ ,$$

and $\mu' = \mu\zeta(Kn)$ where $\mu = (0.2909)\rho\bar{v}\lambda$.

8.4. Using the full-range moment equations given by:

$$\frac{d}{d\tilde{x}} \int v_x Q(\mathbf{v}) f d\mathbf{v} = n\Delta\overline{Q(\mathbf{v})} \ ,$$

where $Q(\mathbf{v}) = v_x$ and $v_x v_y$, determine the mean velocity, $u(\tilde{x})$, and the pressure tensor component, P_{xy}, for the Couette flow. Solve this problem for the two-moment approach using the following form of the correction to the distribution function $\Phi^{\pm}(\mathbf{c}, \tilde{x}) = a_0^{\pm}(\tilde{x}) c_y$. Derive an expression for the apparent viscosity coefficient, μ', that is determined by:

$$\mu' \frac{u_w}{d} = P_{xy} \ .$$

Use the same flow geometry as in Problem 8.1 (see Fig. 8-2).

Solution: The moment equations are obtained by multiplying the Boltzmann equation, Eq. (8-56), by the terms, c_y and $c_x c_y$, and integrating over all velocities. The solution of the moment system then yields:

$$u(\tilde{x}) = \frac{u_w}{d}\tilde{x}\zeta(Kn) ,$$

$$\zeta(Kn) = \left(1 - \frac{2-\alpha_\tau}{\alpha_\tau}\frac{\pi^{3/2}}{2I_2 d}\right)^{-1} ,$$

$$P_{xy} = \tfrac{1}{2}\frac{p\pi}{I_2}\left(\frac{m}{2kT}\right)^{1/2}\frac{u_w}{d}\zeta(Kn) ,$$

$$I_2 = -\frac{3\pi+8}{32}\sqrt{\frac{2}{\pi}}\frac{\pi}{\lambda} ,$$

and $\mu' = \mu\zeta(Kn)$ where $\mu = (0.5100)\rho\bar{v}\lambda$. In this solution, molecules have been assumed to be rigid spheres.

8.5. Determine the temperature distribution between parallel plates by the Maxwell method. Derive an expression for the apparent thermal conductivity coefficient κ', that is given by:

$$\kappa'\frac{2\Delta T}{d} = \kappa\frac{\partial T}{\partial \tilde{x}} .$$

Use the flow geometry in Fig. 8-3. The correction, ΔT, to the temperature, T_0, is assumed to be small such that the relationship $\Delta T \ll T_0$ is satisfied.

Solution: A distribution function should be sought in the form:

$$f = f^{(0)}\left\{1 + c_x S_{3/2}^{(1)}(c^2)\tfrac{15}{16}\sqrt{\pi}\lambda A + \left(c^2 - \tfrac{5}{2}\right)A\tilde{x}\right\} .$$

Using the boundary conditions obtained in Problem 6.5 and conservation of the heat flux as given by:

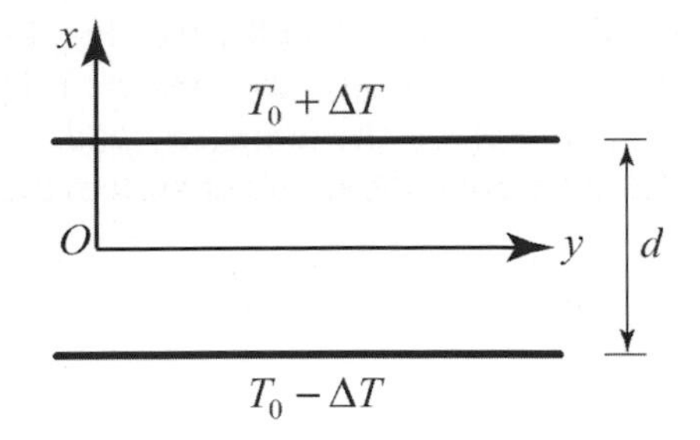

Figure 8-3. The geometry to be used in Problem 8.5 in determining the temperature distribution and apparent thermal conductivity coefficient between parallel plates having different temperatures.

$$\left(c_x c^2\,,\varPhi\left(\mathbf{c},-\tfrac{1}{2}d\right)\right)=0\ ,$$

one can obtain $\tau\left(\tilde{x}\right)=q\tilde{x}\zeta_T\left(Kn\right)$,

$$q=\frac{2\Delta T}{T_0 d}\ ,$$

$$\zeta_T=\left(1+\tfrac{75}{64}\pi\frac{2-\alpha_\tau\alpha_T}{\alpha_\tau\alpha_T}\frac{\lambda}{d}\right)^{-1}\ ,$$

and $\kappa'=\kappa\zeta_T\left(Kn\right)$ where $\tau\left(\tilde{x}\right)$ is a correction to the temperature.

8.6. Derive analytical formulas for the boundary temperature and pressure jumps for a planar surface in a monatomic non-condensable gas using Loyalka's method. Use the second-order Chapman-Enskog solution for rigid-sphere molecules. Assume that molecules reflect diffusely at the surface.

Solution: The following distribution function should be used to describe this boundary problem:

$$f=f^{(0)}\left[1+\left(c^2-\tfrac{5}{2}\right)\tilde{x}q_x-c_x A\left(c\right)q_x+\varPhi^{\pm}\left(\mathbf{c},\tilde{x}\right)\right]\ .$$

The correction, $\varPhi^{\pm}\left(\mathbf{c},\tilde{x}\right)$, should be taken in the form:

$$\Phi^{\pm}(\mathbf{c},\tilde{x}) = a^{\pm}(\tilde{x}) + \left(c^2 - \tfrac{5}{2}\right) b^{\pm}(\tilde{x}) \ .$$

The limiting values of the functions introduced here are given by:

$$\lim_{\tilde{x}\to\infty} a^{\pm}(\tilde{x}) = a(\infty) \ , \quad \lim_{\tilde{x}\to\infty} b^{\pm}(\tilde{x}) = b(\infty) \ ,$$

$$a^{-}(0) = a_w \ , \text{ and } b^{-}(0) = b_w \ .$$

These limiting values are specified by the boundary-jump effects. The unknown constants, $a(\infty)$, $b(\infty)$, a_w, and b_w, may be determined from the conservation laws of Eq. (8-9) which are given by:

$$\left(c_x \, , \Phi^{\pm}(\mathbf{c},0)\right) = 0 \ , \quad \left(c_x^2 \, , \left[\Phi^{\pm}(\mathbf{c},0) - \Phi^{\pm}(\mathbf{c},\infty)\right]\right) = 0 \ ,$$

$$\left(c_x c^2 \, , \Phi^{\pm}(\mathbf{c},0)\right) = 0 \ , \text{ and } \left(c_x^2 A(c), \left[\Phi^{\pm}(\mathbf{c},0) - \Phi^{\pm}(\mathbf{c},\infty)\right]\right) = 0 \ .$$

The distribution function of reflected molecules (see Problem 6.5) is given by:

$$\Phi^{+}(\mathbf{c},0) = c_x A(c) q_x + (1-\alpha_T)\left[(2-c^2)\left(\frac{\delta_2}{\pi} - \frac{2\delta_1}{\pi}\right)\right] - \frac{2\delta_1}{\pi} \ ,$$

where:

$$\delta_1 = -\tfrac{1}{2}\pi\left(a_w - \tfrac{1}{2}b_w\right) \ , \text{ and } \delta_2 = \pi\left(\tfrac{5}{8}\sqrt{\pi}\, a_1^{(2)} q_x - a_w - \tfrac{1}{2}b_w\right) \ .$$

This formulation yields:

$$\frac{T_{asy}(0) - T_0}{T_0} = c_T \lambda_\kappa q_x \ , \text{ and } \frac{p_{asy}(0) - p_0}{p_0} = c_p \lambda_\kappa q_x \ .$$

The jump coefficients are then given by:

$$c_T = \tfrac{75}{128}\pi \frac{2-\alpha_T}{\alpha_T}\left[1+\frac{\alpha_T}{2-\alpha_T}\left(\tfrac{52}{25}\frac{\varepsilon_T}{\pi}-\tfrac{1}{2}\right)\right],$$

and:

$$c_p = \tfrac{75}{256}\pi \frac{2-\alpha_T}{\alpha_T}\left[1+\frac{\alpha_T}{2-\alpha_T}\left(\tfrac{8}{5}\frac{\varepsilon_1}{\pi}-\tfrac{1}{2}\right)\right],$$

where $\varepsilon_T = 1-\tfrac{23}{26}a_2^{(2)*}+\tfrac{433}{208}\left(a_2^{(2)*}\right)^2$, $\varepsilon_1 = 1+\tfrac{1}{4}a_2^{(2)*}$, and $a_2^{(2)*} = (0.08889)$ for rigid-sphere molecules.

8.7. Derive analytical expressions for the boundary temperature and pressure jumps in a monatomic vapor over a planar, liquid surface using Loyalka's method. Use the second-order Chapman-Enskog solution for rigid-sphere molecules. Consider the case of an arbitrary evaporation coefficient.

Solution: The distribution function used should be:

$$f = f^{(0)}\left[1+\left(c^2-\tfrac{5}{2}\right)\tilde{x}q_x + 2c_x u^* - c_x A(c)q_x + \Phi^{\pm}(\mathbf{c},\tilde{x})\right],$$

where $u^* = (m/2kT_0)^{1/2}u_x$. The correction, $\Phi^{\pm}(\mathbf{c},\tilde{x})$, should be taken in the form $\Phi^{\pm}(\mathbf{c},\tilde{x}) = a^{\pm}(\tilde{x})+\left(c^2-\tfrac{5}{2}\right)b^{\pm}(\tilde{x})$. The limiting values of the functions introduced here are given by:

$$\lim_{\tilde{x}\to\infty} a^{\pm}(\tilde{x}) = a(\infty), \quad \lim_{\tilde{x}\to\infty} b^{\pm}(\tilde{x}) = b(\infty),$$

$$a^{-}(0) = a_w, \text{ and } b^{-}(0) = b_w .$$

These limiting values are specified by the boundary-jump effects. The unknown constants, $a(\infty)$, $b(\infty)$, a_w, and b_w, may be determined from the conservation laws of Eq. (8-9) which are given by:

$$\left(c_x, \Phi^{\pm}(\mathbf{c},0)\right) = 0, \quad \left(c_x^2, \left[\Phi^{\pm}(\mathbf{c},0)-\Phi^{\pm}(\mathbf{c},\infty)\right]\right) = 0,$$

$$\left(c_x c^2, \Phi^{\pm}(\mathbf{c},0)\right) = 0, \text{ and } \left(c_x^2 A(c), \left[\Phi^{\pm}(\mathbf{c},0)-\Phi^{\pm}(\mathbf{c},\infty)\right]\right) = 0 .$$

The boundary conditions (see Problem 6.6) for the reflected distribution function may be expressed as:

$$\Phi^{+}(\mathbf{c},0)=c_x A(c)q_x+(1-\alpha_m)$$
$$\times\left\{(2-c^2)\left[(1-\alpha_T)\left(\frac{\delta_2}{\pi}-\frac{2\delta_1}{\pi}\right)\right]-\frac{2\delta_1}{\pi}\right\},$$

where:

$$\delta_1=-\tfrac{1}{2}\pi\left(a_w-\tfrac{1}{2}b_w-\sqrt{\pi}u^*\right)$$

and:

$$\delta_2=\pi\left(\tfrac{5}{8}\sqrt{\pi}a_1^{(2)}q_x-a_w-\tfrac{1}{2}b_w+\tfrac{5}{4}\sqrt{\pi}\right).$$

This formulation yields:

$$\frac{T_{asy}(0)-T_0}{T_0}=c_T^{(T)}\{\alpha_m\}\lambda_\kappa q_x+c_T^{(u)}\{\alpha_m\}u^*,$$

and:

$$\frac{p_{asy}(0)-p_0}{p_0}=c_p^{(T)}\{\alpha_m\}\lambda_\kappa q_x+c_p^{(u)}\{\alpha_m\}u^*,$$

where the jump coefficients are given by:

$$c_T^{(T)}\{\alpha_m\}=\tfrac{75}{128}\pi\frac{2-\alpha}{\alpha}\left[1+\frac{\alpha}{2-\alpha}\left(\tfrac{52}{25}\frac{\varepsilon_T}{\pi}-\tfrac{1}{2}\right)\right],$$

$$c_T^{(u)}\{\alpha_m\}=-\tfrac{1}{8}\sqrt{\pi}\left[\frac{4-3\alpha}{\alpha}+\tfrac{16}{5}\frac{\varepsilon_1}{\pi}\right],$$

$$c_p^{(T)}\{\alpha_m\} = \tfrac{75}{256}\pi \frac{2-\alpha}{\alpha}\left[1+\frac{\alpha}{2-\alpha}\left(\tfrac{8}{5}\frac{\varepsilon_1}{\pi}-\tfrac{1}{2}\right)\right],$$

and:

$$c_p^{(u)}\{\alpha_m\} = -\tfrac{1}{2}\sqrt{\pi}\left[\frac{4-3\alpha_m}{\alpha_m}+\tfrac{1}{8}\left(\frac{4-3\alpha}{\alpha}\right)\right].$$

Here, $\alpha = \alpha_m + \alpha_T(1-\alpha_m)$, $\varepsilon_T = 1-\tfrac{23}{26}a_2^{(2)*}+\tfrac{433}{208}\left(a_2^{(2)*}\right)^2$, $\varepsilon_1 = 1+\tfrac{1}{4}a_2^{(2)*}$, and $a_2^{(2)*} = (0.08889)$ for rigid-sphere molecules.

REFERENCES

1. Maxwell, J.C., "On Stresses in Rarefied Gases Arising from Inequalities of Temperature," *Phil. Trans. Roy. Soc. (London)* **170**, 231 (1879).
2. Loyalka, S.K., "Approximate Method in the Kinetic Theory," *Phys. Fluids* **14(11)**, 2291-2294 (1971).
3. Wang-Chang, C.S. and Uhlenbeck, G.E., *Transport Phenomena in Very Dilute Gases* (VMH-3-F, University of Michigan, 1949).
4. Gross, E.P., Jackson, E.A., and Ziering, S., "Boundary Value Problems in Kinetic Theory of Gases," *Ann. Phys.* **1(2)**, 141-167 (1957).
5. Gross, E.P. and Ziering, S., "Kinetic Theory of Linear Shear Flow," *Phys. Fluids* **1(3)**, 215-224 (1958).
6. Gross, E.P. and Jackson, E.A., "Kinetic Theory of the Impulsive Motion of an Infinite Plane," *Phys. Fluids* **1(4)**, 318-328 (1958).
7. Gross, E.P. and Ziering, S., "Heat Flow Between Parallel Plates," *Phys. Fluids* **2(6)**, 701-712 (1959).
8. Bakanov, S.P. and Derjaguin, B.V., "On the State of a Gas Moving Near a Solid Surface," *Dokl. AN SSSR* **139(1)**, 71-74 (1961).
9. Porodnov, B.T. and Suetin, P.E., "Rarefied Gas Flow Between Two Parallel Plates," *Izv. AN SSSR, M. Zh. G.* **(6)**, 93-98 (1967).
10. Ivchenko, I.N. and Yalamov, Yu.I., "The Kinetic Theory of a Gas Flow Over a Solid Wall in a Velocity Gradient Field," *Izv. AN SSSR, M. Zh. G.* **(6)**, 139-143 (1968).
11. Rolduguin, V.I., *Application of the Non-Equilibrium Thermodynamics Method in Boundary Problems of the Kinetic Theory of Gases* (M.S. Thesis, Moscow, 1979).
12. Savkov, S.A., *Slip Boundary Conditions for Non-Uniform Binary Gas Mixtures and their Application to Aerosol Dynamics* (M.S. Thesis, Moscow, 1987).
13. Yalamov, Yu.I. and Ivchenko, I.N., "Isothermal Heat Flux in the Knudsen Layer," *J. Phys. Chem. (Russian)* **43(10)**, 2622-2624 (1969).
14. Derjaguin, B.V., Ivchenko, I.N., and Yalamov, Yu.I., "About Construction of Solutions of the Boltzmann Equation in the Knudsen Layer," *Izv. AN SSSR, M. Zh. G.* **(4)**, 167-171 (1968).
15. Derjaguin, B.V. and Yalamov, Yu.I., "The Theory of Thermophoresis and Diffusiophoresis of Aerosol Particles and Their Experimental Testing," In *Topics in*

Current Aerosol Research, vol. 3, part 2, edited by Hidy, G.M. (Pergamon Press, Oxford, 1972).

16. Loyalka, S.K. and Lang, H., "On Variational Principles in the Kinetic Theory," in *Proceedings of the Seventh International Symposium on Rarefied Gas Dynamics* (Editrice Tecnico Scientifica, Pisa, Italy, 1970).
17. Loyalka, S.K., "Slip and Jump Coefficients for Rarefied Gas Flows: Variational Results for Lennard-Jones and n(r)-6 Potentials," *Physica A* **163**, 813-821 (1990).

Chapter 9

THE VARIATIONAL METHOD FOR THE PLANAR GEOMETRY

1. ANOTHER FORM OF THE BOLTZMANN EQUATION.

In the previous chapter the basic analytical methods for the planar boundary transport problems were considered. All of these schemes are considered to be moment methods. It is very important to note that the simple analysis of some general properties of the Boltzmann equation related to the conservation of moments results in sufficiently accurate expressions for the velocity-slip and temperature-jump coefficients.

The Maxwell method and its generalization proposed by Loyalka allow one to analyze some integral properties of a gas, for which one can expect reliable results. The physical basis of these methods is not questionable. But within the framework of these methods, as has already been noted, it is impossible to describe the gas behavior in the Knudsen layer. Nevertheless, Loyalka's method of taking into account two conserving moments may be used to calculate the gas parameters both at large distances from the wall and directly on its surface.

In this chapter a simple and accurate method for dealing with planar boundary transport problems will be described by applying a variational technique to the linearized Boltzmann equation. The analysis will be confined to a consideration of the slip and thermal-creep problems which are the classical problems used to analyze the main features and accuracy of this approach. Since the main interest is in linearized transport problems, the distribution function can be written as:

$$f = f^{(0)}\left[1 + \Psi(\mathbf{c}, \tilde{\mathbf{r}}) + \Phi^{\pm}(\mathbf{c}, \tilde{x})\right] , \tag{9-1}$$

where $\Psi(\mathbf{c}, \tilde{\mathbf{r}})$ is the Chapman-Enskog correction for the appropriate problems and $\tilde{x}$ is the normal coordinate. The correction to the distribution function, $\Phi^{\pm}(\mathbf{c}, \tilde{x})$, taking into account the influence of the wall, may be specified from the linearized Boltzmann equation that is given by Eq. (8-9).

The specific features of the variational method are related to the integral form of Boltzmann's equation. To derive the integral equation for the distribution function it is convenient to introduce another form of Boltzmann's equation that, for stationary problems, is defined by:

$$\mathbf{v} \cdot \frac{\partial f}{\partial \tilde{\mathbf{r}}} = J^{(1)} - J^{(2)} , \tag{9-2}$$

where the collision operator is presented as a sum of two separate terms which are given by:

$$J^{(1)} = \int bdbd\varepsilon \int gf'f_1'd\mathbf{v}_1 , \tag{9-3}$$

$$J^{(2)} = f \int bdbd\varepsilon \int gf_1 d\mathbf{v}_1 . \tag{9-4}$$

For molecules of infinite range of interaction these integrals diverge, since they include among the colliding molecules some which interact at arbitrarily large distance with infinitely small changes of state [1]. These integrals converge for the rigid-sphere molecules, and, therefore, can be calculated separately. If the interaction potential falls off rapidly enough, distant collisions may be neglected by means of the use of the definite model for a 'truncated' potential having a finite interaction range.

For planar, stationary, linearized transport problems, Eq. (9-2) may be expressed in the form:

$$c_x \frac{\partial}{\partial \tilde{x}} \Phi(\mathbf{c}, \tilde{x}) + \sigma(c)\Phi(\mathbf{c}, \tilde{x}) = H\Phi(\mathbf{c}, \tilde{x}) . \tag{9-5}$$

Here, $\sigma(c)$ is a function depending only on the magnitude of a molecular velocity vector and $H = H(\mathbf{c}, \mathbf{c}')$ is an operator defined by the relation:

$$H\Phi(\mathbf{c}, \tilde{x}) = \varepsilon \int d\mathbf{c}' \exp(-c'^2) K(\mathbf{c}, \mathbf{c}') \Phi(\mathbf{c}', \tilde{x}) , \tag{9-6}$$

in which $K(\mathbf{c},\mathbf{c}')$ is a symmetric kernel. The parameter ε is defined as:

$$\varepsilon = \frac{1}{\sqrt{2\pi}}\frac{1}{\lambda} \ . \tag{9-7}$$

It is very important to emphasize that both the function, $\sigma(c)$, and the kernel, $K(\mathbf{c},\mathbf{c}')$, can be defined by analytical expressions if molecules are assumed to be rigid spheres. A very simple integration in Eq. (9-4) gives:

$$\sigma(c) = \varepsilon\nu(c) = \frac{1}{\lambda\sqrt{2\pi}}\left[\exp\left(-c^2\right) + \left(2c + \frac{1}{c}\right)\tfrac{1}{2}\sqrt{\pi}\,\mathrm{erf}(c)\right] . \tag{9-8}$$

One can find an analytical expression for $K(\mathbf{c},\mathbf{c}')$ in [2, 3]. This expression for the kernel, $K(\mathbf{c},\mathbf{c}')$, is not utilized in the variational analysis and, thus, it is sufficient to know only the general definition for the operator, H .

2. THE VARIATIONAL TECHNIQUE FOR THE SLIP-FLOW PROBLEM.

In this section the classical slip-flow problem will be analyzed by means of a variational technique that was described by a consideration of the planar boundary transport problems in Loyalka's works [4-7]. The statement of this problem is given in Sec. 8.1 and hence, here, the main focus will be on investigating the features and the accuracy of the variational technique.

Let the distribution function be described by:

$$f^{\pm} = f^{(0)}\left\{1 + 2c_y\tilde{x}h + c_x c_y\phi_u(c)h + \Phi^{\pm}(\mathbf{c},\tilde{x})\right\} . \tag{9-9}$$

where $\phi_u(c) = -2(2kT/m)^{1/2}B(c)$. The correction, $\Phi^{\pm}(\mathbf{c},\tilde{x})$, to the distribution function, which takes into account the perturbation of a gas flow by the wall, has the following asymptotic behavior:

$$\lim_{\tilde{x}\to\infty}\Phi^{\pm}(\mathbf{c},\tilde{x}) = 2c_y u_0^* \ , \tag{9-10}$$

where u_0^* is a dimensionless constant. Then, one uses the Maxwellian boundary model for a description of the gas-surface interaction. This boundary model is given by:

$$\Phi^{+}(\mathbf{c},0)=-(2-\alpha_{\tau})c_x c_y \phi_u(c)h+(1-\alpha_{\tau})\Phi^{-}(-c_x,c_y,c_z,0) \; . \qquad (9\text{-}11)$$

Now, one can use the form of the Boltzmann equation given in Eq. (9-5) to derive the integral equation for $\Phi^{\pm}(\mathbf{c},\tilde{x})$. Taking into account Eqs. (9-10) and (9-11) and performing a formal integration of this linear non-homogeneous differential equation of the first-order, one obtains the following integral equation:

$$\Phi(\mathbf{c},\tilde{x})=L\Phi(\mathbf{c},\tilde{x})+p_u(\mathbf{c},\tilde{x}) \; , \qquad (9\text{-}12a)$$

where L is an integral operator and:

$$p_u(\mathbf{c},\tilde{x})=-\eta(c_x)(2-\alpha_{\tau})c_x c_y \phi_u(c)h\exp\left(-\frac{\sigma(c)}{c_x}\tilde{x}\right) \; , \qquad (9\text{-}12b)$$

$$\eta(x)=\begin{cases}1 & ; \; x>0 \; , \\ 0 & ; \; x<0 \; .\end{cases}$$

The operator, L, is defined by the relation:

$$\begin{aligned} L\,\Phi(\mathbf{c},\tilde{x})=\eta(c_x)&\left\{\frac{1}{c_x}\int_0^{\tilde{x}} H\Phi(\mathbf{c},x')\exp\left(\frac{\sigma(c)}{c_x}(x'-\tilde{x})\right)dx'\right.\\ &\left.+(1-\alpha_{\tau})\frac{1}{c_x}\int_0^{\infty} \mathrm{H}\,\Phi(\mathbf{c},x')\exp\left(-\frac{\sigma(c)}{c_x}(x'+\tilde{x})\right)dx'\right\} \qquad (9\text{-}13)\\ -\eta(-c_x)&\left\{\frac{1}{c_x}\int_{\tilde{x}}^{\infty} H\Phi(\mathbf{c},x')\exp\left(\frac{\sigma(c)}{c_x}(x'-\tilde{x})\right)dx'\right\} \; , \end{aligned}$$

where $\mathrm{H}(\mathbf{c},\mathbf{c}')=RH=H(-c_x,c_y,c_z,\mathbf{c}')$ and R is the reflection operator. If the operator, H, is applied to the function, $\Phi(\mathbf{c},\tilde{x})$, one obtains another function, $\mathrm{H}\,\Phi(\mathbf{c},\tilde{x})$, that is given by:

$$\mathrm{H}\,\Phi(\mathbf{c},\tilde{x})=\varepsilon\int d\mathbf{c}'\exp(-c'^2)\mathrm{K}(\mathbf{c},\mathbf{c}')\Phi(\mathbf{c}',\tilde{x}) \; , \qquad (9\text{-}14)$$

where the following notation is introduced in this integrand:

$$\mathrm{K}(\mathbf{c},\mathbf{c}') = RK(\mathbf{c},\mathbf{c}') \ .$$

Since $K(\mathbf{c},\mathbf{c}')$ is a symmetric function of $(\mathbf{c},\mathbf{c}')$ one can conclude that:

$$\mathrm{K}(\mathbf{c},\mathbf{c}') = \mathrm{K}(\mathbf{c}',\mathbf{c}) \ . \tag{9-15}$$

From both the symmetry of this kernel and Eq. (9-14), it is shown [8] that the operator, H, is self-adjoint, i.e. $\mathrm{H}^* = \mathrm{H}$, in the spaces defined by the scalar products:

$$\left(\zeta_1(\mathbf{c},x), \zeta_2(\mathbf{c},x)\right) = \int d\mathbf{c}\zeta_1 \exp\left(-c^2\right)\zeta_2 \ , \tag{9-16}$$

$$\left[\left(\zeta_1(\mathbf{c},x), \zeta_2(\mathbf{c},x)\right)\right] = \int_0^\infty dx(\zeta_1 , \zeta_2) \ . \tag{9-17}$$

Now, in order to construct a variational principle, one must first express u_0^* as an integral of $\Phi(\mathbf{c},\tilde{x})$. For this purpose one employs the conservation of moments of the Boltzmann equation as was done in the previous analysis. These properties of the Boltzmann equation can be expressed as:

$$\left(c_x c_y , \Phi(\mathbf{c},\tilde{x})\right) = 0 \ , \tag{9-18}$$

$$\left(c_x^2 c_y \phi_u(c) h , \Phi(\mathbf{c},0)\right) = \left(c_x^2 c_y \phi_u(c) h , \Phi(\mathbf{c},\infty)\right) \ . \tag{9-19}$$

Using the asymptotic solution and Eq. (9-19) one can obtain:

$$u_0^* = \frac{1}{2\left(c_x^2 c_y \phi_u(c) h , c_y\right)} \times\left\{-(2-\alpha_\tau)\left(c_x^3 c_y^2 \phi_u^2(c) h^2 , \eta(c_x)\right) - \left[\left(\mathrm{H}\,\Phi(\mathbf{c},\tilde{x}), p_u(\mathbf{c},\tilde{x})\right)\right]\right\} \ . \tag{9-20}$$

To obtain an integral relation for the dimensionless velocity, the first-order Chapman-Enskog solution for $\phi_u(c)$ is used. This simplifies the algebraic calculations in Eq. (9-20) and results in the following integral relation:

$$u_0^* = (2-\alpha_\tau)\tfrac{5}{8}\lambda h + \frac{8}{5\pi^2 \lambda h}\int \exp\left(-c^2\right) d\mathbf{c} \times \int_0^\infty \Phi^*\left(\mathbf{c},x'\right) p_u\left(\mathbf{c},x'\right) dx' \ , \tag{9-21}$$

where $\Phi^*\left(\mathbf{c},\tilde{x}\right) = \mathrm{H}\,\Phi\left(\mathbf{c},\tilde{x}\right)$.

Now, one may construct the functional:

$$I\left(\tilde{\Phi}\right) = \left[\left(\tilde{\Phi}^*, \tilde{\Phi} - L\tilde{\Phi} - 2p_u\right)\right] = F\left(\tilde{\Phi}^*, \tilde{\Phi}\right) .$$

As is shown in Appendix D, this functional is stationary for $\tilde{\Phi} = \Phi$, where Φ is a solution of Eq. (9-12a). If the stationary value of $I\left(\tilde{\Phi}\right)$ is introduced in Eq. (9-21) one can obtain:

$$u_0^* = (2-\alpha_\tau)\tfrac{5}{8}\lambda h - \frac{8}{5\pi^2 \lambda h} I_{st}\left(\tilde{\Phi}\right) = c_m \lambda h \ . \tag{9-22}$$

Consider this functional in detail. It does not involve the real function, $\Phi\left(\mathbf{c},x\right)$, and, therefore, the variational principle can be immediately employed to determine the best value of a given trial function. Since the main purpose is to obtain a solution for the mean velocity, u_0^*, which is directly related to the stationary value taken by $I\left(\tilde{\Phi}\right)$, one may expect that a sufficiently accurate approximation for u_0^* can be determined by considering even some simple trial functions. One simple trial function that takes into account the asymptotic behavior of $\tilde{\Phi}\left(\mathbf{c},\tilde{x}\right)$ is:

$$\tilde{\Phi}\left(\mathbf{c},\tilde{x}\right) = \alpha c_y \ , \tag{9-23}$$

$$\tilde{\Phi}^*\left(\mathbf{c},\tilde{x}\right) = \alpha\sigma\left(c\right) c_y \ , \tag{9-24}$$

where α is a ‘so-called’ variational parameter. Eq. (9-24) can be obtained directly when the trial function, αc_y, is substituted into Eq. (9-5). Having inserted this trial function into the functional and performed a very simple integration, one can obtain the following expression:

$$I\left(\alpha\right) = \tfrac{1}{4}\pi\left[\alpha^2\alpha_\tau - \alpha\left(2-\alpha_\tau\right)\tfrac{5}{4}\pi\lambda h\right] . \tag{9-25}$$

The stationary value for $I(\alpha)$ occurs if $dI/d\alpha = 0$, so that:

$$\alpha = \frac{2-\alpha_\tau}{\alpha_\tau} \tfrac{5}{8} \pi \lambda h \ . \tag{9-26}$$

From Eq. (9-22) one can obtain the expression for the slip coefficient that may be written as:

$$c_m = \frac{2-\alpha_\tau}{\alpha_\tau} \left(1 + \frac{4-\pi}{2\pi} \alpha_\tau \right) \tfrac{5}{16} \pi \ . \tag{9-27}$$

It is very important to note that this formula is exactly the same as the expression derived by Loyalka's method.

Now, consider how one might describe the gas flow in the Knudsen layer. It should be emphasized that the variational method, even employed in its simplest form, permits one to perform this analysis. The main interest here is to investigate the velocity defect expression that is derived by taking a moment of the distribution function. The correction to the distribution function, $\tilde{\Phi}(\mathbf{c}, \tilde{x})$, obtained from Eq. (9-12) has the following form for a given trial function:

$$\begin{aligned} \Phi(\mathbf{c}, \tilde{x}) = \eta(c_x) \Bigg\{ & -(2-\alpha_\tau) c_x c_y \phi_u(c) h \exp\left(-\frac{\sigma(c)}{c_x} \tilde{x} \right) \\ & + \alpha c_y \left[1 - \alpha_\tau \exp\left(-\frac{\sigma(c)}{c_x} \tilde{x} \right) \right] \Bigg\} + \eta(-c_x) \{ \alpha \, c_y \} \ . \end{aligned} \tag{9-28}$$

To obtain the right asymptotic form for this correction, insert $\alpha = 2u_0^*$ into Eq. (9-28) instead of the value of α given by Eq. (9-26). This gives:

$$\begin{aligned} \Phi(\mathbf{c}, \tilde{x}) = 2u_0^* c_y + \eta(c_x) \exp\left(-\frac{\sigma(c)}{c_x} \tilde{x} \right) \\ \times \left[(2-\alpha_\tau) c_x c_y \tfrac{5}{4} \sqrt{\pi} \lambda h - 2u_0^* c_y \alpha_\tau \right] . \end{aligned} \tag{9-29}$$

Allowing for Eq. (9-29), one can obtain the following expressions for both the mean velocity and the velocity defect of a gas:

$$u(\tilde{x}) = \left(\frac{\partial u}{\partial \tilde{x}}\right)_\infty \lambda \left[c_m + \frac{\tilde{x}}{\lambda} - \alpha_\tau \frac{2c_m}{\pi^{3/2}} \mathrm{J}_1(\tilde{x}) + (2 - \alpha_\tau) \frac{5}{4\pi} \mathrm{J}_2(\tilde{x}) \right], \quad \text{(9-30)}$$

$$u_d(\tilde{x}) = \left(\frac{\partial u}{\partial \tilde{x}}\right)_\infty \lambda \left[\alpha_\tau \frac{2c_m}{\pi^{3/2}} \mathrm{J}_1(\tilde{x}) - (2 - \alpha_\tau) \tfrac{5}{4} \pi^{-1} \mathrm{J}_2(\tilde{x}) \right]. \quad \text{(9-31)}$$

Here, the functions, $\mathrm{J}_1(\tilde{x})$ and $\mathrm{J}_2(\tilde{x})$, are defined by:

$$\mathrm{J}_1(\tilde{x}) = \pi \int_0^\infty c^4 \exp(-c^2) \left[E_2\left(\frac{\sigma(c)}{c}\tilde{x}\right) - E_4\left(\frac{\sigma(c)}{c}\tilde{x}\right) \right] dc\ , \quad \text{(9-32)}$$

$$\mathrm{J}_2(\tilde{x}) = \pi \int_0^\infty c^5 \exp(-c^2) \left[E_3\left(\frac{\sigma(c)}{c}\tilde{x}\right) - E_5\left(\frac{\sigma(c)}{c}\tilde{x}\right) \right] dc\ , \quad \text{(9-33)}$$

where the functions, $E_n(z)$, are the exponential integrals that can be expressed in the form:

$$E_n(z) = \int_0^1 t^{n-2} \exp\left(-\frac{z}{t}\right) dt\ . \quad \text{(9-34)}$$

As can be easily seen, the variational method allows one to investigate the Knudsen layer. The accuracies of this method and the other methods described previously will be considered in the next section.

3. DISCUSSION OF THE SLIP-FLOW RESULTS.

There are several expressions for the mean velocity and slip coefficient that have been obtained here by employing various methods and using various boundary models. To establish the accuracy of these methods it is necessary to compare the approximate analytical results with direct numerical results. For this purpose, we have found it convenient to use the numerical results of Loyalka and Hickey [9]. The analytical results that we are comparing are those that take into account the second-order corrections to the Chapman-Enskog distribution functions and, moreover, the molecules are assumed in the analysis to be rigid spheres.

It is easy to see that both Loyalka's method and the variational method give the same results for the slip coefficient if the trial function given by Eq. (9-23) is employed in the variational analysis. These methods result in the following relation:

$$c_m = \frac{2-\alpha_\tau}{\alpha_\tau}\tfrac{5}{16}\pi\left[1+(0.1086)\alpha_\tau\right]. \tag{9-35}$$

The velocity defect cannot be obtained by Loyalka's method with the exception of one point for which $\tilde{x}=0$. Since the expression for a_w given by Eq. (8-34) is not altered, if the second-order Chapman-Enskog solution is used, one can obtain the velocity defect at the wall given by:

$$u_d^*(0) = \tfrac{5}{32}\pi(2-\alpha_\tau)\left[\frac{4\omega_1}{\pi}-\frac{2}{\pi}\left(1+\tfrac{1}{2}b_2^{(2)*}\right)\right] = (0.2759)(2-\alpha_\tau). \tag{9-36}$$

where $\omega_1 = 1-b_2^{(2)*}+\tfrac{17}{4}\left(b_2^{(2)*}\right)^2$.

The half-range moment method (four-moment approach) gives different results for various boundary models. The boundary model proposed by Ivchenko, Loyalka, and Tompson [10] (the ILT boundary model) results in the following expressions for c_m and $u_d^*(\tilde{x})$:

$$c_m = \frac{2-\alpha_\tau}{\alpha_\tau}\tfrac{5}{8}\sqrt{\pi}\left[\frac{(0.6690)+(0.1775)\alpha_\tau}{(0.7549)+(0.09714)\alpha_\tau}\right], \tag{9-37}$$

$$u_d^*(\tilde{x}) = (1.4229)\frac{(0.3413)-(0.1706)\alpha_\tau}{(0.7549)+(0.09714)\alpha_\tau}\exp\left(-(2.2015)\frac{\tilde{x}}{\lambda}\right). \tag{9-38}$$

The same method for the Maxwellian boundary model gives:

$$c_m = \frac{2-\alpha_\tau}{\alpha_\tau}\tfrac{5}{8}\sqrt{\pi}\left[\frac{(3.1929)+\alpha_\tau}{(3.6028)+(0.5241)\alpha_\tau}\right], \tag{9-39}$$

$$u_d^*(\tilde{x}) = (1.4229)\frac{2-\alpha_\tau}{(3.6028)+(0.5241)\alpha_\tau}\exp\left(-(2.2015)\frac{\tilde{x}}{\lambda}\right). \tag{9-40}$$

Table 9-1. The slip-flow coefficient, c_m.

		Four-Moment Approach			
	Numerical Values [9]	ILT Boundary Model [10]		Maxwellian Boundary Model [10]	
α_τ	c_m	c_m	$\delta\%$	c_m	$\delta\%$ †
0.2	9.0458	9.0710	0.28	9.1235	0.86
0.4	4.1147	4.1310	0.40	4.1759	1.49
0.6	2.4531	2.4650	0.49	2.5027	2.02
0.8	1.6096	1.6185	0.55	1.6496	2.48
1.0	1.0940	1.1006	0.60	1.1254	2.87

† $\delta\% = \left(\left|a - a_{num}\right| / a_{num}\right) \times 100$; $a = c_m$

Table 9-2. The slip-flow coefficient, c_m.

	Numerical Values [9]	Loyalka Method & Variational Results, Eq. (9-35)	
α_τ	c_m	c_m	$\delta\%$ †
0.2	9.0458	9.0277	0.20
0.4	4.1147	4.0976	0.42
0.6	2.4531	2.4400	0.53
0.8	1.6096	1.6006	0.56
1.0	1.0940	1.0884	0.51

† $\delta\% = \left(\left|a - a_{num}\right| / a_{num}\right) \times 100$; $a = c_m$

To compare the accuracy of the various expressions for the slip coefficient the analytical and numerical results are presented together in Tables 9-1 and 9-2. It can be seen that a very good agreement exists between analytical and numerical values of c_m for the ILT boundary model. The discrepancy of these results is less than 0.6% for all α_τ. The maximum deviation of the other analytical data amounts to 0.56% and 2.87% for the variational method and the Maxwellian boundary model, respectively, and, therefore, Eqs. (9-35) and (9-37) are the most accurate analytical formulas for the slip coefficient.

Another macroscopic parameter which has a significant influence on the description of the gas flow in the Knudsen layer is the velocity defect at the wall. The values of this parameter are presented in Tables 9-3 and 9.4. For this parameter the best agreement with numerical values in the full region of α_τ is obtained from the ILT boundary model. The maximum difference for this model, for all α_τ, is about 5.5%, while the discrepancies associated with the Loyalka and variational methods are 8.4% and 18.5% respectively for $\alpha_\tau = 1.0$, and range up to 16.9% and 42.2% respectively for $\alpha_\tau = 0.2$. The numerical and analytical data for the dimensionless velocity defect are presented in Table 9-5.

Table 9-3. The velocity defect at the wall, $u_d^*(0)$.

		Four-Moment Approach			
	Numerical Values [9]	ILT Boundary Model [10]		Maxwellian Boundary Model [10]	
α_τ	$u_d^*(0)$	$\delta\,\%$ †	$u_d^*(0)$	$\delta\,\%$	$u_d^*(0)$
0.2	0.5976	0.5645	5.5	0.6908	15.6
0.4	0.5180	0.4895	5.5	0.5972	15.3
0.6	0.4422	0.4181	5.5	0.5085	15.0
0.8	0.3701	0.3500	5.4	0.4245	14.7
1.0	0.3013	0.2851	5.4	0.3448	14.5

† $\delta\% = \left(\left|a - a_{num}\right| / a_{num}\right) \times 100$; $a = u_d^*(0)$

Table 9-4. The velocity defect at the wall, $u_d^*(0)$.

	Numerical Values [9]	Loyalka Method Eq. (9-36)		Variational Method Eq. (9-31)	
α_τ	$u_d^*(0)$	$\delta\,\%$ †	$u_d^*(0)$	$\delta\,\%$	$u_d^*(0)$
0.2	0.5976	0.4967	16.9	0.3452	42.2
0.4	0.5180	0.4414	14.8	0.3283	36.6
0.6	0.4422	0.3863	12.6	0.3061	30.8
0.8	0.3701	0.3311	10.5	0.2784	24.8
1.0	0.3013	0.2759	8.4	0.2455	18.5

† $\delta\% = \left(\left|a - a_{num}\right| / a_{num}\right) \times 100$; $a = u_d^*(0)$

An analysis of Table 9-5 shows that for the ILT boundary model a satisfactory agreement with the numerical results occurs if $0 < x' < 0.5$. The maximum deviation from the numerical values for this domain of x' is about 11.5%. When $x' > 0.5$ the velocity defect decreases more rapidly than the corresponding numerical value. In this region the discrepancy of the results is very large. This occurs because the asymptotic velocity defect behavior for this approach is described by one exponential term which does not characterize a real dependence on the normal coordinate [11]. This dependence is more complicated than a simple exponential drop, as one can see from Eq. (9-12). Moreover, when $x' > 0.5$, the velocity defect values become extremely small so that the moment approach employed does not have a sufficient accuracy to describe this very subtle quantity. The variational method does not accurately describe the velocity defect because a very simple trial function has been used.

The results for the mean velocity profile are shown in Table 9-6. For the ILT model the maximum deviation is 2.92%. For the Maxwellian boundary model this difference is 3.77%.

Table 9-5. The dimensionless velocity defect, $u_d^*(x')$, ($\alpha_\tau = 1$).

Dimensionless Normal Coordinate	Numerical Values [9]	Four-Moment Approach: ILT Boundary Model [10]	Four-Moment Approach: Maxwellian Boundary Model [10]
x' †	$u_d^*(x')$	$u_d^*(x')$	$u_d^*(x')$
0.0	0.3013	0.2851	0.3448
0.1	0.1997	0.2226	0.2692
0.2	0.1560	0.1738	0.2102
0.5	0.0888	0.0827	0.1000
1.0	0.0430	0.0240	0.0290
1.5	0.0233	0.0070	0.0084
2.0	0.0133	0.0020	0.0024

† $\tilde{x} = \frac{5}{8}\sqrt{\pi}\lambda_\mu x' = (1.12423)\lambda x'$

Table 9-6. The mean velocity, $u^*(x')$, ($\alpha_\tau = 1$).

	Numerical Values [9]	Four-Moment Approach: ILT Boundary Model [10]		Four-Moment Approach: Maxwellian Boundary Model [10]	
x' †	$u^*(x')$	$u^*(x')$	$\delta\%$ #	$u^*(x')$	$\delta\%$
0.0	0.7924	0.8155	2.92	0.7806	1.49
0.1	1.0049	0.9888	1.60	0.9670	3.77
0.2	1.1594	1.1484	0.95	1.1368	1.95
0.5	1.5589	1.5718	0.83	1.5793	1.31
1.0	2.1586	2.1844	2.58	2.2042	2.11
1.5	2.7322	2.7553	2.31	2.7787	1.70
2.0	3.2961	3.3142	0.55	3.3386	1.29

† $\tilde{x} = \frac{5}{8}\sqrt{\pi}\lambda_\mu x' = (1.12423)\lambda x'$

$\delta\% = \left(|a - a_{num}|/a_{num}\right)\times 100$; $a = u^*(x')$

4. THE VARIATIONAL SOLUTION FOR THE THERMAL-CREEP PROBLEM.

The thermal-creep problem has been stated in detail in the previous chapter, and therefore, only a brief outline will be given here. Let the distribution function be described by:

$$f = f^{(0)}\left\{1 + \left(c^2 - \tfrac{5}{2}\right)\tilde{y}q_y + c_y\phi_T(c)q_y + \Phi^{\pm}(\mathbf{c},\tilde{x})\right\} . \tag{9-41a}$$

Here, $\phi_T(c) = -A(c)$ and may be written as:

$$\phi_T(c) = -A(c) = \tfrac{15}{16}\sqrt{\pi}\lambda_\kappa\left[S_{3/2}^{(1)}(c^2) + a_2^{(2)*}S_{3/2}^{(2)}(c^2)\right], \tag{9-41b}$$

where $\lambda_\kappa = \frac{45}{44}\lambda$ and $a_2^{(2)*} = (0.08889)$.

The correction, $\Phi^\pm(\mathbf{c},0)$, to the distribution function can be specified from Eq. (9-5) and its boundary conditions which can be expressed as:

$$\Phi^+(\mathbf{c},0) = -\alpha_\tau c_y \phi_T(c) q_y + (1-\alpha_\tau)\Phi^-(-c_x, c_y, c_z, 0), \tag{9-42}$$

$$\lim_{\tilde{x}\to\infty} \Phi^\pm(\mathbf{c},\tilde{x}) = 2c_y u_0^* . \tag{9-43}$$

As it is shown in Section 9.2, one can obtain the following integral equation for $\Phi^\pm(\mathbf{c},0)$:

$$\Phi(\mathbf{c},\tilde{x}) = L\Phi(\mathbf{c},\tilde{x}) + p_T(\mathbf{c},\tilde{x}) . \tag{9-44}$$

Here the integral operator, $L\Phi(\mathbf{c},\tilde{x})$, is given by Eq. (9-13) and $p_T(\mathbf{c},\tilde{x})$ is defined by the relation:

$$p_T(\mathbf{c},\tilde{x}) = -\eta(c_x)\left[\alpha_\tau c_y \phi_T(c) q_y \exp\left(-\frac{\sigma(c)}{c_x}\tilde{x}\right)\right]. \tag{9-45}$$

This function represents the external non-uniformity of a gas in Eq. (9-44).

Now, taking into account the conservation of moments of the Boltzmann equation given by Eqs. (9-18) and (9-19) one can derive the following integral equation for the mean velocity:

$$u_0^* = \frac{-\alpha_\tau q_y h\left(c_x^2 c_y^2 \phi_u(c)\phi_T(c), \eta(c_x)\right) + I}{2h\left(c_y, c_x^2 c_y \phi_u(c)\right)}, \tag{9-46}$$

where $I = -\left[\left(\mathrm{H}\,\Phi(\mathbf{c},\tilde{x}), p_u(\mathbf{c},\tilde{x})\right)\right]$ and $p_u(\mathbf{c},\tilde{x})$ is the known function given by Eq. (9-12a) that is proportional to the non-uniformity of a gas for the slip-flow problem. The standard notations given by Eqs. (9-16), (9-17), and (9-14) are used in this expression.

The functional contained in Eq. (9-46) is a linear form with respect to the trial function $\Phi(\mathbf{c},\tilde{x})$ and, therefore, it cannot be immediately used for solution of a variational problem. A new functional must be constructed on the basis of the square form of the trial functions and which has a stationary value equal to I. Let this new functional be the sum of two functionals, the first of which is equal to $-[(\tilde{\Phi}^*, p_u(\mathbf{c},\tilde{x}))]$. Since p_u is proportional to λh and $\tilde{\Phi}^* \sim \sigma(c)\lambda q_y \sim q_y$, the integrand of the second functional must be proportional to $\lambda q_y h$, and moreover this term is equal to zero if the trial function coincides with the real correction, $\Phi(\mathbf{c},\tilde{x})$. This indicates that the integrand of the second functional has to contain the factor $\tilde{\Phi} - L\tilde{\Phi} - p_T$, which is proportional to λq_y. Only the unique square form for this integrand, which is the product of $\tilde{\Phi}_u^*$, where $\tilde{\Phi}_u$ is a trial function for the slip-flow problem, and $\tilde{\Phi} - L\tilde{\Phi} - p_T$, can be constructed so that it should be proportional to $\sigma(c)\lambda^2 hq_y$. This very simple dimensional analysis shows that the following form can be used for this variational problem:

$$F\left(\tilde{\Phi}_u^*,\tilde{\Phi}\right) = -\left[\left(\tilde{\Phi}^*, p_u\right)\right] + \left[\left(\tilde{\Phi}_u^*, \tilde{\Phi} - L\tilde{\Phi} - p_T\right)\right]. \tag{9-47}$$

The stationary value of this functional is equal to I. Now, the asymptotic solutions of Eqs. (9-12) and (9-44) can be chosen as the trial functions for $F\left(\tilde{\Phi}_u^*,\tilde{\Phi}\right)$ which can be expressed in the form:

$$\tilde{\Phi} = \alpha_1 c_y \quad , \quad \tilde{\Phi}^* = \sigma(c)\alpha_1 c_y \ , \tag{9-48a}$$

$$\tilde{\Phi}_u = \alpha_2 c_y \ , \quad \tilde{\Phi}_u^* = \sigma(c)\alpha_2 c_y \ . \tag{9-48b}$$

It is very important to note that different constants, α_1 and α_2, are used for these trial functions as they are asymptotic solutions of different equations. The first-order Chapman-Enskog solutions for $\phi_u(c)$ and $\phi_T(c)$ are assumed to be employed in a later analysis and one can use unity instead of $\phi_u(c)$ as it is easy to see from Eq. (9-46). In accordance with this suggestion one can obtain the following relationship:

$$F(\alpha_1,\alpha_2) = A_1\alpha_1 + A_2\alpha_1\alpha_2 + A_3\alpha_2 \ . \tag{9-49}$$

Here, the following notations are introduced:

$$A_1 = (2-\alpha_\tau)\tfrac{1}{8}\pi^{3/2} \ , \ A_2 = \alpha_\tau \tfrac{1}{4}\pi \ , \ A_3 = -\alpha_\tau \tfrac{15}{128}\pi^{3/2}\lambda q_y \ . \tag{9-50}$$

The desired values of α_1 and α_2 for which this functional is stationary can be defined from the following equations:

$$\frac{\partial F}{\partial \alpha_1}=0\ ,\quad \frac{\partial F}{\partial \alpha_2}=0\ .$$

The stationary value of the functional can be expressed in the form:

$$F_{st}\left(\tilde{\Phi}_u^*,\tilde{\Phi}\right)=\left(2-\alpha_\tau\right)\tfrac{15}{254}\pi^2\lambda q_y\ . \tag{9-51}$$

Using Eq. (9-51) in Eq. (9-46), one can obtain:

$$u_0=c_{Tsl}\nu q_y\ . \tag{9-52}$$

The thermal-creep coefficient in this formula is given by:

$$c_{Tsl}=\tfrac{3}{4}\left(1+\tfrac{1}{2}\alpha_\tau\right)\ . \tag{9-53}$$

This expression coincides exactly with the one obtained by Loyalka's method.

Substituting the asymptotic value of the mean velocity $2u_0^*c_y$ instead of $\alpha_1 c_y$ into an expression of the distribution function, one can obtain the relation for the velocity defect defined by:

$$u_d^*\left(\tilde{x}\right)=\frac{u_d\left(\tilde{x}\right)}{\nu q_y}=\alpha_\tau\left\{\frac{2c_{Tsl}}{\pi^{3/2}}\mathrm{J}_1\left(\tilde{x}\right)+\frac{3}{\pi^{3/2}}\mathrm{J}_3\left(\tilde{x}\right)\right\}\ . \tag{9-54}$$

Here, $\mathrm{J}_1\left(\tilde{x}\right)$ is given by Eq. (9-32) and $\mathrm{J}_3\left(\tilde{x}\right)$ is defined by:

$$\begin{aligned}\mathrm{J}_3\left(\tilde{x}\right)=\pi\int_0^\infty c^4\left(\tfrac{5}{2}-c^2\right)\exp\left(-c^2\right)\\ \times\left[E_2\left(\frac{\sigma(c)}{c}\tilde{x}\right)-E_4\left(\frac{\sigma(c)}{c}\tilde{x}\right)\right]dc\ ,\end{aligned} \tag{9-55}$$

where the notations in Eq. (9-55) are the same as in Eqs. (9-32) and (9-33).

5. DISCUSSION OF THE THERMAL-CREEP RESULTS.

The accuracy of the thermal-creep problem results obtained by various approximate methods will be discussed here. This analysis can be performed by means of a comparison of analytical values with the numerical results obtained by Loyalka [12]. First, the accuracy of the various expressions of the thermal-creep coefficient will be examined. All of the analytical expressions for this coefficient and the velocity defect will be given here in the most convenient form for this analysis.

Loyalka's method and the variational solution give the same results for the thermal-creep coefficient. If the second-order Chapman-Enskog solution is used, these methods both yield:

$$c_{Tsl} = \tfrac{3}{4}\left[\varepsilon_1 + \left(\varepsilon_2 - \tfrac{1}{2}\varepsilon_1\right)\alpha_\tau\right], \tag{9-56}$$

where $\varepsilon_1 = 1 + \frac{1}{4}a_2^{(2)*}$ and $\varepsilon_2 = 1 - \frac{7}{2}b_2^{(2)*} + \frac{7}{2}a_2^{(2)*}b_2^{(2)*}$. The variational velocity defect obtained by means of the simplest trial functions is given by Eq. (9-54). The velocity defect at the wall obtained by Loyalka's method can be written in the form:

$$u_d^*(0) = \tfrac{3}{4}\alpha_\tau\varepsilon_2 \; . \tag{9-57}$$

In the previous chapter the various solutions of the thermal-creep problem were obtained within the framework of the half-range moment method for the different boundary models. The boundary model proposed by Ivchenko, Loyalka, and Tompson [13, 14] results in the following expressions for c_{Tsl} and $u_d^*(\tilde{x})$:

$$c_{Tsl} = \tfrac{3}{2}\frac{(0.4354)+(0.2179)\alpha_\tau}{(0.8518)+(0.1096)\alpha_\tau}, \tag{9-58}$$

$$u_d^*(\tilde{x}) = \frac{(0.5822)\alpha_\tau}{(0.8518)+(0.1096)\alpha_\tau}\exp\left(-(2.2015)\frac{\tilde{x}}{\lambda}\right). \tag{9-59}$$

For the integral boundary model that is usually employed in transport problems, these relations can be written as:

Table 9-7. The thermal-creep coefficient, c_{Tsl}.

		Four-Moment Approach			
	Numerical Values [12] #	ILT Boundary Model [13] Eq. (9-58)		Integral Boundary Model [13] Eq. (9-60)	
α_τ	c_{Tsl}	c_{Tsl}	$\delta\,\%$ †	c_{Tsl}	$\delta\,\%$
0.2	0.8234	0.8223	0.1	0.8395	1.7
0.4	0.8733	0.8752	0.2	0.9240	5.8
0.6	0.9207	0.9255	0.5	1.0041	9.1
0.8	0.9657	0.9735	0.8	1.0799	11.8
1.0	1.0089	1.0193	1.0	1.1519	14.2

† $\delta\% = \left(\left|a - a_{num}\right| / a_{num}\right) \times 100$; $a = c_{Tsl}$
# These additional numerical values have been obtained by Loyalka.

Table 9-8. The thermal-creep coefficient, c_{Tsl}.

	Numerical Values [12] #	Loyalka Method & Variational Results, Eq. (9-56)	
α_τ	c_{Tsl}	c_{Tsl}	$\delta\,\%$ †
0.2	0.8234	0.8120	1.4
0.4	0.8733	0.8573	1.8
0.6	0.9207	0.9027	2.0
0.8	0.9657	0.9480	1.8
1.0	1.0089	0.9933	1.5

† $\delta\% = \left(\left|a - a_{num}\right| / a_{num}\right) \times 100$; $a = c_{Tsl}$
# These additional numerical values have been obtained by Loyalka.

$$c_{Tsl} = \tfrac{3}{2}\frac{(1.5464)+(1.1741)\alpha_\tau}{(3.0927)+(0.4499)\alpha_\tau}\,, \tag{9-60}$$

$$u_d^*(\tilde{x}) = \frac{(3.4149)\alpha_\tau}{(3.0927)+(0.4499)\alpha_\tau}\exp\left(-(2.2015)\frac{\tilde{x}}{\lambda}\right). \tag{9-61}$$

Now, consider the accuracy of the various analytical results for this problem. The numerical and analytical data for the thermal-creep coefficient versus the accommodation coefficient, α_τ, are presented in Tables 9-7 and 9-8. It is easy to see that the most accurate values of c_{Tsl} are predicted by the moment method for the ILT boundary model. In this case, the maximum discrepancy between the numerical and analytical results is about 1.0%, while Loyalka's method and the variational method give results that differ from the numerical results by 1.4% for $\alpha_\tau = 0.2$ up to 2.0% when $\alpha_\tau = 0.6$.

Table 9-9. The velocity defect at the wall, $u_d^*(0)$.

		Four-Moment Approach			
	Numerical Values [12] #	ILT Boundary Model [13] Eq. (9-59)		Integral Boundary Model [13] Eq. (9-61)	
α_τ	$u_d^*(0)$	$u_d^*(0)$	$\delta\%$ †	$u_d^*(0)$	$\delta\%$
0.2	0.1596	0.1333	16.5	0.2146	34.5
0.4	0.3078	0.2600	15.5	0.4174	35.6
0.6	0.4458	0.3807	14.6	0.6093	36.7
0.8	0.5748	0.4958	13.7	0.7913	37.7
1.0	0.6960	0.6056	13.0	0.9640	38.5

† $\delta\% = \left(\left|a - a_{num}\right| / a_{num}\right) \times 100$; $a = u_d^*(0)$

These additional numerical values have been obtained by Loyalka.

Table 9-10. The velocity defect at the wall, $u_d^*(0)$.

	Numerical Values [12] #	Loyalka Method Eq. (9-57)		Variational Method Eq. (9-54)	
α_τ	$u_d^*(0)$	$u_d^*(0)$	$\delta\%$ †	$u_d^*(0)$	$\delta\%$
0.2	0.1596	0.1220	23.6	0.0825	48.3
0.4	0.3078	0.2440	20.7	0.1800	41.5
0.6	0.4458	0.3660	17.9	0.2925	34.4
0.8	0.5748	0.4880	15.1	0.4200	26.9
1.0	0.6960	0.6100	12.8	0.5625	19.2

† $\delta\% = \left(\left|a - a_{num}\right| / a_{num}\right) \times 100$; $a = u_d^*(0)$

These additional numerical values have been obtained by Loyalka.

The integral boundary model values deviate greatly from the numerical results (from 1.7% when $\alpha_\tau = 0.2$ up to 14.2% when $\alpha_\tau = 1.0$).

In Tables 9-9 and 9-10 values of the velocity defect at the wall are presented. A comparison of these velocity defect values shows that the discrepancy between the analytical and numerical results is noteworthy for all of the analytical methods discussed. Loyalka's method and the ILT boundary model have the best agreement with the numerical values. The maximum deviations of these values are 23.6% and 16.5%, respectively, when $\alpha_\tau = 0.2$ and are seen to decrease to only 12.8% and 13.0%, respectively, when $\alpha_\tau = 1.0$.

Values of the velocity defect versus the normal coordinate for $\alpha_\tau = 1.0$ are presented in Table 9-11. It can be seen that the relative deviations of all of the analytical values from the corresponding numerical values are very large for $x' \geq 0.5$. The velocity defect behavior inside the Knudsen layer seems to be more complicated [15] than the behavior predicted by these analytical formulae and, moreover, the accuracy of these methods appears insufficient overall to describe this extremely subtle quantity.

Table 9-11. The dimensionless velocity defect, $u_d^*(x')$, ($\alpha_\tau = 1$).

Dimensionless Normal Coordinate	Numerical Values [12]	Four-Moment Approach: ILT Boundary Model [13] Eq. (9-59)	Four-Moment Approach: Integral Boundary. Model [13] Eq. (9-61)
x' †	$u_d^*(x')$	$u_d^*(x')$	$u_d^*(x')$
0.00	0.6960	0.6056	0.9640
0.25	0.3840	0.3052	0.3783
0.50	0.2539	0.0933	0.1485
0.75	0.1747	0.0366	0.0583
1.00	0.1229	0.0143	0.0229
1.50	0.0632	0.0022	0.0035
2.00	0.0349	0.0003	0.0005

† $\tilde{x} = \frac{15}{16}\sqrt{\pi}\lambda_\kappa x' = (1.69944)\lambda x'$

Table 9-12. Numerical values of the velocity defect, $u_d^*(x')$.

x' †	Velocity Defect, $u_d^*(x')$: α_τ=0.2	α_τ=0.4	α_τ=0.6	α_τ=0.8	α_τ=1.0
0.00	0.1596	0.3078	0.4458	0.5748	0.6960
0.25	0.0840	0.1639	0.2400	0.3126	0.3840
0.50	0.05499	0.1076	0.1579	0.2062	0.2539
0.75	0.03768	0.0738	0.1085	0.1419	0.1747
1.00	0.02645	0.05184	0.07629	0.09984	0.1229
1.50	0.01355	0.02659	0.03915	0.05127	0.0632
2.00	0.007017	0.01404	0.02069	0.02712	0.0349

† $\tilde{x} = \frac{15}{16}\sqrt{\pi}\lambda_\kappa x' = (1.69944)\lambda x'$

The results for the mean velocity profile are shown in Fig. 9-1. For the ILT model the maximum deviation of the mean velocity is approximately 13% which occurs when $x' = 0$, while for the integral model this discrepancy is 40%. It is clearly necessary to use even more exact approaches if one is to succeed in obtaining a good overall agreement between the analytical and numerical results for the mean velocity profile.

In Table 9-12 the numerical results for the velocity defect are presented. These data have been obtained by Loyalka by the use of the numerical technique reported in [12]. The accuracy of any approximate method can be examined by using these data.

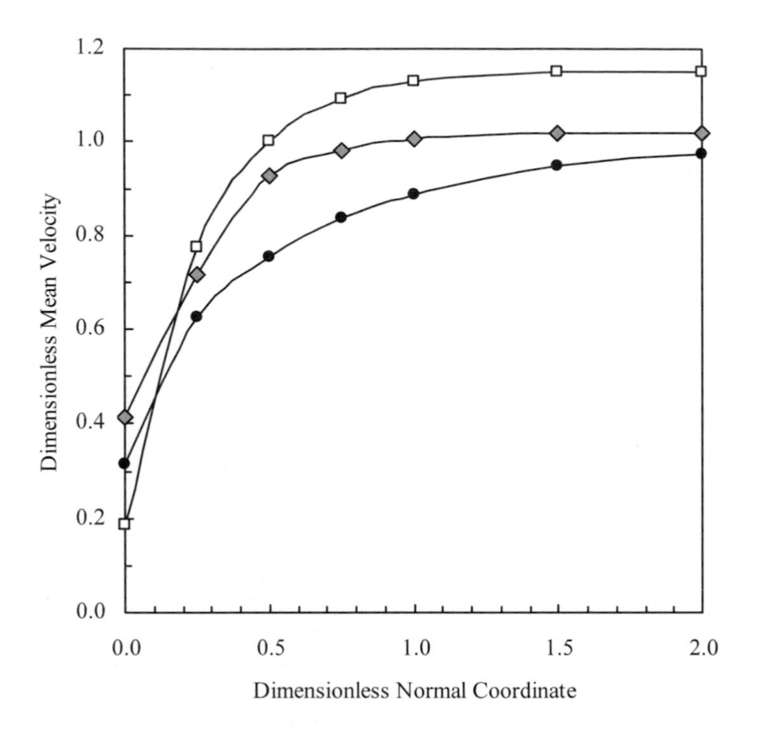

Figure 9-1. The dimensionless, mean velocity profile, $u^*(x') = u(x')/\nu q$ as a function of the dimensionless normal coordinate, x', where $\tilde{x} = (1.69944)\lambda x'$. Numerical values [12] are shown as black circles, ILT Boundary Model results [13] are shown as gray diamonds, and Integral Boundary Model results [13] are shown as open squares. Curves are shown for continuity only in order to help the reader visualize the shape of the profile.

6. SLIP-FLOW AND TEMPERATURE-JUMP COEFFICIENTS FOR THE LENNARD-JONES (6-12) POTENTIAL MODEL.

In the framework of the variational and generalized Maxwellian methods all the slip and temperature-jump coefficients depend upon a potential model if one uses the second- and higher-order Chapman-Enskog approximations to describe the gas state beyond the wall. For the second-order Chapman-Enskog approximation both these methods yield:

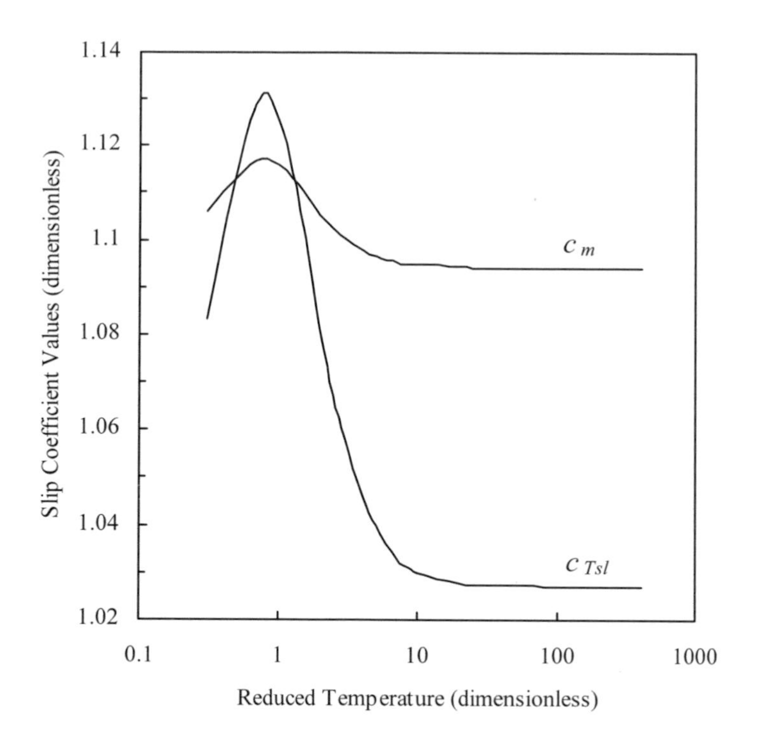

Figure 9-2. The dependence of the slip coefficients, c_m and c_{Tsl}, on the reduced temperature, T^*, for the Lennard-Jones (6-12) potential model)(assuming that $\alpha_\tau = 1$).

$$c_m = \frac{2-\alpha_\tau}{\alpha_\tau}\tfrac{5}{16}\pi\left(1+\frac{4\omega_1-\pi}{2\pi}\alpha_\tau\right), \tag{9-62}$$

$$c_{Tsl} = \tfrac{3}{4}\left[\varepsilon_1 + \left(\varepsilon_2 - \tfrac{1}{2}\varepsilon_1\right)\alpha_\tau\right], \tag{9-63}$$

$$c_T = \frac{2-\alpha_\tau\alpha_T}{\alpha_\tau\alpha_T}\tfrac{75}{128}\pi\left[1+\left(\tfrac{52}{25}\frac{\varepsilon_T}{\pi}-\tfrac{1}{2}\right)\frac{\alpha_\tau\alpha_T\left(2-\alpha_\tau\right)}{2-\alpha_\tau\alpha_T}\right] \tag{9-64}$$

where the following notations have been introduced:

$$\omega_1 = 1 - b_2^{(2)*} + \tfrac{17}{4}\left(b_2^{(2)*}\right)^2, \tag{9-65}$$

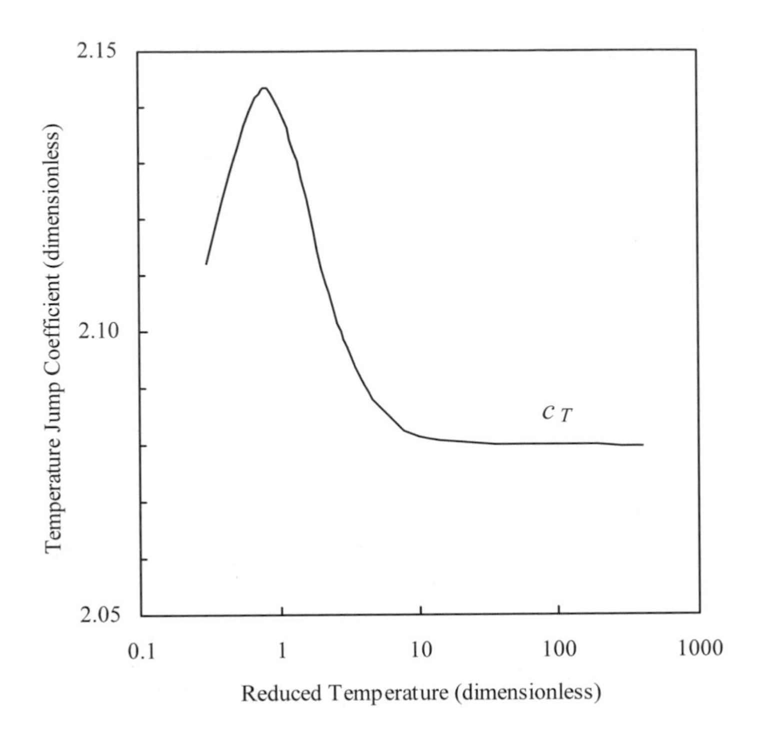

Figure 9-3. The dependence of the temperature-jump coefficient, c_T, on the reduced temperature, T^*, for the Lennard-Jones (6-12) potential model (assuming that $\alpha_\tau = \alpha_T = 1$).

$$\varepsilon_1 = 1 + \tfrac{1}{4} a_2^{(2)*} \,, \tag{9-66}$$

$$\varepsilon_2 = 1 - \tfrac{7}{2} b_2^{(2)*} + \tfrac{7}{2} a_2^{(2)*} b_2^{(2)*} \,, \tag{9-67}$$

$$\varepsilon_T = 1 - \tfrac{23}{26} a_2^{(2)*} + \tfrac{433}{208} \left(a_2^{(2)*} \right)^2 . \tag{9-68}$$

The slip and jump coefficients depend upon the reduced temperature (see Section 5.9), T^*, through their dependence on $a_2^{(2)*}$ and $b_2^{(2)*}$. Using the data given in Tab. 5-1, one can calculate the appropriate values of c_m, c_{Tsl}, and c_T for the Lennard-Jones (6-12) potential model.

The variation of c_m, c_{Tsl}, and c_T with the reduced temperature, T^*, is illustrated in Figs. 9-2 and 9-3. All the curves have a maximum when T^* is about unity. The relative deviations of the maximum values from those, when T^* is large, are about 10%, 3% and 2% for c_{Tsl}, c_T, and c_m, respectively. This analysis shows that the thermal-creep coefficient which is dependent on the cross effects (viscosity and thermal conductivity) is quite sensitive to the parameters of the Lennard-Jones (6-12) potential.

Now, examine dependence of the slip and jump coefficients on the intermolecular potential model. The analysis is restricted to consideration of the rigid-sphere and Lennard-Jones (6-12) models. For rigid-sphere molecules the potential well depth, $\varepsilon = kT/T^*$, is equal to zero and, therefore, all results depending on intermolecular interactions are very close to those for the Lennard-Jones (6-12) model in which $T^* \to \infty$. This shows that only the thermal-creep coefficient is fairly sensitive to the potential model. The maximum relative discrepancy between c_{Tsl} for these two models is about 12% if $\alpha_\tau = 1$ and T^* is about unity. It is very important to note that c_{Tsl} is independent of the temperature, T, for the rigid-sphere model.

PROBLEMS

9.1. Determine the slip and thermal-creep coefficients using the Maxwellian method for the second-order Chapman-Enskog solution.

Solution: The distribution function may be written as:

$$f = f_{eq}\left\{1 + c_x c_y \phi_u(c) h + c_y \phi_T(c) q_y + \Phi^{\pm}(\mathbf{c}, \tilde{x})\right\} ,$$

where $\Phi^-(\mathbf{c},0) = \Phi^{\pm}(\mathbf{c},\infty) = a_0 c_y$ and the functions $\phi_u(c)$ and $\phi_T(c)$ are given by Eqs. (8-75) and (9-41b), respectively. The constant, a_0, may be found from the relation $\left(c_x c_y, \eta(c_x) A\Phi^-(\mathbf{c},0) + \eta(-c_x)\Phi^-(\mathbf{c},0)\right) = 0$. Having performed some simple integrations, one can obtain:

$$c_m = \frac{2-\alpha_\tau}{\alpha_\tau} \tfrac{5}{16}\pi ,$$

and $c_{Tsl} = \frac{3}{4}\varepsilon_1$, where $\varepsilon_1 = \left(1 + \frac{1}{4} a_2^{(2)*}\right) = (1.02222)$. Thus, for a Maxwellian analysis, the usage of the second-order Chapman-Enskog solution results in an increase of the thermal-creep coefficient by 2.2% while the slip-flow coefficient is not altered.

9.2. Determine the temperature-jump at the wall using the Maxwellian method for the second-order Chapman-Enskog solution. Consider the case when $\alpha_\tau = 1$.

Solution: The distribution function is given by:

$$f = f^{(0)}\left\{1 + \tilde{x}q_x\left(c^2 - \tfrac{5}{2}\right) + c_x\phi_T\left(c\right)q_x + \Phi^{\pm}\left(\mathbf{c}, \tilde{x}\right)\right\} ,$$

where $\phi_T\left(c\right) = \frac{15}{16}\sqrt{\pi}\lambda_\kappa\left(S_{3/2}^{(1)}\left(c^2\right) + a_2^{(2)*}S_{3/2}^{(2)}\left(c^2\right)\right)$ and:

$$\Phi^{\pm}\left(\mathbf{c}, \infty\right) = \Phi^{-}\left(\mathbf{c}, 0\right) = a_1\left(c^2 - \tfrac{5}{2}\right) .$$

The temperature-jump may be calculated by employing the general moment solution of the Boltzmann equation given by $\left(c_x c^2, \Phi^{\pm}\left(\mathbf{c}, 0\right)\right) = 0$. Since:

$$\int_+ c_x^2 c^2 S_{3/2}^{(k)}\left(c^2\right)\exp\left(-c^2\right)d\mathbf{c} = 0 \quad ; \quad k \geq 2 ,$$

the temperature-jump is not altered for arbitrary orders of the Chapman-Enskog solution and may be expressed in the form:

$$\Delta T = \frac{2 - \alpha_T}{\alpha_T}\tfrac{75}{128}\pi\lambda_\kappa\left(\frac{\partial T}{\partial x}\right)_\infty \quad ; \quad \lambda_\kappa = \tfrac{45}{44}\lambda .$$

9.3. Determine the temperature-jump coefficient using the Maxwellian method and the general form of the boundary conditions given by Eqs. (6-12) and (6-15). Use the second-order Chapman-Enskog solution.

Solution: Following the solution of Problem 9.2, one obtains:

$$c_T = \frac{2 - \alpha_\tau\alpha_T}{\alpha_\tau\alpha_T}\tfrac{75}{128}\pi .$$

9.4. Derive expressions for the velocity defect at the wall for both the slip-flow and thermal-creep problems by using the Maxwellian method.

Solution: The velocity defect at the wall for these problems is given by:

$$\frac{u_d\left(0\right)}{\lambda_\mu\left(\partial u/\partial\tilde{x}\right)_\infty} = c_m - \frac{u\left(0\right)}{\lambda_\mu\left(\partial u/\partial\tilde{x}\right)_\infty} ,$$

and:

$$\frac{u_d(0)}{\nu^* q_y} = c_{Tsl} - \frac{u(0)}{\nu^* q_y} \quad ; \quad \nu^* = \nu \frac{\lambda_\kappa}{\lambda_\mu} \ .$$

Using Maxwellian boundary conditions to calculate the mean velocity at the wall, $u(0)$, one can obtain for the slip and thermal-creep problems:

$$\frac{u_d(0)}{\lambda_\mu (\partial u/\partial \tilde{x})_\infty} = \tfrac{5}{32}\pi(2-\alpha_\tau)\left[1 - \frac{2}{\pi}\left(1 + \tfrac{1}{2} b_2^{(2)*}\right)\right] ,$$

and:

$$\frac{u_d(0)}{\nu^* q_y} = \tfrac{3}{8}\alpha_\tau \varepsilon_1 \ ,$$

respectively, where:

$$\varepsilon_1 = \left(1 + \tfrac{1}{4} a_2^{(2)*}\right)$$

9.5. Determine an expression for the temperature defect at the wall by using the Maxwellian method for the first- and second-order Chapman-Enskog solutions. Obtain the result for $\alpha_\tau = 1$ and compare with the values from [12].

Solution: The temperature defect at the wall is given by:

$$T_d(0) = T_{asy}(0) - T(0) = c_T \lambda_\kappa \left(\frac{\partial T}{\partial \tilde{x}}\right)_\infty - T_0 \tau(0) \ ,$$

where $\tau(0) = \tfrac{2}{3}\pi^{-3/2} \int \left(c^2 - \tfrac{3}{2}\right)\Phi^\pm(\mathbf{c},0)\exp\left(-c^2\right)d\mathbf{c}$. Having performed the necessary integrations and algebraic transformations, one obtains:

$$T_d^*(0) = \frac{T_d(0)}{\lambda_\kappa (\partial T/\partial \tilde{x})_\infty} = \tfrac{75}{256}\pi(2-\alpha_\tau)\left[1 - \tfrac{28}{15}\pi^{-1}\left(1 + \tfrac{13}{28} a_2^{(2)*}\right)\right] .$$

If $\alpha_\tau = 1$ in this expression, then $T_d^*(0) = (0.3509)$, while the value reported in [12] is $T_{d[12]}^*(0) = (0.6374)$. The temperature defect for the first-order

Chapman-Enskog approximation may be formally obtained from the general expression if $a_2^{(2)*}=0$. This then results in $T_d^*(0)=(0.3735)$ which is an improvement in agreement with the value reported in [12].

9.6. Derive a formula for the temperature-jump coefficient by Loyalka's method for the second-order Chapman-Enskog solution and using the general form of the boundary conditions given by Eqs. (6-12) and (6-15).

Solution: The correction to the distribution function is given by:

$$\Phi^{\pm}(\mathbf{c},\infty)=a_1\left(c^2-\tfrac{5}{2}\right)\ ,\ \text{and}\ \Phi^{-}(\mathbf{c},0)=a_w\left(c^2-\tfrac{5}{2}\right)\ ,$$

where $a_1 \neq a_w$. The two constants, a_1 and a_w, may be found from the two conservation laws given by Eqs. (8-42) and (8-43). Boundary conditions are described by Eqs. (6-12) and (6-15). Integration yields the following expression:

$$c_T=\tfrac{75}{128}\pi\frac{2-\alpha_\tau\alpha_T}{\alpha_\tau\alpha_T}\left[1+\frac{\alpha_\tau\alpha_T(2-\alpha_\tau)}{2-\alpha_\tau\alpha_T}\left(\tfrac{52}{25}\pi^{-1}\varepsilon_T-\tfrac{1}{2}\right)\right]\ ,$$

where $\varepsilon_T=1-\tfrac{23}{26}a_2^{(2)*}+\tfrac{433}{208}\left(a_2^{(2)*}\right)^2=(0.9378)$ for rigid-sphere molecules.

9.7. Generalize Loyalka's expression for the temperature defect at the wall using two accommodation coefficients and the second-order Chapman-Enskog solution. Compare your analytical results with the numerical results [12] for the first- and second-order Chapman-Enskog approximations when $\alpha_\tau=1$.

Solution: Employing the same techniques as in Problem 9.5, one can obtain:

$$T_d^*(0)=\tfrac{75}{256}\pi(2-\alpha_\tau)\left[\tfrac{104}{25}\pi^{-1}\varepsilon_T-\tfrac{28}{15}\pi^{-1}\left(1+\tfrac{13}{28}a_2^{(2)*}\right)\right]\ ,$$

where $\varepsilon_T=(0.9378)$. If $\alpha_\tau=1$, then the values obtained for the first- and second-order Chapman-Enskog approximations are $T_d^*(0)=(0.6719)$ and $T_d^*(0)=(0.5734)$, respectively. These values represent deviations from the numerical value reported in [12] of 5.4% and 10%, respectively.

Table 9-13. Velocity defect data obtained during the solution of Problem 9.8.

α_τ	1	0.8	0.6	0.4	0.2
$u^*_{d[9]}(0)$	0.3013	0.3701	0.4422	0.5180	0.5976
$u^*_d(0)$	0.2850	0.3420	0.3990	0.4560	0.5130
deviation, %	5.4	7.6	9.8	12.0	14.2

9.8. Using the integral representation of the distribution function given by Eq. (9-12a), derive an expression for the velocity defect at the wall for the slip-flow problem if a trial function containing two independent constants is chosen in the form:

$$\tilde{\Phi} = \eta(c_x)\left[\alpha c_y\right] + \eta(-c_x)\left[\alpha c_y + \beta c_y \exp(\sigma\tilde{x}/c_x)\right] .$$

For different values of α_τ, compare these results with those obtained in [9]. Use the second-order Chapman-Enskog solution.

Solution: To obtain an approximate solution for the distribution function one can employ the approximate method described in [16,17]. Having substituted the trial function into Eq. (9-12a), one can obtain:

$$\begin{aligned}\Phi(\mathbf{c},\tilde{x}) = \eta(c_x)\big[&-(2-\alpha_\tau)c_x c_y \phi_u(c) h \exp(-\sigma\tilde{x}/c_x) \\ &+ \alpha c_y\left(1-\alpha_\tau \exp(-\sigma\tilde{x}/c_x)\right) + (1-\alpha_\tau)\beta c_y \exp(-\sigma\tilde{x}/c_x)\big] \\ &+\eta(-c_x)\left[\alpha c_y + \beta c_y \exp(\sigma\tilde{x}/c_x)\right] .\end{aligned}$$

The velocity defect at the wall may then be written as:

$$\frac{u_d(0)}{\lambda_\mu\left(\partial u/\partial\tilde{x}\right)_\infty} = \tfrac{5}{8}(2-\alpha_\tau)\left(\omega_1 - \tfrac{1}{2}\right) ,$$

where:

$$\omega_1 = 1 - b_2^{(2)*} + \tfrac{17}{4}\left(b_2^{(2)*}\right)^2 = (0.95602) .$$

For various values of α_τ, the velocity defect data are shown in Table 9-13.

9.9. For Problem 9.8, derive an expression for the velocity defect.

Solution: The mean velocity is given by:

$$u(\tilde{x}) = \left(\frac{2kT}{m}\right)^{1/2} \pi^{-3/2} \int c_y \Phi(\mathbf{c}, \tilde{x}) \exp(-c^2) d\mathbf{c} .$$

After integration of the distribution function presented in Problem 9.8, one can obtain:

$$\frac{u_d(\tilde{x})}{\lambda_\mu (\partial u/\partial \tilde{x})_\infty} = \tfrac{5}{4}(2-\alpha_\tau)\left[\frac{\mathrm{J}_1(\tilde{x})}{\pi^{3/2}} 2\omega_1 - \frac{\mathrm{J}_2(\tilde{x})}{\pi}\right],$$

where $\omega_1 = (0.95602)$ and the functions, $\mathrm{J}_1(\tilde{x})$ and $\mathrm{J}_2(\tilde{x})$, are given by Eqs. (9-32) and (9-33), respectively.

9.10. Using the same trial function as in Problem 9.8, derive an expression for the velocity defect for the thermal-creep problem. In particular, determine the velocity defect at the wall.

Solution: Integration of the distribution function given in Problem 9.8 results in:

$$u_d^*(\tilde{x}) = \frac{u_d(\tilde{x})}{v^* q} = 3\alpha_\tau \left[\frac{\mathrm{J}_3(\tilde{x})}{\pi^{3/2}} + \varepsilon_2 \frac{\mathrm{J}_1(\tilde{x})}{\pi^{3/2}}\right],$$

where the functions, $\mathrm{J}_1(\tilde{x})$ and $\mathrm{J}_3(\tilde{x})$, are given by Eqs. (9-32) and (9-55), respectively. The velocity defect at the wall may then be written as $u_d^*(0) = \tfrac{3}{4}\alpha_\tau \varepsilon_2$ where $\varepsilon_2 = 1 - \tfrac{7}{2} b_2^{(2)*} + \tfrac{7}{2} a_2^{(2)*} b_2^{(2)*} = (0.81332)$. This expression is exactly the same as that obtained via Loyalka's method.

REFERENCES

1. Kogan, M.N., *Rarefied Gas Dynamics* (Plenum Press, NY, 1969).
2. Ivchenko, I.N., Loyalka, S.K., and Tompson, R.V., "On the Collision Kernels for Gas Mixtures," *Ann. Nucl. Energy* **23(18)**, 1489-1495 (1996).
3. Loyalka, S.K. and Hickey, K.A., "Plane Poiseuille Flow: Near Continuum Results for a Rigid Sphere Gas," *Physica* **A160**, 395-408 (1989).
4. Loyalka, S.K., "Momentum and Temperature-Slip Coefficients with Arbitrary Accommodation at the Surface," *J. Chem. Phys.* **48(12)**, 5432-5436 (1968).
5. Loyalka, S.K. and Ferziger, J.H., "Model Dependence of the Temperature Slip Coefficient," *Phys. Fluids* **11(8)**, 1668-1671 (1968).

6. Loyalka, S.K., "Slip in the Thermal Creep Flow," *Phys. Fluids* **14(1)**, 21-24 (1971).
7. Loyalka, S.K., "The Slip Problems for a Simple Gas," *Z. Naturforsch.* **26a**, 964-972 (1971).
8. Morse, P.M. and Feshbach, H., *Methods of Theoretical Physics* (McGraw-Hill, NY, 1953).
9. Loyalka, S.K. and Hickey, K.A., "The Kramers Problem: Velocity Slip and Defect for a Hard Sphere Gas with Arbitrary Accommodation," *ZAMP* **41**, 246-253 (1990).
10. Ivchenko, I.N., Loyalka, S.K., and Tompson, R.V., "The Precision of Boundary Models in the Gas Slip Problem," *High Temperature* **31(1)**, 127-129 (1993).
11. Loyalka, S.K., "Velocity Profile in the Knudsen Layer for the Kramer's Problem," *Phys. Fluids* **18(12)**, 1666-1669 (1975).
12. Loyalka, S.K., "Temperature Jump and Thermal Creep Slip: Rigid Sphere Gas," *Phys. Fluids A* **1(2)**, 403-408 (1989).
13. Ivchenko, I.N., Loyalka, S.K., and Tompson, R.V., "A Boundary Model for the Thermal Creep Problem," *Fluid Dynamics* **28(6)**, 876-878 (1993).
14. Ivchenko, I.N., Loyalka, S.K., and Tompson, R.V., "On the Use of Conservation Laws in Plane Slip Problems," *Teplofizika Vysokikh Temperatur* (Russian) **33(1)**, 66-72 (1995).
15. Loyalka, S.K., "Velocity Profile in the Thermal Creep Slip Problem," *Phys. Fluids* **19(10)**, 1641-1642 (1976).
16. Loyalka, S.K., "An Approximate Method in Transport Theory," *Phys. Fluids* **24(10)**, 1912-1914 (1981).
17. Loyalka, S.K. and Cipolla, J.W.Jr., "On Choice of Trial Functions in Integro-Differential Variational Principles of Transport Theory," *Nucl. Sci. Engineering* **99**, 118-122 (1988).

Chapter 10

THE SLIP-FLOW REGIME

1. BASIC EQUATIONS.

In this chapter, the description of a particle in a rarefied gas is considered under conditions where the Knudsen number is small. The analysis is restricted to the usual conditions assumed for aerosol particle motion in non-uniform gases. These conditions will be discussed later in detail. The classical sphere drag and thermal force problems are solved as important practical applications of the theory and techniques described here.

In this regime, an aerosol particle disturbs the surrounding gas out to a distance on the order of the particle radius. The behavior of this particle can be described using the Navier-Stokes continuum equations except for the thin Knudsen layer in the neighborhood of the particle where the disturbance is greatest. Since, in the slip-flow regime, the typical dimension of the system is much greater than the mean free path, the continuum equations are assumed to be suitable for all distances. The local features of the gas in the Knudsen layer are then taken into account in the boundary conditions which have a special form discussed in Section 6.4.

The closed system of continuum equations derived in Chapters 2 and 5 may be written as:

$$\frac{Dn}{Dt}+n\frac{\partial u_i}{\partial \tilde{x}_i}=0 \ , \qquad (10\text{-}1)$$

$$\frac{\partial P_{ij}}{\partial \tilde{x}_j} - \rho\left(F_i - \frac{Du_i}{Dt} \right) = 0 \ , \tag{10-2}$$

$$\frac{DT}{Dt} = -\frac{2}{Nk\,n}\left(P_{ij}\frac{\partial u_i}{\partial \tilde{x}_j} + \frac{\partial q_i}{\partial \tilde{x}_i} \right) , \tag{10-3}$$

$$P_{ij} = p\delta_{ij} - \mu\left(\frac{\partial u_i}{\partial \tilde{x}_j} + \frac{\partial u_j}{\partial \tilde{x}_i} - \tfrac{2}{3}\delta_{ij}\frac{\partial u_k}{\partial \tilde{x}_k} \right) , \tag{10-4}$$

$$Q_i = -\kappa\frac{\partial T}{\partial \tilde{x}_i} \ , \tag{10-5}$$

where μ and κ are the viscosity and thermal conductivity coefficients that are given by Eqs. (5-53) and (5-52), respectively, D/Dt is a time-derivative following the motion, as in hydrodynamics, and N is the number of degrees of freedom of a molecule. For monatomic gases, $N = 3$, while for other gases, $N > 3$.

Each of these equations is a non-linear partial differential equation, and therefore, there are many difficulties in solving this system in the most general form. Thus, one considers possible methods that might be used to linearize the equations. In this analysis, the non-linear and linear terms are estimated using the technique described in [1]. The force on a stationary particle is proportional to the external parameters of the surrounding non-uniform gas flow. Such parameters are the mean gas velocity for the sphere drag problem and the temperature gradient for the thermal force problem The following conditions are assumed to be satisfied in the statement of the above mentioned problems:

1. the gas mean velocity is much less than the heat velocity of a molecule;
2. the relative alteration of the temperature is much less than unity for a distance of the same order as the mean free path;
3. the relative temperature variation is small for a typical dimension distance, R.

These conditions governing the external parameters can be expressed mathematically as:

$$\frac{U}{\overline{V}} \ll 1 \,, \quad \frac{\lambda\left|(\nabla T)_\infty\right|}{T} \ll 1 \,, \quad \frac{R\left|(\nabla T)_\infty\right|}{T} \ll 1 \,. \tag{10-6}$$

As mentioned above, to evaluate the order of the partial derivatives contained in Eqs. (10-1)-(10-3) the technique described in [1] is used. It is necessary to note that the typical distance of a variation of all the disturbed parameters is the characteristic dimension of the suspended particle, R. The disturbance of the pressure in the sphere drag problem, and that of the local thermal flow velocity in the thermal force problem, can be evaluated by dimensional analysis which gives:

$$\Delta p \sim \rho U^2 \,, \quad U_{th} \sim \frac{\mu}{\rho T}\left|(\nabla T)_\infty\right| \,. \tag{10-7}$$

The ratio of the terms in the continuity equation for each problem can be expressed as:

$$\left|\frac{u_i\left(\partial n/\partial \tilde{x}_i\right)}{n\left(\partial u_i/\partial \tilde{x}_i\right)}\right| \sim \frac{U\Delta p}{kTR}\frac{R}{nU} \sim \frac{U^2}{\overline{V}^2} \ll 1 \,,$$

and:

$$\left|\frac{u_i\left(\partial n/\partial \tilde{x}_i\right)}{n\left(\partial u_i/\partial \tilde{x}_i\right)}\right| \sim \frac{U_{th} p_0 \left|(\nabla T)_\infty\right|}{kT^2}\frac{kTR}{p_0 U_{th}} \sim \frac{R\left|(\nabla T)_\infty\right|}{T} \ll 1 \,,$$

where the pressure in the thermal force problem is assumed to be constant. By analogy, one obtains the following for the appropriate ratio in Eq. (10-2):

$$\frac{\left|u_i\left(\partial u_i/\partial \tilde{x}_i\right)\right|}{\nu\left|\nabla^2 u_i\right|} \sim \frac{U^2}{R}\frac{R^2}{\nu U} \sim \frac{UR}{\nu} = Re \ll 1 \,,$$

where ν is the kinematic viscosity and Re is the Reynolds number.

For the thermal force problem, one needs to use the third moment equation which contains two non-linear terms. The first ratio of the non-linear to linear terms can be evaluated by the relation:

$$\frac{\left|n\,k u_i\left(\partial T/\partial \tilde{x}_i\right)\right|}{\kappa\left|\nabla^2 T\right|} \sim \frac{nkTU_{th}}{R}\frac{R^2}{\kappa T} \sim \frac{U_{th}R}{\overline{V}\lambda} \sim \frac{R\left|(\nabla T)_\infty\right|}{T} \ll 1 \,,$$

where $\kappa \sim \rho \bar{V} \lambda c_v \sim nk\bar{V}\lambda$ and $U_{th} \sim \bar{V}\lambda \left|(\nabla T)_\infty\right| / T$.
For the other non-linear term the appropriate evaluation is given by:

$$\frac{\left|P_{ij}\left(\partial u_i / \partial \tilde{x}_j\right)\right|}{\kappa\left|\nabla^2 T\right|} \sim \frac{nkTU_{th}}{R} \frac{R^2}{nk\bar{V}\lambda T} \sim \frac{R\left|(\nabla T)_\infty\right|}{T} \ll 1 .$$

As one can see, if the basic assumptions defined by Eq. (10-6) are satisfied, the ratio of the non-linear to the linear terms is much less than unity for all the moment equations and, therefore, this system can be linearized by the simple expedient of neglecting the non-linear terms. The stationary linearized form of this system can be expressed as:

$$\nabla \cdot \mathbf{u} = 0 , \tag{10-8}$$

$$\mu \nabla^2 \mathbf{u} = \nabla p , \tag{10-9}$$

$$\nabla^2 T = 0 , \tag{10-10}$$

where the external force, F_i, is assumed to be equal to zero. Having obtained the basic equations describing the gas flow over the particle, one is now ready to consider, in detail, the various transfer problems of the slip-flow regime.

2. THE SPHERICAL DRAG PROBLEM.

An understanding of translational motion of small single particles in a gas is required in disciplines as diverse as nano-phase materials synthesis, environmental physics, clean-room technology, cloud physics, and nuclear reactor safety [2,3]. While these particles, excepting liquid drops, are generally non-spherical, studies of spherical particles are necessary for the intrinsic fundamentals involved as well as for applied reasons. The drag problem is of great significance and, since the work of Millikan [4], there has been a substantial body of related experimental [5] and theoretical work [6-9] generated.

Consider the short statement of the spherical drag problem. The origin of a spherical coordinate system is assumed to be at the center of a stationary spherical particle and the direction of the polar axis is the same as that of the mean velocity of the uniform flow at large distances from the particle, **U** (Fig. 7-2). The gas state can be described by Eqs. (10-8) and (10-9) which, for the spherical geometry and azimuthally symmetric problems, are given by [1]:

$$\frac{1}{r}\frac{\partial^2(rv_r)}{\partial r^2}+\frac{1}{r^2}\frac{\partial^2 v_r}{\partial\theta^2}+\frac{\cot(\theta)}{r^2}\frac{\partial v_r}{\partial\theta}-\frac{2v_r}{r^2} -\frac{2}{r^2}\left(\frac{\partial v_\theta}{\partial\theta}+v_\theta\cot(\theta)\right)=\frac{R}{\mu}\frac{\partial p}{\partial r}\ , \tag{10-11}$$

$$\frac{1}{r}\frac{\partial^2(r\,v_\theta)}{\partial r^2}+\frac{1}{r^2}\frac{\partial^2 v_\theta}{\partial\theta^2}+\frac{\cot(\theta)}{r^2}\frac{\partial v_\theta}{\partial\theta} +\frac{2}{r^2}\frac{\partial v_r}{\partial\theta}-\frac{v_\theta}{r^2\sin^2(\theta)}=\frac{R}{\mu}\frac{1}{r}\frac{\partial p}{\partial\theta}\ , \tag{10-12}$$

$$\frac{\partial v_r}{\partial r}+\frac{2v_r}{r}+\frac{1}{r}\frac{\partial v_\theta}{\partial\theta}+\frac{v_\theta}{r}\cot(\theta)=0\ . \tag{10-13}$$

The special features of the problem associated with the slip-flow regime are accounted for in the boundary conditions specified at the particle surface. For large distances from the particle, the radial and tangential components of the mean velocity of the gas are given by:

$$u_r(\infty)=U\cos(\theta)\ \text{and}\ u_\theta(\infty)=-U\sin(\theta)\ . \tag{10-14}$$

At the sphere surface, however, the tangential component of the mean velocity is not equal to zero as it is for the continuum regime. This implies that the gas is slipping along the surface. Since this slipping is the consequence of momentum conservation at the surface, the slip boundary conditions take the form:

$$v_\theta=-c_m\frac{\lambda}{\mu}P_{r\theta}\ , \tag{10-15}$$

$$v_r=0\ , \tag{10-16}$$

where c_m is the isothermal-slip coefficient. This coefficient, from a physical point of view, is the relative rate of communication of the tangential momentum, per unit area of the particle surface, to the surrounding gas.

For this regime the force on the particle is given by:

$$F_u = -\int_S \left(P_{rr} \cos(\theta) - P_{r\theta} \sin(\theta) \right) n_r dS \ ; \quad \tilde{r} = R \ , \tag{10-17}$$

where n_r is a unit normal vector at the surface of the sphere which points into the gas. The pressure tensor components, P_{rr} and $P_{r\theta}$, are defined by [1]:

$$P_{rr} = p_0 - 2\mu \frac{\partial v_r}{\partial \tilde{r}} \ , \tag{10-18a}$$

$$P_{r\theta} = -\mu \left(\frac{1}{\tilde{r}} \frac{\partial v_r}{\partial \theta} + \frac{\partial v_\theta}{\partial \tilde{r}} - \frac{v_\theta}{\tilde{r}} \right) . \tag{10-18b}$$

Again, the standard method involving separation of variables [1] is used to obtain the following solution for the system of Eqs. (10-11)-(10-13):

$$v_r = \left(\frac{A}{\tilde{r}^3} + \frac{B}{\tilde{r}} + U \right) \cos(\theta) \ , \tag{10-19}$$

$$v_\theta = \left(\frac{A}{2\tilde{r}^3} - \frac{B}{2\tilde{r}} - U \right) \sin(\theta) \ , \tag{10-20}$$

$$p = p_0 + \mu \frac{B}{\tilde{r}^2} \cos(\theta) \ , \tag{10-21}$$

where the two constants, A and B, may be found from boundary conditions given by Eqs. (10-15) and (10-16). Simple calculations result in the following expression for the force on the particle:

$$F_u = 6\pi\mu R\, U \frac{1 + 2c_m Kn}{1 + 3c_m Kn} \ , \tag{10-22}$$

where $Kn = \lambda_\mu / R$ and λ_μ is given by Eq. (8-16a).

Table 10-1. The isothermal-slip coefficient, c_m, for different values of α_τ.

α_τ	1.00	0.95	0.90	0.85	0.80
c_m	1.1006	1.2106	1.3321	1.4673	1.6185

The most reliable analytical expression for the slip coefficient is given by Eq. (8-79) that can be presented in the form:

$$c_m = \frac{2-\alpha_\tau}{\alpha_\tau}\tfrac{5}{8}\sqrt{\pi}\,\frac{(0.6690)+(0.1775)\alpha_\tau}{(0.7549)+(0.09714)\alpha_\tau}\,. \tag{10-23}$$

Table 10.1 gives the slip coefficient, c_m, for a number of values of the tangential momentum accommodation coefficient, α_τ.

Compare the analytical expression given by Eq. (10-22) with the Millikan experimental data for this regime. It is convenient to use the 'so-called' drag ratio value, D, which is given by:

$$D \equiv \frac{F_u(Kn)}{F_{fm}}\,,$$

where F_{fm} is the drag force in the free-molecular regime that is defined by Eq. (7-13). From Eq. (10-22) one can obtain:

$$D_{th} = \tfrac{45}{8}\frac{\pi Kn}{(8+\alpha_\tau\pi)}\frac{1+2c_m Kn}{1+3c_m Kn}\,. \tag{10-24}$$

The Millikan data for the full range of Kn can be described by means of the use of the Cunningham slip correction factor the most exact form of which has been proposed by Buckley and Loyalka [8]. The use of this factor gives:

$$D_{\exp} = \frac{(1.617)Kn}{C_C(Kn)}\,, \tag{10-25}$$

where $C_C(Kn)$ is defined by [8]:

$$C_C(Kn) = 1 + Kn\left[(1.099)+(0.518)\exp\left(-\frac{(0.425)}{Kn}\right)\right]. \tag{10-26}$$

Table 10-2. Comparison of experimental and theoretical values of drag on a sphere.

		D_{th}				
Kn^{-1}	$D_{\exp}$	α_τ=1.00	α_τ=0.95	α_τ=0.90	α_τ=0.85	α_τ=0.80
1	0.6633	1.1803	1.1883	1.1970	1.2063	1.2163
2	0.4870	0.6284	0.6315	0.6348	0.6383	0.6420
4	0.3113	0.3368	0.3384	0.3401	0.3417	0.3434
6	0.2265	0.2331	0.2344	0.2358	0.2371	0.2384
8	0.1774	0.1790	0.1802	0.1814	0.1825	0.1837
10	0.1456	0.1455	0.1466	0.1477	0.1487	0.1498
12	0.1234	0.1227	0.1237	0.1247	0.1257	0.1266
14	0.1071	0.1061	0.1070	0.1079	0.1089	0.1098
16	0.0946	0.0935	0.0943	0.0952	0.0961	0.0969
18	0.0847	0.0836	0.0844	0.0852	0.0860	0.0868
20	0.0766	0.0756	0.0763	0.0771	0.0778	0.0786

The Millikan experimental data and analytical results are presented in Table 10-2. A comparison of experimental results with the theory for the slip-flow regime is illustrated by Fig. 10-1. It can be easily seen that the best agreement between the theoretical and experimental data for this regime is reached when $\alpha_\tau = 0.95$. For this case the discrepancy of results is less than 0.5% for $Kn < 0.1$. This excellent agreement of data is an additional confirmation of the sufficiently high accuracy of Eq. (10-23).

3. THE THERMAL FORCE PROBLEM.

In Section 7.4 the problem of thermal forces (thermophoresis) on aerosol particles in the free-molecular regime was considered. Here, this problem will be examined in the slip-flow regime. A quantitative knowledge of thermophoresis is of great importance in many areas such as optical fiber fabrication, nuclear reactor safety, micro-contamination control, etc.

Thermophoresis has been studied extensively, both experimentally [10-18] and theoretically [19-29]. In spite of the numerous articles associated with this problem, both the experimental data and the theory remain controversial. The former due to the difficulties in measuring the relatively small effect because of competing phenomena, and the latter because of the approximations employed in the theoretical efforts. None of the previous work, with the exception of Loyalka [29], has attempted to solve this problem in its general form. Loyalka's work contains the most general numerical analysis that may be used to obtain results for all Knudsen numbers.

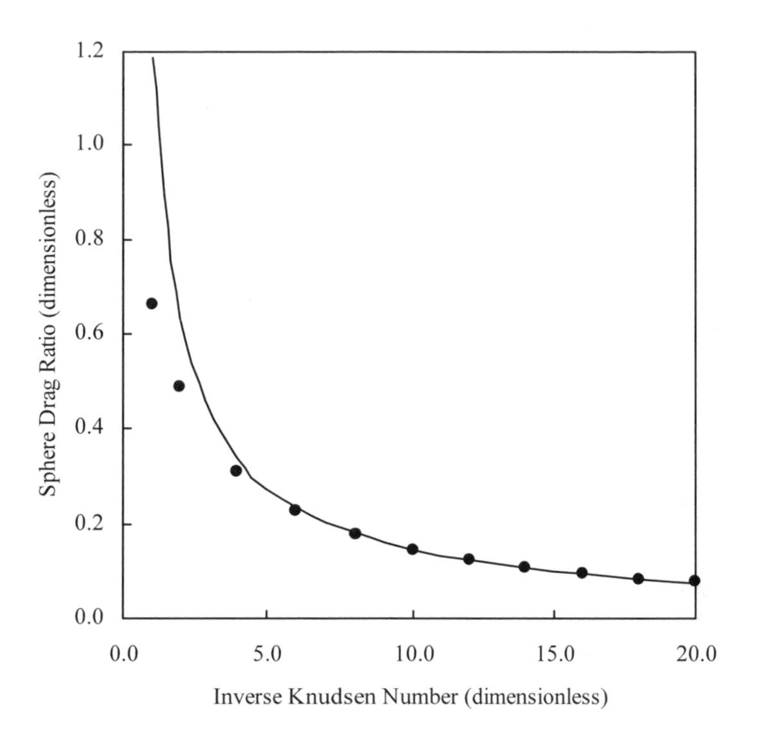

Figure 10-1. Comparison of the experimental and theoretical values of the drag on a sphere. Theoretical values are given by the solid line. Experimental points are obtained from the interpolation formula of Buckley and Loyalka [8] for $\alpha_\tau = 0.95$.

Consider the various aspects of this problem within the slip-flow framework. Let a single spherical particle of radius, R, be located in an infinite gas with a temperature gradient ∇T, which is constant at large distances from the particle. All the other conditions have been specified in Section 7.4. For the slip-flow regime, it is shown here that the Navier-Stokes equations may be applied, with boundary conditions appropriate for $Kn \ll 1$, to the quantitative description of the thermal force. The momentum and continuity equations are given by Eqs. (10-11)-(10-13). The temperature distribution, both inside and outside of the particle, is described by the thermal conductivity equation which may be written as:

$$\frac{1}{\tilde{r}^2}\frac{\partial}{\partial \tilde{r}}\left(\tilde{r}^2\frac{\partial T}{\partial \tilde{r}}\right)+\frac{1}{\tilde{r}^2\sin(\theta)}\frac{\partial}{\partial \theta}\left(\sin(\theta)\frac{\partial T}{\partial \theta}\right)=0\ . \qquad (10\text{-}27)$$

The proper slip-flow boundary conditions for the motion of the gas at the particle surface are defined by the relations:

$$v_r = 0 \; ; \quad \tilde{r} = R \; , \tag{10-28}$$

$$v_\theta = c_m \lambda_\mu \left[\tilde{r} \frac{\partial}{\partial \tilde{r}} \left(\frac{v_\theta}{\tilde{r}} \right) + \frac{1}{\tilde{r}} \frac{\partial v_r}{\partial \theta} \right] + c_{Tsl} \nu^* \frac{1}{RT_0} \left(\frac{\partial T}{\partial \theta} \right) \; ; \; \tilde{r} = R \; , \tag{10-29}$$

where c_{Tsl} is the thermal-creep coefficient. The most exact analytical expression for c_{Tsl} is given by Eq. (8-88). At large distances from the particle we have:

$$v_r = v_\theta = 0 \; . \tag{10-30}$$

The appropriate boundary conditions for the thermal conductivity equation have the form:

$$\kappa_g \frac{\partial T_g}{\partial \tilde{r}} = \kappa_p \frac{\partial T_p}{\partial \tilde{r}} \; ; \; \tilde{r} = R \; , \tag{10-31}$$

$$T_g - T_p = c_T \lambda_\kappa \frac{\partial T_g}{\partial \tilde{r}} \; ; \; \tilde{r} = R \; , \tag{10-32}$$

$$T_g \to \left| (\nabla T)_\infty \right| \tilde{r} \cos(\theta) \; ; \; \tilde{r} \to \infty \; , \tag{10-33}$$

where c_T is the temperature-jump coefficient. The most exact expression for c_T is given by Loyalka's formula, Eq. (8-50). The solutions of Eqs. (10-11)-(10-13) and (10-27) can be presented in the form:

$$v_r = \left(\frac{A}{\tilde{r}^3} + \frac{B}{\tilde{r}} \right) \cos(\theta) \; ,$$

$$v_\theta = \left(\frac{A}{2\tilde{r}^3} - \frac{B}{2\tilde{r}} \right) \sin(\theta) \; ,$$

$$p = p_0 + \mu \frac{B}{\tilde{r}^2} \cos(\theta) \ ,$$

$$T_g = T_0 + \left|(\nabla T)_\infty\right| \tilde{r} \cos(\theta) + \frac{C_1}{\tilde{r}^2} \cos(\theta) \ ,$$

$$T_p = T_0 + C_2 \tilde{r} \cos(\theta) \ ,$$

where the constants A, B, C_1, and C_2 can be found from boundary conditions. By means of Eq. (10-17), the following expression is obtained for the thermal force:

$$\mathbf{F} = -12\pi \frac{\mu^2}{\rho} \frac{R(\nabla T)_\infty}{T_0} c_{Tsl} \zeta_m \zeta_T \ , \tag{10-34}$$

where the quantities ζ_m and ζ_T are defined by:

$$\zeta_m = \frac{1}{1 + 3c_m Kn} \ ,$$

$$\zeta_T = \frac{(\kappa_g / \kappa_p) + c_T Kn}{1 + 2(\kappa_g / \kappa_p) + 2c_T Kn} \ .$$

The trustworthiness of the results obtained here may be analyzed by comparing the thermal force values given by Eq. (10-34) with Loyalka's numerical values [29]. For this comparison, the thermal force expression of Eq. (10-34) must be transformed into the form used by Loyalka. It is convenient to introduce a dimensionless thermal force, F_T^*, by means of the relation:

$$|\mathbf{F}| = p_0 R^2 \frac{\lambda\left|(\nabla T)_\infty\right|}{T_0} F_T^* \ , \tag{10-35}$$

where, F_T^*, may be written as:

$$F_T^* = \tfrac{75}{32} \pi^2 Kn c_{Tsl} \zeta_m \zeta_T \ , \tag{10-36}$$

Table 10-3. The reduced thermal force, F_T^*.

	F_T^*			
	$\kappa_p/\kappa_g = 10.0$		$\kappa_p/\kappa_g = 100.0$	
Kn^{-1}	Eq. (10-36)	Loyalka [29]	Eq. (10-36)	Loyalka [29]
1.2463	2.0175	2.2056	1.9936	2.1378
1.6617	1.7303	1.7965	1.6972	1.7099

Table 10-4. The reduced thermal force on NaCl aerosol particles.

	F_T^*			F_T^*	
Kn^{-1}	experimental [15]	theoretical Eq. (10-36)	Kn^{-1}	experimental [15]	theoretical Eq. (10-36)
1.00	2.4300	2.2059	2.50	1.1828	1.2654
1.25	2.1552	1.9879	2.75	1.0490	1.1690
1.50	1.9115	1.8006	3.00	0.9304	1.0832
1.75	1.6935	1.6387	3.25	0.8252	1.0066
2.00	1.5036	1.4977	3.50	0.7319	0.9378
2.25	1.3336	1.3742	3.75	0.6491	0.8759

and molecules are assumed to be rigid spheres. It is important to note that F_T^* depends on the nature of the aerosol particle through the accommodation coefficients, the Knudsen number, and the ratio of the thermal conductivity coefficients of the gas and the particle. Let the accommodation coefficients, $\alpha_\tau = \alpha_T = 1$. For this case, the slip and jump coefficients are specified by:

$$c_{Tsl} = (1.0193) \;, \quad c_m = (1.1006) \;, \quad c_T = (2.0633) \;.$$

A comparison of the results obtained from Eq. (10-36) with Loyalka's numerical results [29] is given in Table 10-3. It should be noted that the results reported here as Loyalka's were transformed in a simple fashion to conform to the current notation. One can see that the discrepancy between these results is very small.

One may further compare the theoretical results with the experimental values for F_T^* that have been reported by Jacobsen and Brock [15]. The experimental results for sodium chloride aerosols in argon can be fitted, for $0 < Kn^{-1} < 4$, by the expression:

$$\left(F_T^*\right)_{\text{exp}} = \tfrac{5}{4}\pi \exp\left(-\frac{\tau}{Kn}\right), \tag{10-37}$$

where the constant $\tau = (0.48)$. The theoretical and experimental results are presented in Table 10-4. A graphical comparison of these values is shown in

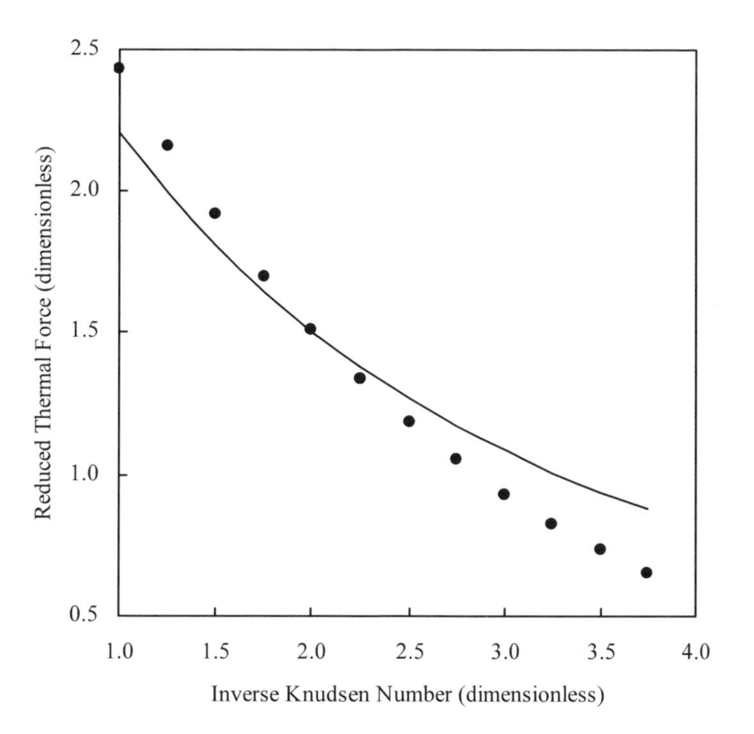

Figure 10-2. A comparison of experimental and theoretical reduced thermal forces for NaCl. Theoretical values from Eq. (10-36) are shown as a solid line and experimental values reported in [15] are shown as solid circles.

Fig. 10-2. One can easily see that the agreement of the theory with the experiment is within 10% for the very narrow region of inverse Knudsen number, $0 < Kn^{-1} < 3$. A substantial discrepancy occurs outside this region, however, and further experimental and theoretical work is clearly necessary before an acceptable understanding of the whole slip-flow region is possible.

One contributing factor to the discrepancy between theory and experiment is that the theory does not take into account all of the first-order boundary effects. A strict theory must account for corrections that are proportional to Kn for the thermal-creep. These additional terms occur due to surface curvature and to Barnett's thermal stresses [30]. The influence of all the corrections that are proportional to Kn on the thermophoresis of aerosol particles has been analyzed in [30]. Unfortunately, the accuracy of the additional terms was not sufficiently discussed and, moreover, a kinetic theory analysis of these phenomena leads to very complicated calculations that are of the same order of difficulty as an appropriate solution for all Knudsen numbers. Consequently, it stands to reason that the most reliable theory must ultimately be based on the direct solution of the Boltzmann equation by use of different moment methods.

PROBLEMS

10.1. A sphere suspended in a gas has radius, R, and temperature, $T_0 + \Delta T \left(\Delta T \ll T_0\right)$. Determine the heat flux and the temperature distribution for the surrounding gas if its number density and temperature far from the sphere are n_0 and T_0, respectively. Solve this problem for the slip-flow regime.

Solution: The temperature distribution can be determined from the Laplace equation:

$$\frac{1}{\tilde{r}^2}\frac{\partial}{\partial r}\left(\tilde{r}^2\frac{\partial T}{\partial \tilde{r}}\right)=0 \ .$$

The boundary conditions are:

$$T=\left(T_0+\Delta T\right)+c_T\lambda\left.\frac{\partial T}{\partial \tilde{r}}\right|_{\tilde{r}=R} \quad \text{and } T=T_0 \ ; \quad \tilde{r}\to\infty \ .$$

The solution of this boundary problem may then be expressed as:

$$T(\tilde{r})=T_0+\Delta T\zeta\frac{R}{\tilde{r}} \ ; \quad \zeta=\left(1+c_T\frac{\lambda}{R}\right)^{-1} \ .$$

The heat flux then takes the form:

$$Q=-4\pi R^2\kappa\left.\frac{\partial T}{\partial \tilde{r}}\right|_{\tilde{r}=R}=4\pi R\kappa\Delta T\zeta \ .$$

10.2. To maintain the constant surface temperature, $T_0 + \Delta T \ \left(\Delta T \ll T_0\right)$, of a sphere (radius R), there is a heat source of power, Q, located at the center of the sphere. Determine the thermal conductivity coefficient of the surrounding gas if its number density and temperature far from the sphere are n_0 and T_0, respectively. Solve this problem for the slip-flow regime.

Solution: Using the expression for the heat flux obtained in Problem 10.1, one has that:

$$\kappa = \frac{Q}{4\pi R \Delta T \zeta} \ ; \quad \zeta = \left(1 + c_T \frac{\lambda}{R}\right)^{-1} .$$

10.3. One method of measuring the thermal conductivity of a gas is to determine the loss of heat through the gas from a hot sphere of radius, R. For the same conditions used in Problem 10.2, determine a calculation error for the thermal conductivity coefficient of the gas if the temperature-jump at the surface of the sphere is not taken into account.

Solution: In the continuum regime, the thermal conductivity coefficient is given by:

$$\kappa_0 = \frac{Q}{4\pi R \Delta T} .$$

Using the result of Problem 10.2, one can obtain:

$$\delta = \frac{|\kappa - \kappa_0|}{\kappa} = \left|1 - \frac{1}{\zeta}\right| = c_T \frac{\lambda}{R} .$$

10.4. Determine the heat flux and temperature distribution of a gas between two concentric spheres the radii and temperatures of which are $R_1, T_0 - \Delta T$ and $R_2, T_0 + \Delta T$, respectively. Assume $\Delta T \ll T_0$. The equilibrium number density of the gas is n_0. Solve this problem for the slip-flow regime.

Solution: The temperature distribution is given by:

$$T(\tilde{r}) = \frac{A}{\tilde{r}} + B .$$

The two constants, A and B, are determined from the boundary conditions:

$$T(R_1) = (T_0 - \Delta T) + c_T \lambda \frac{\partial T}{\partial \tilde{r}} , \text{ and:}$$

$$T(R_2) = (T_0 + \Delta T) - c_T \lambda \frac{\partial T}{\partial \tilde{r}} .$$

The temperature distribution and the thermal flux may be expressed as:

$$T(\tilde{r}) = T_0 \left\{1 + \frac{\Delta T}{T_0} \zeta \left[-\frac{2R_1}{\tilde{r}} + \left(1 + c_T \frac{\lambda}{R_1}\right) + \frac{R_1}{R_2}\left(1 - c_T \frac{\lambda}{R_2}\right)\right]\right\} ,$$

$$Q=-8\pi R_1\kappa\Delta T\zeta \ ,$$

$$\zeta=\left[\left(1+c_T\frac{\lambda}{R_1}\right)-\frac{R_1}{R_2}\left(1-c_T\frac{\lambda}{R_2}\right)\right]^{-1} .$$

10.5. Determine the heat flux and temperature distribution of a gas between two coaxial cylinders the radii and temperatures of which are $R_1, T_0+\Delta T$ and R_2, T_0 for the inside and outside cylinders, respectively. Assume $\Delta T \ll T_0$. Solve this problem for the slip-flow regime.

Solution: The temperature distribution may be found from the following boundary problem:

$$\frac{1}{\tilde{r}}\frac{\partial}{\partial\tilde{r}}\left(\tilde{r}\frac{\partial T}{\partial\tilde{r}}\right)=0 \ ,$$

$$T(R_1)=(T_0+\Delta T)+c_T\lambda\frac{\partial T}{\partial\tilde{r}} \ ,$$

$$T(R_2)=T_0-c_T\lambda\frac{\partial T}{\partial\tilde{r}} \ .$$

After some calculations, one can obtain:

$$T(\tilde{r})=T_0\left\{1-\frac{\Delta T}{T_0}\frac{\ln(\tilde{r}/R_2)-c_T(\lambda/R_2)}{\ln(R_2/R_1)+c_T\lambda(1/R_1+1/R_2)}\right\} \ , \text{ and:}$$

$$Q=2\pi\kappa\Delta T\left[\ln\left(\frac{R_2}{R_1}\right)+c_T\lambda\left(\frac{1}{R_1}+\frac{1}{R_2}\right)\right]^{-1} .$$

10.6. One method of measuring the thermal conductivity of a gas is to determine the loss of heat through the gas from a long hot wire enclosed by a cylinder having a radius much greater than that of the wire. If the thermal power generated per unit length of the wire, Q, and the temperature difference, ΔT, are measured during the experiment, determine the thermal conductivity coefficient of the ambient gas. Assume that the temperature of the external cylinder is T_0 and that slip-flow conditions are satisfied during this experiment.

Solution: Using the expression for the heat flux from Problem 10.5 and allowing for $R \gg \tilde{r}$, one can obtain:

$$\kappa = \frac{Q}{2\pi \Delta T}\left[\ln\left(\frac{R}{\tilde{r}}\right) + c_T \frac{\lambda}{\tilde{r}}\right].$$

10.7. For the conditions given in Problem 10.6, determine the relative error of the predicted values of the thermal conductivity coefficient of the ambient gas if the temperature-jump at the surface of the wire is neglected.

Solution: In the continuum regime, the thermal conductivity coefficient is given by:

$$\kappa_0 = \frac{Q}{2\pi \Delta T}\ln\left(\frac{R}{\tilde{r}}\right).$$

From this one may generate the following relative error:

$$\delta = \left|\frac{\kappa - \kappa_0}{\kappa}\right| = \frac{c_T\left(\lambda/\tilde{r}\right)}{\ln\left(R/\tilde{r}\right) + c_T\left(\lambda/\tilde{r}\right)}.$$

10.8. Two flat parallel round disks of equal radius, R, are separated by a small distance, $h \ll R$. While the lower disk is stationary, the upper disk moves toward it with a velocity, $\mathbf{u}$. Determine the resistance force on the upper disk for the slip-flow regime. Use the geometry in Fig. 10-3.

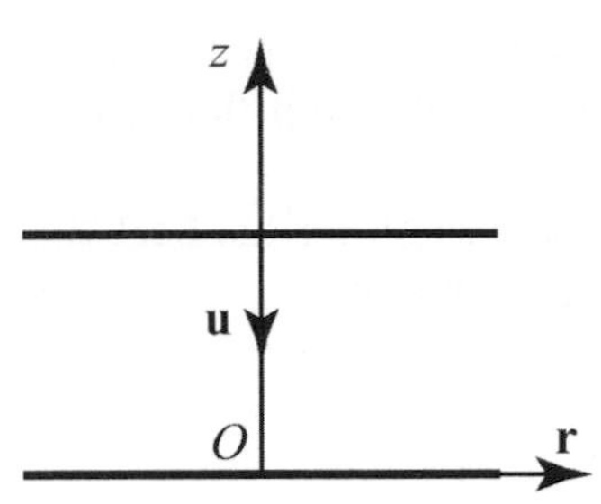

Figure 10-3. The geometry to be used in Problem 10.8 in determining the resistance force on a flat, round disk in the slip-flow regime due to its approach to an identical, parallel, stationary disk.

Solution: The motion of the gas in this problem is axially symmetric such that $u_\phi = 0$ and, therefore, the Navier-Stokes system of equations in cylindrical coordinates may be written as [1]:

$$\frac{1}{\tilde{r}}\frac{\partial}{\partial \tilde{r}}(\tilde{r}u_r) + \frac{\partial u_z}{\partial \tilde{z}} = 0 \ ,$$

$$\mu\left(\Delta u_r - \frac{u_r}{\tilde{r}^2}\right) = \frac{\partial p}{\partial \tilde{r}} \ ,$$

$$\mu \Delta u_z = \frac{\partial p}{\partial \tilde{z}} \ , \text{ where:}$$

$$\Delta f = \frac{1}{\tilde{r}}\frac{\partial}{\partial \tilde{r}}\left(\tilde{r}\frac{\partial f}{\partial \tilde{r}}\right) + \frac{\partial^2 f}{\partial \tilde{z}^2} \ .$$

It may readily be shown that $u_z \sim u$ and $u_r \sim (R/h)u$ and, therefore, that $u_r \gg u_z$. For the partial derivatives, one can obtain the following estimates:

$$\frac{\partial u_r}{\partial \tilde{r}} \sim \frac{Ru}{Rh} = \frac{u}{h} \ ,$$

$$\frac{\partial u_r}{\partial \tilde{z}} \sim \frac{Ru}{hh} = \frac{R}{h^2}u \ ,$$

$$\mu\Delta u_z \sim \rho\bar{V}\lambda\frac{u}{h^2} ,$$

$$\frac{\partial p}{\partial\tilde{z}} \sim \frac{\rho kT}{mh} \sim \frac{\rho\bar{V}^2}{h} .$$

Consequently:

$$\frac{\mu\Delta u_z}{\partial p/\partial\tilde{z}} \sim \frac{u}{\bar{V}}\frac{\lambda}{h} \ll 1 \text{, and } \frac{\partial u_r/\partial\tilde{r}}{\partial u_r/\partial\tilde{z}} \sim \frac{h}{R} \ll 1 .$$

Taking into consideration these estimates, one can reduce the Navier-Stokes system to the form:

$$\frac{1}{\tilde{r}}\frac{\partial}{\partial\tilde{r}}\left(\tilde{r}u_r\right) + \frac{\partial u_z}{\partial\tilde{z}} = 0 , \quad \text{(P-1)}$$

$$\mu\frac{\partial^2 u_r}{\partial\tilde{z}^2} = \frac{\partial p}{\partial\tilde{r}} , \quad \text{(P-2)}$$

$$\frac{\partial p}{\partial\tilde{z}} = 0 . \quad \text{(P-3)}$$

The boundary conditions may be expressed as:

$$\tilde{z} = 0 , \quad u_z = 0 , \quad u_r = -c_m\frac{\lambda}{\mu}P_{rz} = c_m\lambda\frac{\partial u_r}{\partial\tilde{z}} ,$$

$$\tilde{z} = h , \quad u_z = -u , \quad u_r = c_m\frac{\lambda}{\mu}P_{rz} = -c_m\lambda\frac{\partial u_r}{\partial\tilde{z}} ,$$

$$\tilde{r} = R , \quad p = p_0 .$$

From Eqs. (P-2) and (P-3) and the boundary conditions, one can obtain:

$$u_r = \frac{1}{2\mu}\frac{dp}{d\tilde{r}}\left[\tilde{z}\left(\tilde{z}-h\right)-c_m\lambda h\right] .$$

Integration of Eq. (P-1) yields:

$$u = \frac{1}{\tilde{r}}\frac{\partial}{\partial \tilde{r}}\int_0^h \tilde{r}u_r d\tilde{z} = -\frac{h^3}{12\mu\tilde{r}}\frac{d}{d\tilde{r}}\left(\tilde{r}\frac{dp}{d\tilde{r}}\right)\left(1+6c_m\frac{\lambda}{h}\right) .$$

From this relation, the pressure may be expressed in the form:

$$p = p_0 + \frac{3\mu u}{h^3\left(1+6c_m\lambda/h\right)}\left(R^2-\tilde{r}^2\right) .$$

The resistance force on the upper disk is then given by:

$$F_z = \tfrac{3}{2}\pi\frac{\mu R^4 \zeta u}{h^3} \quad \text{where } \zeta = \left(1+6c_m\frac{\lambda}{h}\right)^{-1} .$$

10.9. Determine the velocity distribution and the net mass transport of a gas flowing along a cylindrical tube owing to a pressure gradient, $dp/d\tilde{x}$. The tube radius is R.

Solution: The gas velocity has only an x-component which depends only on the radial coordinate, $\tilde{r}$. The Navier-Stokes system may be written as:

$$\frac{1}{\tilde{r}}\frac{d}{d\tilde{r}}\left(\tilde{r}\frac{du}{d\tilde{r}}\right) = \frac{1}{\mu}\frac{\partial p}{\partial \tilde{x}} ,$$

$$\frac{\partial p}{\partial \tilde{r}} = 0 .$$

The slip-flow boundary condition is given by:

$$u\left(R\right) = -c_m\lambda\frac{\partial u}{\partial \tilde{r}} .$$

The solution of this boundary problem can be written as:

$$u(\tilde{r}) = -\frac{1}{4\mu}\frac{dp}{d\tilde{x}}\left[R^2\left(1+2c_m\frac{\lambda}{R}\right)-\tilde{r}^2\right], \text{ with } \frac{dp}{d\tilde{x}} = \text{const} .$$

The net mass transfer is given by:

$$M = 2\pi\rho\int_0^R \tilde{r}ud\tilde{r} = -\frac{\pi\rho R^4}{8\mu}\frac{dp}{d\tilde{x}}\left(1+4c_m\frac{\lambda}{R}\right). \qquad \text{(P-4)}$$

10.10. Determine the pressure difference at the ends of a cylindrical tube of radius, R, and length, L, if the net mass transfer through the tube is M. The gas flow is assumed to be isothermal.

Solution: The pressure gradient may be expressed in the form (see Problem 10.9):

$$\frac{dp}{d\tilde{x}} = -\frac{8\mu kTM}{\pi mR^4\left(p+4c_m p_1\lambda_1/R\right)},$$

where p_1 and λ_1 are the pressure and mean free path at the entrance of the tube. Integrating this expression, one can obtain:

$$p_1 - p_2 = p_1\left(1+4c_m\lambda_1/R\right)\left\{1-\left[1-\frac{16\mu kTML}{\pi mR^4 p_1^2\left(1+4c_m\lambda_1/R\right)^2}\right]^{1/2}\right\}.$$

10.11. Two vessels containing the same gas at the different temperatures, T_1 and T_2, are connected by a long cylindrical tube of radius, R. There exists a pressure difference between the vessels owing to the phenomenon of thermal-creep. Determine this pressure difference and the velocity distribution of the gas in the tube.

Solution: The net mass transfer per unit time, ignoring transfer in the Knudsen layer, may be written as (see Problem 10.9):

$$M = -\tfrac{1}{8}\pi\frac{\rho R^4}{\mu}\left(1+4c_m\frac{\lambda}{R}\right)\frac{dp}{d\tilde{x}} + c_{Tsl}\nu\rho\pi R^2\frac{1}{T}\frac{dT}{d\tilde{x}} .$$

The mechanical equilibrium condition, $M = 0$, gives the following relation:

$$p\left(1+4c_m\frac{\lambda}{R}\right)\frac{dp}{d\tilde{x}} = \frac{8k\mu^2 c_{Tsl}}{mR^2}\frac{dT}{d\tilde{x}} . \qquad \text{(P-5)}$$

If one takes into consideration that $\lambda \sim 1/p$ and $\mu \sim \sqrt{T}$, this equation may be transformed to:

$$\left(p + 4c_m p_1 \frac{\lambda_1}{R}\right)\frac{dp}{d\tilde{x}} = \frac{8k\mu_1^2 c_{Tsl}}{mR^2 T_1} T \frac{dT}{d\tilde{x}} .$$

Integrating this expression, one obtains:

$$p_2 - p_1 = p_1 \zeta^{-1}\left(\sqrt{1 + A\left(T_2^2 - T_1^2\right)} - 1\right) \text{ where:}$$

$$\zeta = \left(1 + 4c_m \frac{\lambda_1}{R}\right)^{-1} \text{ and } A = \frac{8k\mu_1^2 \zeta^2 c_{Tsl}}{mT_1 p_1^2 R^2} .$$

The velocity distribution is then given by:

$$u = -\frac{1}{4\mu}\left[R^2\left(1 + 2c_m \frac{\lambda}{R}\right) - \tilde{r}^2\right]\frac{dp}{d\tilde{x}} + c_{Tsl} \nu \frac{1}{T}\frac{dT}{d\tilde{x}} .$$

Substituting $dp/d\tilde{x}$ into this relation from Eq. (P-4), one obtains:

$$u(\tilde{r}) = c_{Tsl} \nu \zeta \left(\frac{2\tilde{r}^2}{R^2} - 1\right)\frac{1}{T}\frac{dT}{d\tilde{x}} .$$

For $\tilde{r} = 0$ and $\tilde{r} = R$, the gas velocity has opposite signs.

10.12. For the same conditions used in Problem 10.11, determine the pressure difference if the variation in the mean velocity due to thermal-creep in the Knudsen layer is taken into account.

Solution: The mean velocity due to the temperature gradient is given by Eq. (8-87) and may be expressed in the form:

$$u_{th}(\tilde{r}) = c_{Tsl} \nu \left[1 + \psi \exp\left(-\alpha\left(R - \tilde{r}\right)\right)\right]\frac{1}{T}\frac{dT}{d\tilde{x}} \text{ where:}$$

$$\psi = -\frac{(0.3882)\alpha_\tau}{(0.4354) + (0.2179)\alpha_\tau} ,$$

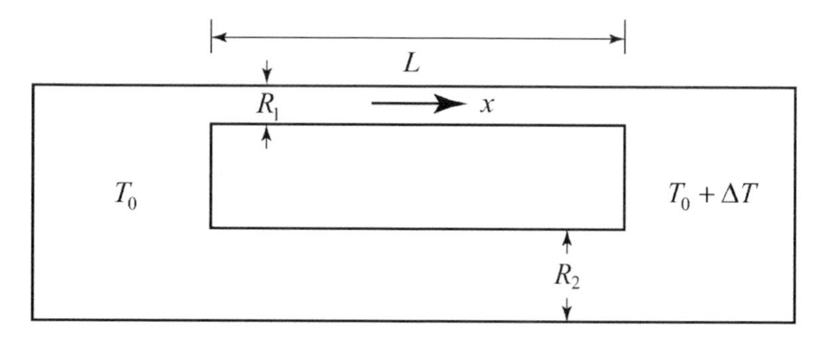

Figure 10-4. The geometry to be used for Problem 10.13 in determining the net mass transfer in a closed capillary loop with different radii and end temperatures.

and $\alpha = (2.2015)\lambda^{-1}$. Taking into account the mass transport in the Knudsen layer, Eq. (P-5) from Problem 10.11 may be cast in the form:

$$p\left(1+4c_m\frac{\lambda}{R}\right)\left(1+\frac{2\psi}{\alpha R}\right)^{-1}\frac{dp}{d\tilde{x}} = \frac{8k\mu^2 c_{Tsl}}{mR^2}\frac{dT}{d\tilde{x}}.$$

The pressure difference is given by:

$$p_2 - p_1 = p_1\zeta_1^{-1}\left(\sqrt{1+A_1\left(T_2^2 - T_1^2\right)} - 1\right) \text{ where:}$$

$$\zeta_1 = \left(1+\left(4c_m - \frac{2\psi}{\alpha\lambda}\right)\frac{\lambda_1}{R}\right)^{-1} \text{ and } A_1 = \frac{8k\mu_1^2\zeta_1^2 c_{Tsl}}{mT_1 p_1^2 R^2}.$$

10.13. Two capillary tubes having an equal length, L, but different radii, R_1 and R_2 where $R_1 < R_2$, are connected at each end to form a closed loop (see the figure). The ends of the capillary tubes are maintained at the temperatures, T_0 and $T_0 + \Delta T$ where $\Delta T \ll T_0$, as shown in Fig. 10-4. Determine the net mass transfer through any cross section of either tube that is due to a circular motion of the gas. The gas parameters, μ and ρ, are assumed to be constant and the isothermal end connections are assumed to make no contribution to the flow.

Solution: The solution for the net mass transfer from Problem 10.9 may be presented in the form:

$$\frac{M\zeta_i}{R_i^4} = -\frac{\pi\rho}{8\mu}\frac{dp}{d\tilde{x}} + \frac{c_{Tsl}\pi\mu\zeta_i}{R_i^2}\frac{1}{T_0}\frac{dT}{d\tilde{x}} \text{ where:}$$

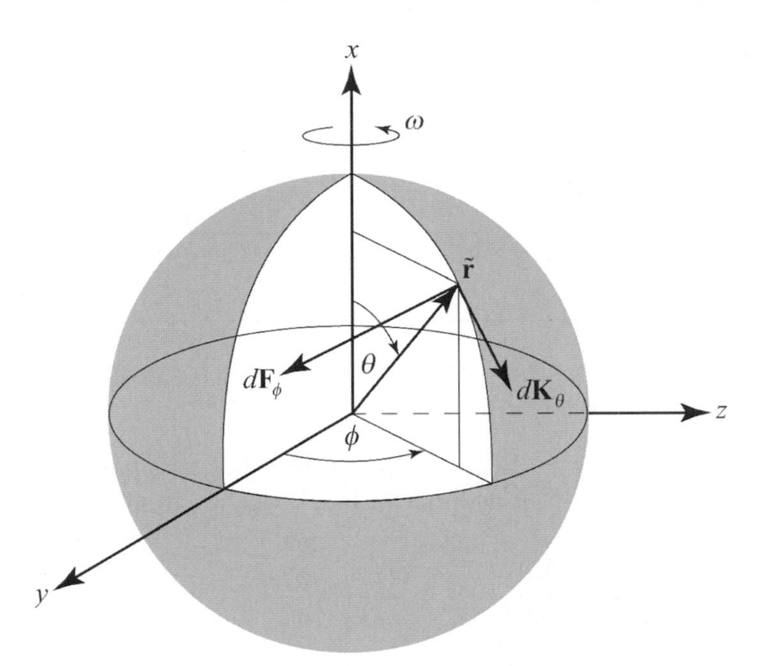

Figure 10-5. The geometry to be used in Problem 10.14 in determining the torque on a rotating sphere in the slip-flow regime.

$$\zeta_i = \left(1 + 4c_m \frac{\lambda}{R_i}\right)^{-1} .$$

Integrating over all values of $\tilde{x}$ along a closed path, one can obtain:

$$M = c_{Tsl} \mu \frac{\Delta T}{T_0} \frac{\pi R_1^2 R_2^2}{L} \frac{\zeta_1 R_2^2 - \zeta_2 R_1^2}{\zeta_1 R_2^4 + \zeta_2 R_1^4} .$$

10.14. A sphere of radius, R, is rotating with an angular velocity, ω, such that $\omega R \ll (2kT/m)^{1/2}$. Determine the torque, $\mathbf{K}$, on the sphere in the slip-flow regime. Use the geometry in Fig. 10-5.

Solution: Owing to the axial symmetry of this problem, the mean velocity of the gas has only one component, $u_\phi(\tilde{r},\theta) = u(\tilde{r},\theta)$. Additionally, one also has that $p = \text{const}$. The Navier-Stokes system of equations thus reduces to a single equation given by [1]:

$$\frac{1}{\tilde{r}^2}\frac{\partial}{\partial\tilde{r}}\left(\tilde{r}^2\frac{\partial u}{\partial\tilde{r}}\right)+\frac{1}{\tilde{r}^2\sin(\theta)}\frac{\partial}{\partial\theta}\left(\sin(\theta)\frac{\partial u}{\partial\theta}\right)-\frac{u}{\tilde{r}^2\sin^2(\theta)}=0 \ .$$

The boundary conditions may be expressed as:

$$u(R)=\omega R\sin(\theta)-c_m\frac{\lambda}{\mu}P_{r\phi}(R) \text{ and } u(\infty)=0 \text{ where:}$$

$$P_{r\phi}(R)=-\mu\left(\frac{\partial u}{\partial\tilde{r}}-\frac{u}{\tilde{r}}\right)\Bigg|_{\tilde{r}=R} \ .$$

Let $u(\tilde{r},\theta)$ be expressed in the following form $u(\tilde{r},\theta)=f(\tilde{r})\sin(\theta)$. Then, the function, $f(\tilde{r})$, may be determined from the equation:

$$\frac{d}{d\tilde{r}}\left(\tilde{r}^2\frac{df}{d\tilde{r}}\right)-2f=0 \ .$$

The solution of the above boundary problem is then:

$$u(\tilde{r},\theta)=\frac{\omega R^3\zeta}{\tilde{r}^2}\sin(\theta) \ ,$$

$$P_{r\phi}(R)=3\mu\omega\zeta\sin(\theta) \ ,$$

$$dF_\phi=-3\mu\omega R^2\zeta\sin^2(\theta)d\theta d\phi \ ,$$

where $\zeta=(1+3c_m\lambda/R)^{-1}$. The net torque, K_x, is then:

$$K_x=-R\int_0^{2\pi}d\phi\int_0^{\pi}\sin(\theta)\left|dF_\phi\right|d\theta=-8\pi\mu\omega R^3\zeta \ .$$

10.15. An infinite cylinder of radius, R, is rotating with an angular velocity, ω, such that $\omega R \ll (2kT/m)^{1/2}$. Determine the torque per unit length in the slip-flow regime.

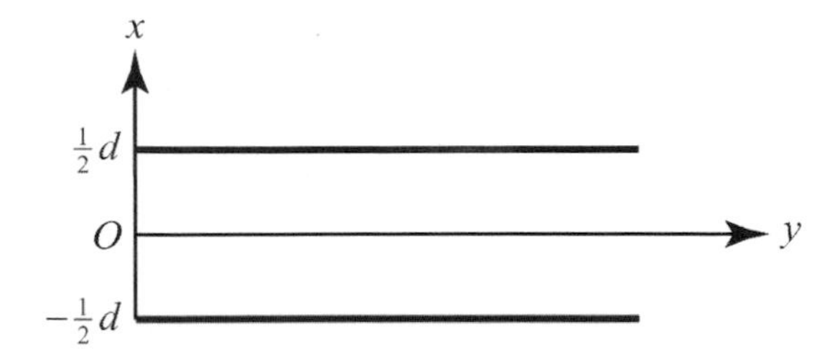

Figure 10-6. The geometry to be used for Problem 10.16 in determining the average speed of a steady gas flow between two, parallel, planar surfaces in the slip-flow regime.

Solution: The mean velocity of the gas possesses only one component, $u_\phi = u(\tilde{r})$. Additionally, $p = \text{const}$. The velocity distribution may be found from the following boundary problem:

$$\frac{1}{\tilde{r}}\frac{\partial}{\partial \tilde{r}}\left(\tilde{r}\frac{\partial u}{\partial \tilde{r}}\right) - \frac{u}{\tilde{r}^2} = 0 \ ,$$

$$u(R) = \omega R - c_m \frac{\lambda}{\mu} P_{r\phi}(R) \text{ and } u(\infty) = 0 \ , \text{ where:}$$

$$P_{r\phi}(R) = -\mu\left(\frac{\partial u}{\partial \tilde{r}} - \frac{u}{\tilde{r}}\right) .$$

The solution of this problem may be expressed as:

$$u = \frac{\omega R^2 \zeta}{\tilde{r}} \text{ and } P_{r\phi}(R) = 2\mu\omega\zeta \text{ where } \zeta = 1 + 2c_m \frac{\lambda}{R} .$$

The torque is then given by $K_x = -4\pi\mu\omega R^2 \zeta$ where the direction of the angular velocity vector, $\boldsymbol{\omega}$, is in the direction of the x-axis.

10.16. In the steady flow of a gas between two parallel planar surfaces, if $d/\lambda \gg 1$ (the slip-flow regime) determine the average gas speed, $\bar{u}$, owing to a constant pressure gradient, $dp/d\tilde{y} = \text{const}$, where the velocity is assumed 1-dimensional in the y-direction. Use the geometry in Fig. 10-6.

Solution: The gas flow may be described by Eqs. (10-8) and (10-9) which, for this particular case, take the form:

$$\frac{\partial u_y}{\partial \tilde{y}} = 0, \ \mu\left(\frac{\partial^2 u}{\partial \tilde{x}^2}\right) = \frac{\partial p}{\partial \tilde{y}}, \ \partial p/\partial \tilde{x} = 0, \text{ and } u_y = u(\tilde{x}) .$$

The boundary conditions are given by $u(-d/2)=c_m\lambda(\partial u/\partial\tilde{x})$ and $u(d/2)=-c_m\lambda(\partial u/\partial\tilde{x})$. This statement of the problem then yields:

$$\bar{u}=d^{-1}\int_{-d/2}^{d/2}u(\tilde{x})\,d\tilde{x}=-\tfrac{1}{12}\frac{d^2}{\mu}\left(1+6c_m\frac{\lambda}{d}\right)\frac{dp}{d\tilde{y}}.$$

REFERENCES

1. Landau, L.D. and Lifshitz, E.M., *Fluid Mechanics* (Pergamon Press, Oxford, 1987).
2. Hidy, G.M. and Brock, J.R., *Topics in Aerosol Research*, vols. 1-3 (Pergamon Press, Oxford, 1971-1973).
3. Williams, M.M.R. and Loyalka, S.K., *Aerosol Science: Theory and Practice, with Special Applications to Nuclear Industry* (Pergamon Press, Oxford, 1991).
4. Allen, M.D. and Raabe, O.G., "Re-evaluation of Millikan's Oil Drop Data for the Motion of Small Particles in Air," *J. Aerosol Sci.* **13(6)**, 537-547 (1982).
5. Cercignani, C., Pagani, C.D., and Bassanini, P., "Flow of Rarefied Gas Past an Axisymmetric Body. II. Case of a Sphere," *Phys. Fluids* **11(7)**, 1399-1403 (1968).
6. Phillips, W.F., "Drag on a Small Sphere Moving Through a Gas," *Phys. Fluids* **18(9)**, 1083-1089 (1975).
7. Lea, K.C. and Loyalka, S.K., "Motion of a Sphere in Rarefied Gas," *Phys. Fluids* **25(9)**, 1550-1557 (1982).
8. Buckley, R.L. and Loyalka, S.K., "Cunningham Correction Factor and Accommodation Coefficient: Interpretation of Millikan's Data," *J. Aerosol Sci.* **20(3)**, 347-349 (1989).
9. Loyalka, S.K., "Motion of a Sphere in a Gas: Numerical Solution of the Linearized Boltzmann Equation," *Phys. Fluids A* **4(5)**, 1049-1056 (1992).
10. Rosenblatt, P. and La Mer, V.K., "Motion of a Particle in a Temperature Gradient. Thermal Repulsion as a Radiometer Phenomenon," *Phys. Rev.* **70(5-6)**, 385-395 (1946).
11. Saxton, R.L. and Ranz, W.E., "Thermal Force on an Aerosol Particle in a Temperature Gradient," *J. Appl. Phys.* **23(8)**, 917-923 (1952).
12. Schadt, C.F. and Cadle, R.D., "Thermal Forces on Aerosol Particles in a Thermal Precipitator," *J. Colloid Sci.* **12**, 356-362 (1957).
13. Schadt, C.F. and Cadle, R.D., "Thermal Forces on Aerosol Particles," *J. Phys. Chem.* **65(10)**, 1689-1694 (1961).
14. Schmitt, K.H., "Untersuchungen an Schwebstoffteilchen im Temperaturfeld," *Z. Naturforschg.* **14a**, 870-881 (1959).
15. Jacobsen, S. and Brock, J.R., The Thermal Force on Spherical Sodium Chloride Aerosols," *J. Colloid Sci.* **20(6)**, 544-556 (1965).
16. Waldmann, L. and Schmitt, K.H., "Thermophoresis and Diffusiophoresis of Aerosols," in *Aerosol Science*, edited by Davies, C.N. (Academic Press, London, 1966).
17. Derjaguin, B.V., Storozhilova, A.I., and Rabinovich, Ya.I., "Experimental Verification of the Theory of Thermophoresis of Aerosol Particles," *J. Colloid and Interface Sci.* **21(1)**, 35-58 (1966).
18. Keng, E.Y.H. and Orr, C.Jr., "Thermal Precipitation and Particle Conductivity," *J. Colloid and Interface Sci.* **22**, 107-116 (1966).
19. Epstein, P.S., "Zur Theorie des Radiometers," *Z. Physik* **54(4)**, 537-563 (1929).

20. Brock, J.R., "On the Theory of Thermal Forces Acting on Aerosol Particles," *J. Colloid Sci.* **17**, 768-780 (1962).
21. Dwyer, H.A., "Thirteen-Moment Theory of the Thermal Force on a Spherical Particle," *Phys. Fluids* **10(5)**, 976984 (1967).
22. Ivchenko, I.N. and Yalamov, Yu.I., "The Hydrodynamic Calculation Method of the Thermophoresis of Sufficiently Large Nonvolatile Aerosol Particles," *J. Phys. Chem. (Russia)* **45(3)**, 577-582 (1971) (in Russian).
23. Derjaguin, B.V. and Yalamov, Yu.I., "The Theory of Thermophoresis and Diffusiophoresis of Aerosol Particles and their Experimental Testing," in *Topics in Current Aerosol Research*, vol. 3, part 2, edited by Hidy, G.M. (Pergamon Press, Oxford, 1972).
24. Sone, Y. and Aoki, K., "Negative Thermophoresis: Thermal Stress Slip Flow Around a Spherical Particle in a Rarefied Gas," in *Rarefied Gas Dynamics*, vol. 74, part I, edited by Fisher, S.S. (American Institute of Aeronautics and Astronautics, 1981).
25. Yamamoto, K. and Ishihara, Y., "Thermophoresis of a Spherical Particle in a Rarefied Gas of a Transition Regime," *Phys. Fluids* **31(12)**, 3618-3624 (1988).
26. Bakanov, S.P., "Thermophoresis in Gases at Small Knudsen Numbers," *Aerosol Sci. Tech.* **15(1)**, 77-92 (1991).
27. Loyalka, S.K., "Mechanics of Aerosols in Nuclear Reactor Safety: A Review," *Prog. Nucl. Energy* **12(1)**, 1-56 (1983).
28. Loyalka, S.K., "Rarefied Gas Dynamics Problems in Environmental Sciences," in *Rarefied Gas Dynamics, XVI Symposium*, edited by Boffi, V. and Cercignani, C. (Teubner, Stuttgart, 1986).
29. Loyalka, S.K., "Thermophoretic Force on a Single Particle-I. Numerical Solution of the Linearized Boltzmann Equation," *J. Aerosol Sci.* **23(3)**, 291-300 (1992).
30. Poddoskin, A.B., Yushkanov, A.A., and Yalamov, Yu.I., "The Thermophoresis Theory for Sufficiently Large Aerosol Particles," *J. Tech. Phys. (Russian)* **52(11)**, 2253-2261 (1982) (in Russian).

Chapter 11

BOUNDARY VALUE PROBLEMS FOR ALL KNUDSEN NUMBERS

1. THE MOMENT EQUATIONS IN ARBITRARY CURVILINEAR COORDINATES.

The most advanced analysis possible for boundary value transport problems involving arbitrary Knudsen number and for various molecular interaction laws must be founded on the solution of the Boltzmann equation. In the transition regime, the effects of intermolecular collisions become dominant and their influence on the gas behavior is accurately described only by the full Boltzmann form of the collision operator found in the Boltzmann equation. For this reason, different model forms of the collision integral and model equations will not be considered here.

There are two kinds of difficulties associated with the solution of the Boltzmann integro-differential equation [1]. The first difficulty is connected with the direct solution of this equation in which the distribution function depends, in the general case, on seven independent variables. It is well known that the difficulty in solving differential equations increases very rapidly with the number of independent variables. The second difficulty is associated with the very complicated non-linear structure of the collision term in the Boltzmann equation.

The first difficulty may be overcome by replacing the Boltzmann equation with a system of equations for macroscopic values (moments). Such a moment system involves, in general, functions of only four independent variables and, thus, the difficulty of solution does not increase to the same level as it would with the Boltzmann equation due to this lesser number of independent variables. Additionally, for specific transport problems involving stationary or symmetry conditions, the number of independent variables may even be less than four. The second difficulty may be overcome, for some applied transport problems, through the process of linearization. This may be done, however, only if the appropriate linearization conditions are physically satisfied.

In order to investigate boundary value transport problems for all Knudsen numbers a moment method is considered. Instead of seeking exact solutions of the Boltzmann equation, this equation may be satisfied in a certain average sense. For boundary value transport problems, one is often not particularly interested in the distribution function itself since only certain lower moments of this function have the practical significance of being connected with the macroscopic values that describe the gas. Since one is mainly interested in mean flow quantities rather than the distribution function itself, the Maxwell integral transport equations, or moment equations, will be taken as the basis of the theory of transport phenomena for arbitrary Knudsen number.

The general moment equation can be obtained by multiplying the Boltzmann equation by the function, $Q(\mathbf{v})$, of the components of the molecular velocity and integrating over the velocity space. The general form of this equation was derived in [2]. The transport moment equation of the molecular property, $Q(\mathbf{v})$, in arbitrary, orthogonal, curvilinear coordinates, has the following form:

$$\begin{aligned} &\frac{\partial}{\partial t}\left(\int fQ(\mathbf{v})d\mathbf{v}\right)+\frac{1}{h_1h_2h_3}\sum_{i=1}^{3}\frac{\partial}{\partial\alpha_i}\left(h_jh_k\int fv_iQ(\mathbf{v})d\mathbf{v}\right)\\ &-\int f\left(\sum_{i=1}^{3}\frac{F_i}{m}\frac{\partial}{\partial v_i}Q(\mathbf{v})\right)d\mathbf{v}-\int f\sum_{i=1}^{3}\left\{\frac{1}{h_ih_j}\left(v_j^2\frac{\partial h_j}{\partial\alpha_i}-v_iv_j\frac{\partial h_i}{\partial\alpha_j}\right)\right.\\ &\left.+\frac{1}{h_ih_k}\left(v_k^2\frac{\partial h_k}{\partial\alpha_i}-v_iv_k\frac{\partial h_i}{\partial\alpha_k}\right)\right\}\frac{\partial}{\partial v_i}Q(\mathbf{v})d\mathbf{v}=n\Delta\overline{Q(\mathbf{v})}\ , \end{aligned} \tag{11-1}$$

where $\alpha_1,\alpha_2,\alpha_3$ are the curvilinear coordinates, $h_i(\alpha_1,\alpha_2,\alpha_3)$ denotes the metric coefficient such that $dl_i=h_id\alpha_i$, F_i is the component of the external force acting on a molecule, and the moment of the collision operator has been expressed in standard notation as $n\Delta\overline{Q(\mathbf{v})}$. For the spherical and cylindrical geometries these metric coefficients can be written as:

$$\alpha_1=\tilde{r}\qquad \alpha_2=\theta\qquad \alpha_3=\phi\ ,$$

$$h_1=1\qquad h_2=\tilde{r}\qquad h_3=\tilde{r}\sin(\theta)\ ,$$

and:

$$\alpha_1 = \tilde{r} \qquad \alpha_2 = \theta \qquad \alpha_3 = \tilde{z} \ ,$$
$$h_1 = 1 \qquad h_2 = \tilde{r} \qquad h_3 = 1 \ .$$

For the stationary problem with radial symmetry, in the absence of external forces, the general moment equation, Eq. (11-1), takes the following form [2,3]:

$$\begin{aligned}
&\frac{1}{r^j}\frac{d}{dr}\left(r^j \int f v_r Q(\mathbf{v})d\mathbf{v}\right) - \frac{1}{r}\int f \left\{\left[v_\theta^2 + (j-1)v_\phi^2\right]\frac{\partial}{\partial v_r}Q(\mathbf{v})\right. \\
&+\left[(j-1)\cot(\theta)v_\phi^2 - v_\theta v_r\right]\frac{\partial}{\partial v_\theta}Q(\mathbf{v}) \qquad\qquad (11\text{-}2)\\
&\left. -(j-1)\left[v_\phi v_r + \cot(\theta)v_\phi v_\theta\right]\frac{\partial}{\partial v_\phi}Q(\mathbf{v})\right\}d\mathbf{v} = Rn\Delta\overline{Q(\mathbf{v})} \ ,
\end{aligned}$$

where, for the cylindrical problem, $j=1$, for the spherical problem, $j=2$, $\mathbf{r} = \tilde{\mathbf{r}}/R$ is a dimensionless radial vector and R is the radius of the sphere or cylinder.

If $Q(\mathbf{v})$ is successively selected to be the collisional invariants, m, mv_r, and $\frac{1}{2}mv^2$, for which $n\Delta Q(\mathbf{v}) = 0$, one can obtain the ordinary continuity equation, and the radial momentum and energy transport equations. These moment equations are usually employed in all boundary value transport problems. Other equations can be obtained by using the arbitrary functions, $Q(\mathbf{v})$. It is necessary to note, however, that such a system of moment equations will never be closed as the number of the unknown moments is always greater than the number of moment equations. A common feature of all moment methods is their inclusion of some technique for closing the moment system. Typically, the technique for closing the moment system consists of making a special choice of the distribution functions so that they contain a certain number of unknown quantities that depend upon the various coordinates and time. The number of these unknown quantities must be equal to the number of moment equations and thus is a characteristic of the order of a given moment approximation.

Two ways may be distinguished by which to close the moment system. The first is based on the use of polynomial expansions in velocity space for the distribution function. The coefficients in this expansion are unknown functions of time and space but may be determined from the moment system if one employs a final number of expansion terms and uses an equal number of moment equations [4-7].

The second method of closing the system of moment equations involves making a special choice for the distribution function that is pertinent to the specific transport problem under consideration [2,3,8-12]. Simple approximate functions may be chosen if one takes into account conditions specific to the problem. However, one must insure that the distribution function chosen has the features normally characteristic of both the free-molecular and continuum regimes.

Mott-Smith [13], to describe planar shock waves, was the first to suggest the use of the discontinuous Maxwellian distribution function. This idea was then generalized by Lees and Liu [2,3] for arbitrary curvilinear geometries. At present, the distribution functions proposed by Mott-Smith and Lees are considered to be the most suitable for the solution of boundary value transport problems. The Lees method will be discussed in the next section.

2. THE TWO-SIDED MAXWELLIAN DISTRIBUTION FUNCTIONS.

As mentioned above, for arbitrary Knudsen number it is most convenient to use a moment method. Any moment method is an integral method. Just as in any other integral method, the distribution function that is employed is not necessarily an exact solution of the original Boltzmann equation. Nevertheless, this function must have features common to both the free-molecular and continuum distributions. It is also desirable that the distribution function be adaptable in such a fashion as to satisfy the boundary conditions. The best distribution function satisfying these conditions for the Maxwellian boundary model is that proposed by Lees and Liu [2,3]. In the Lees method the distribution function is assumed to be two-sided Maxwellians which are discontinuous on the surface of the 'so-called' 'cone of influence' in the velocity space. Consider the arbitrary point, $M(\mathbf{r})$, where $\mathbf{r}$ is a dimensionless position vector. For each specified point there exists a cone of influence the surface of which is formed by tangents to the particle surface passing through the point, $M(\mathbf{r})$. At this point it is necessary to distinguish two regions in the molecular velocity space. In the body coordinates all the outwardly directed molecular velocity vectors lying within the cone of influence are considered to belong to the second region in the velocity space. All other molecules belong to the first velocity set, denoted as region 1. For these velocity regions, the simplest distribution function having a discontinuous structure and being capable of exhibiting a smooth transition between the free-molecular and continuum regimes is the two-sided Maxwellian distribution given by:

$$f_1 = n_1(\mathbf{r},t)\left[\frac{m}{2\pi k T_1(\mathbf{r},t)}\right]^{3/2} \exp\left(-\frac{m(\mathbf{v}-\mathbf{u}_1(\mathbf{r},t))^2}{2kT_1(\mathbf{r},t)}\right), \quad (11\text{-}3a)$$

and:

$$f_2 = n_2(\mathbf{r},t)\left[\frac{m}{2\pi k T_2(\mathbf{r},t)}\right]^{3/2} \exp\left(-\frac{m(\mathbf{v}-\mathbf{u}_2(\mathbf{r},t))^2}{2kT_2(\mathbf{r},t)}\right), \quad (11\text{-}3b)$$

where $n_i(\mathbf{r},t)$, $T_i(\mathbf{r},t)$, and $\mathbf{u}_i(\mathbf{r},t)$ are ten unknown functions of position and time. These 'so-called' characteristic densities, temperatures, and mean velocities form the basic set of parameters over which all of the moments of the distribution function and, therefore, all of the macroscopic quantities for a gas, may be calculated. As one can see, the two-sided Maxwellian distribution function is a straightforward generalization of the distribution function for free-molecular flow.

Such a choice for the distribution function gives a chance to close the moment system but, unfortunately, it limits the number of moment equations and consequently the accuracy of this analysis. An improvement in the accuracy can be achieved by the use of the 'so-called' modified two-stream Maxwellians proposed by Krook [14] and described in detail by Kogan [15]. For various regions of the velocity space these modified two-stream Maxwellians can be represented by:

$$f_i = n_i\left[\frac{m}{2\pi k T_i}\right]^{3/2} \exp\left(-\frac{m(\mathbf{v}-\mathbf{u}_i)^2}{2kT_i}\right)\left(1 + A_k^{(i)} v_k + A_{kl}^{(i)} v_k v_l + \cdots\right), \quad (11\text{-}4)$$

where, $i=1$ for the first region and $i=2$ for the second. The functions n_i, $\mathbf{u}_i$, T_i, $A_k^{(i)}$, and $A_{kl}^{(i)}$ depend on r and t in the general case.

To facilitate the employment of the discontinuous distribution functions given by Eqs. (11-3a), (11-3b), and (11-4), one should attempt to represent these functions by analytical formulas that might be suitable for the entire velocity space [16]. For two specific geometries of interest, this type of representation will be discussed in detail.

For the spherical geometry, at each space point, $M(\mathbf{r})$, there exists a cone of influence, the surface of which is described by the equation $\chi = \chi_0$, $\chi_0 = \arcsin(R/\tilde{r})$, where R is the radius of the particle, $\tilde{\mathbf{r}} = R\mathbf{r}$, and χ is

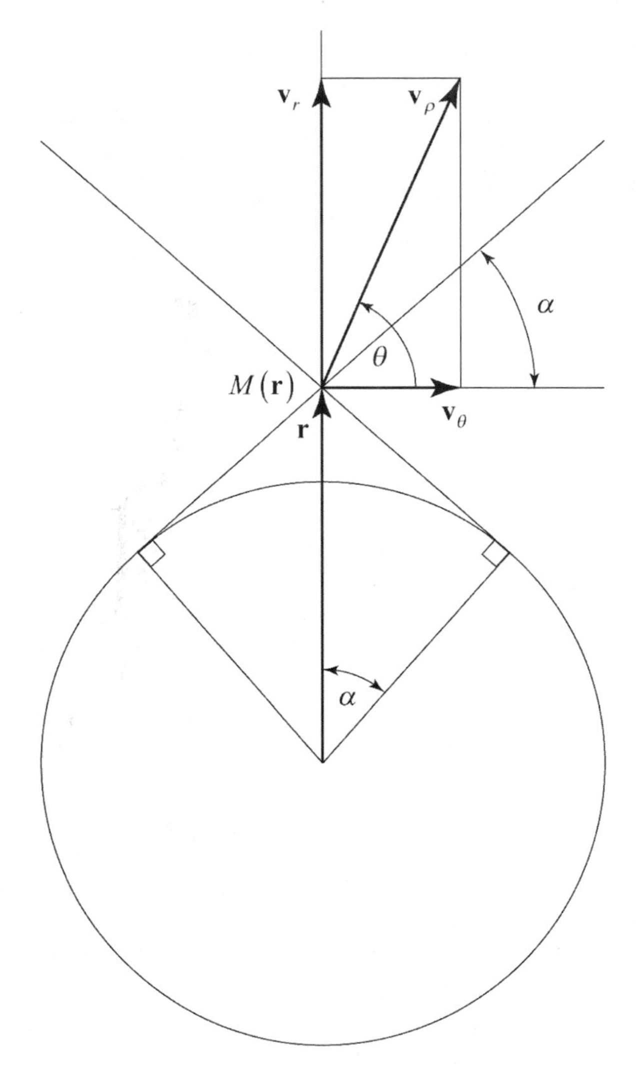

Figure 11-1. The cone of influence in cylindrical coordinates.

the angle between vectors **v** and **r** (Fig. 7.1). In velocity space, the cone of influence divides the molecules into two groups with distribution functions f_1 and f_2 (f_2 describes the molecules for which $v_r > v_r^*, v_r^* = v\cos(\chi_0)$; f_1 describes all other molecules). In the full velocity space the distribution function may be expressed as [16]:

$$f = \tfrac{1}{2}(f_1 + f_2) + \tfrac{1}{2}(f_2 - f_1)\operatorname{sign}(v_r - v_r^*) . \tag{11-5}$$

This is a representation of the distribution function that is very convenient to use in a moment system.

Now, consider the corresponding representation for the cylindrical geometry shown in Fig 11-1. While still traditionally termed a 'cone of influence,' the region is actually tent-shaped and is bounded by the two planes passing through the point $M(\mathbf{r})$ and lying tangent to the surface of the cylinder. The cylinder, which nominally has a dimensional radius of R, is shown in Fig. 11-1 in dimensionless form where $\mathbf{r} = \tilde{\mathbf{r}}/R$ and thus the cylinder has a dimensionless radius of unity. Let v_θ, v_r, v_z and v_ρ, θ, v_z be Cartesian and cylindrical coordinates in velocity space, respectively. Now consider an arbitrary velocity, $\mathbf{v} = \mathbf{v}_\theta + \mathbf{v}_r + \mathbf{v}_z$. From Fig. 11-1 one can see that $\mathbf{v}_\rho = \mathbf{v}_\theta + \mathbf{v}_r$ with $v_r = v_\rho \sin(\theta)$ and $v_\theta = v_\rho \cos(\theta)$. The velocity component, $\mathbf{v}_z$, which is the same in both coordinate systems, is not shown in Fig. 11-1 which is only a two-dimensional representation, but it may be envisioned as projecting normally upward from the plane of the page perpendicular to the components, $\mathbf{v}_\theta$ and $\mathbf{v}_r$. If one defines $v_r^* = v_\rho \sin(\alpha)$ for the special case when $\mathbf{v}_\rho$ and $\mathbf{v}$ both lie on the surface of the cone of influence, i.e. when $\theta = \alpha = \arccos(1/r)$, then from Fig. 11-1 one can conclude that all molecules for which $v_r > v_r^*$ belong to the second region of the velocity space in which the angle, θ, varies over the domain $\alpha \le \theta \le \pi - \alpha$. For all of the remaining molecules, $v_r < v_r^*$ and the molecules belong to the first region for which $\pi - \alpha < \theta < 2\pi + \alpha$. Eliminating the angular dependence, v_r^* is given in terms of v_ρ and r by:

$$v_r^* = v_\rho \left(1 - r^{-2}\right)^{1/2} . \tag{11-6}$$

This analysis allows one to write an expression for the distribution function in the full velocity space in the same form as that given by Eq. (11-5) which is a form of the distribution function that allows one to find all of the moments necessary to construct the moment system. The next section contains an analytical method for the calculation of these moments for both the spherical and cylindrical geometries.

3. MOMENTS OF DISCONTINUOUS DISTRIBUTION FUNCTIONS.

Let value $Q(\mathbf{v})$ be any function of the molecular velocity, $\mathbf{v}$. Then, the mean value of $Q(\mathbf{v})$, being a moment of the distribution function, can be calculated by the formula:

$$
\begin{aligned}
n\overline{Q(\mathbf{v})} &= \int_{(1)} Q(\mathbf{v}) f_1 d\mathbf{v} + \int_{(2)} Q(\mathbf{v}) f_2 d\mathbf{v} \\
&= \int_{(1+2)} Q(\mathbf{v}) f_1 d\mathbf{v} + \int_{(2)} Q(\mathbf{v}) f_2 d\mathbf{v} - \int_{(2)} Q(\mathbf{v}) f_1 d\mathbf{v} ,
\end{aligned}
\tag{11-7}
$$

where $\int_{(i)}$ and $\int_{(1+2)}$ are interpreted to mean integration over the specified region of the velocity space and the entire velocity space, respectively. For the spherical geometry, integration over velocities lying within the cone of influence proceeds according to the formula:

$$
\int_{(2)} d\mathbf{v} = \int_0^\infty v^2 dv \int_0^{2\pi} d\beta \int_0^{\chi_0} \sin(\chi) d\chi , \tag{11-8}
$$

where $\chi_0 = \arcsin(1/r)$ and r is the dimensionless radial coordinate of the point, $M(\mathbf{r})$. The relationships between spherical and Cartesian coordinates in velocity space are:

$$
v_r = v\cos(\chi) , \ v_\theta = v\sin(\chi)\sin(\beta) , \ v_\phi = v\sin(\chi)\cos(\beta) .
$$

Thus, the same integration for the cylindrical geometry proceeds according to the formula:

$$
\int_{(2)} d\mathbf{v} = \int_0^\infty v_\rho dv_\rho \int_{-\infty}^\infty dv_z \int_\alpha^{\pi-\alpha} d\theta , \tag{11-9}
$$

where $\alpha = \arccos(1/r)$ and $v_r = v_\rho \sin(\theta)$, $v_\theta = v_\rho \cos(\theta)$, and $v_z = v_z$.

For example, one may calculate the gas number density in the four-moment approach for which the Lees distribution function is given by:

$$
f_i = n_i(\mathbf{r},t) \left[\frac{m}{2\pi k T_i(\mathbf{r},t)} \right]^{3/2} \exp\left(-\frac{m\mathbf{v}^2}{2kT_i(\mathbf{r},t)} \right) ; \ (i = 1; 2) , \tag{11-10}
$$

Using Eq. (11-7) with $Q(\mathbf{v}) = 1$, one obtains for the spherical geometry:

$$n(\mathbf{r},t) = n_1(\mathbf{r},t) + \frac{n_2(\mathbf{r},t)}{\pi^{3/2}} \int_0^\infty c_2^2 \exp(-c_2^2) dc_2 \int_0^{2\pi} d\beta \int_0^{\chi_0} \sin(\chi) d\chi$$
$$- \frac{n_1(\mathbf{r},t)}{\pi^{3/2}} \int_0^\infty c_1^2 \exp(-c_1^2) dc_1 \int_0^{2\pi} d\beta \int_0^{\chi_0} \sin(\chi) d\chi \qquad (11\text{-}11)$$
$$= \tfrac{1}{2}(n_1 + n_2) - \tfrac{1}{2}(n_2 - n_1) x ,$$

where $x = \left(1 - r^{-2}\right)^{1/2}$ and $\mathbf{c}_i = \left(m/2kT_i\right)^{1/2} \mathbf{v}$. The appropriate expression for the cylindrical geometry is then given by:

$$n = \tfrac{1}{2}(n_1 + n_2) - \tfrac{1}{2}(n_2 - n_1) \frac{2\alpha}{\pi} , \qquad (11\text{-}12)$$

where $\alpha = \arccos(1/r)$. Other moments can be calculated in the same way.

4. ANALYTICAL EXPRESSIONS FOR THE BRACKET INTEGRALS.

Let this analysis be confined to a description of linearized transport problems in which the four-moment approach is used. The characteristic densities and temperatures in Eqs. (11-3a) and (11-3b), for these linearized problems, may be written as:

$$n_i(\mathbf{r},t) = n_0\left[1 + \nu_i(\mathbf{r},t)\right] , \qquad (11\text{-}13a)$$

$$T_i(\mathbf{r},t) = T_0\left[1 + \tau_i(\mathbf{r},t)\right] . \qquad (11\text{-}13b)$$

The corrections, ν_i and τ_i, are assumed to satisfy the following conditions $\nu_i \ll 1$ and $\tau_i \ll 1$. The two-sided Maxwellian distribution function takes the form:

$$f_i = f^{(0)}\left\{1 + \nu_i(\mathbf{r},t) + \left(c^2 - \tfrac{3}{2}\right)\tau_i(\mathbf{r},t)\right\} , \qquad (11\text{-}14)$$

where:

$$f^{(0)} = n_0 \left(\frac{m}{2\pi k T_0}\right)^{3/2} \exp\left(-c^2\right) .$$

In the full velocity space, for the spherical and cylindrical geometries, this distribution function may be expressed, according to Eq. (11-5), as:

$$f = f^{(0)}\left\{1+\tfrac{1}{2}\nu^{+}+\tfrac{1}{2}\left(c^{2}-\tfrac{3}{2}\right)\tau^{+} + \left[\tfrac{1}{2}\nu^{-}+\tfrac{1}{2}\left(c^{2}-\tfrac{3}{2}\right)\tau^{-}\right]\operatorname{sign}\left(c_r - c_r^{*}\right)\right\}, \tag{11-15}$$

where $\nu^{\pm} = \nu_2 \pm \nu_1$ and $\tau^{\pm} = \tau_2 \pm \tau_1$.

If the specific molecular properties, $\left(1, c_r, c^2, c^2 c_r\right)$, are substituted into Eq. (11-2) instead of $Q(\mathbf{v})$, then four-moment equations will be produced. The right-hand-side of the fourth equation will contain the following bracket integrals:

$$\left[\operatorname{sign}\left(c_r - c_r^{*}\right), c^2 c_r\right], \quad \left[c^2 \operatorname{sign}\left(c_r - c_r^{*}\right), c^2 c_r\right].$$

Consider the first of the above bracket integrals when molecules are assumed to act as rigid spheres. For the spherical geometry, this integral has the following analytical form:

$$\begin{aligned}
\left[\operatorname{sign}\left(c_r - c_r^{*}\right), c^2 c_r\right] &= \tfrac{1}{2}\frac{\sigma^2}{\pi^3}\left(\frac{kT_0}{m}\right)^{1/2} \\
&\times \int_0^{\pi}\sin(\theta)\,d\theta \int_0^{2\pi} d\varepsilon \int_0^{\pi}\sin(\alpha)\,d\alpha \int_0^{2\pi} d\beta \\
&\times \int_{-\infty}^{\infty}\exp\left(-G_r^2\right)dG_r \int_{-\infty}^{\infty}\exp\left(-G_\theta^2\right)dG_\theta \\
&\times \int_{-\infty}^{\infty}\exp\left(-G_\phi^2\right)dG_\phi \int_0^{\infty} g^3\exp\left(-g^2\right)dg \\
&\times \sqrt{2}\left\{G_r\left(g_r^2 - g_r'^2\right) + G_\theta\left(g_r g_\theta - g_r' g_\theta'\right) + G_\phi\left(g_r g_\phi - g_r' g_\phi'\right)\right\} \\
&\times \operatorname{sign}\left(G_r - g_r - \sqrt{(\mathbf{G}-\mathbf{g})^2}\left(1-\frac{1}{r^2}\right)^{1/2}\right).
\end{aligned} \tag{11-16}$$

Here, the relations given in Appendix A are used to specify the components of the relative velocity vector, $\mathbf{g}$, before and after a collision.

From Eq. (11-16) one can see that this integral depends only on the radial coordinate, r. Two limiting values of this integral may easily be found. On the particle surface, when $r=1$, the integrand has the usual form encountered in planar transport problems. This integral may be expressed as [16]:

$$\left[\mathrm{sign}\left(c_r - c_r^*\right), c^2 c_r\right]_{r=1} = \left[\mathrm{sign}\left(c_r\right), c^2 c_r\right] = -\tfrac{2}{3}\left(\frac{2kT_0}{m}\right)^{1/2} \sigma^2 .$$

In the other limit, when $r \to \infty$, $\mathrm{sign}\left(G_r - g_r - |\mathbf{G}-\mathbf{g}|\right) = -1$ and, therefore, the integrand is an odd function with respect to the components of the molecular velocity, G_r, G_θ, G_ϕ. Consequently, in this limit one has:

$$\lim_{r\to\infty}\left[\mathrm{sign}\left(c_r - c_r^*\right), c^2 c_r\right] = 0 .$$

For arbitrary values of r, these multiple integrations result in very cumbersome algebraic calculations. After integration with respect to the variables θ, ε, and g, and after introducing the new variables $G_r = xg$, $G_\theta = yg$, $G_\phi = zg$, and $\cos(\alpha) = t$, this integral becomes:

$$\left[\mathrm{sign}\left(c_r - c_r^*\right), c^2 c_r\right] = \frac{96\sigma^2}{\sqrt{2}\pi^2}\left(\frac{kT_0}{m}\right)^{1/2} I_1^{[1]}(r) , \qquad (11\text{-}17)$$

where:

$$I_1^{[1]}(r) = \int_{-1}^{1} dt \int_{0}^{2\pi} d\beta \int_{-\infty}^{\infty} dy \int_{-\infty}^{\infty} dz \int_{x_0}^{\infty} F(t,\beta,y,z,x)\, dx ,$$

$$F(t,\beta,y,z,x) = \frac{x\left(t^2 - \frac{1}{3}\right) + yt\sqrt{1-t^2}\cos(\beta) + zt\sqrt{1-t^2}\sin(\beta)}{\left(1 + x^2 + y^2 + z^2\right)^5} ,$$

and:

$$x_0 = t + \left[1 - t^2 + y^2 + z^2 - 2y\sqrt{1-t^2}\cos(\beta) - 2z\sqrt{1-t^2}\sin(\beta)\right]^{1/2}\left(r^2 - 1\right)^{1/2} .$$

For the cylindrical geometry the integrand is the same, but it is convenient to perform the first integration over all values of z as the lower limit of the integral with respect to x does not depend on z. In this case, the bracket integral may be written as:

$$\left[\operatorname{sign}\left(c_r - c_r^*\right), c^2 c_r\right] = \frac{96\sigma^2}{\sqrt{2}\pi^2}\left(\frac{kT_0}{m}\right)^{1/2} I_2^{[1]}(r) , \tag{11-18}$$

with:

$$I_2^{[1]}(r) = \int_{-1}^{1} dt \int_{0}^{2\pi} d\beta \int_{-\infty}^{\infty} dy \int_{x_0'}^{\infty} dx \int_{-\infty}^{\infty} F(t,\beta,y,z,x)\, dz ,$$

and:

$$x_0' = t + \left|y - \sqrt{1-t^2}\cos(\beta)\right|\left(r^2 - 1\right)^{1/2} .$$

In the same way, analogous expressions may be obtained for the second bracket integral that is contained in the right-hand-side of the heat flow transport equation. For the spherical geometry this integral may be represented as:

$$\left[c^2\operatorname{sign}\left(c_r - c_r^*\right), c^2 c_r\right] = \frac{240\sigma^2}{\sqrt{2}\pi^2}\left(\frac{kT_0}{m}\right)^{1/2} I_1^{[2]}(r) , \tag{11-19}$$

with:

$$I_1^{[2]}(r) = \int_{-1}^{1} dt \int_{0}^{2\pi} d\beta \int_{-\infty}^{\infty} dy \int_{-\infty}^{\infty} dz \int_{x_0}^{\infty} F(t,\beta,y,z,x)\Phi(t,\beta,y,z,x)\, dx ,$$

and:

$$\Phi(t,\beta,y,z,x) = \frac{1 + x^2 + y^2 + z^2 - 2xt - 2y\sqrt{1-t^2}\cos(\beta) - 2z\sqrt{1-t^2}\sin(\beta)}{\left(1 + x^2 + y^2 + z^2\right)} .$$

For the cylindrical geometry, the appropriate integral is given by:

$$\left[c^2 \operatorname{sign}\left(c_r - c_r^*\right), c^2 c_r\right] = \frac{240\sigma^2}{\sqrt{2}\pi^2}\left(\frac{kT_0}{m}\right)^{1/2} I_2^{[2]}(r) , \tag{11-20}$$

with:

$$I_2^{[2]}(r) = \int_{-1}^{1} dt \int_0^{2\pi} d\beta \int_{-\infty}^{\infty} dy \int_{x_0'}^{\infty} dx \int_{-\infty}^{\infty} F(t,\beta,y,z,x)\Phi(t,\beta,y,z,x)dz ,$$

The numerical values of the functions, $I_i^{[j]}(r)$, are necessary to obtain results of practical interest for arbitrary Knudsen numbers. The application of these functions will be more convenient if one introduces some special functions depending only on a radial coordinate by some normalization procedure. Using the values of the bracket integrals for the planar geometry [16], one can introduce the following notation for these functions:

$$\left[\operatorname{sign}\left(c_r - c_r^*\right), c^2 c_r\right] = -\tfrac{2}{3}\left(\frac{2kT_0}{m}\right)^{1/2} \sigma^2 I_i^{*[1]}(r) , \tag{11-21}$$

$$\left[c^2 \operatorname{sign}\left(c_r - c_r^*\right), c^2 c_r\right] = \left(\tfrac{2}{3} + \tfrac{1}{4}\pi\right)\left(\frac{2kT_0}{m}\right)^{1/2} \sigma^2 I_i^{*[2]}(r) , \tag{11-22}$$

where the normalized functions, $I_i^{*[j]}(r)$, may be written as:

$$I_i^{*[1]}(r) = -\frac{72}{\pi^2} I_i^{[1]}(r) ,$$

and:

$$I_i^{*[2]}(r) = \frac{1440}{\pi^2(8+3\pi)} I_i^{[2]}(r) ,$$

and in which the index, i, assumes the value of $i=1$ for the spherical geometry and assumes the value $i=2$ for the cylindrical geometry. Now, the functions, $I_i^{*[j]}(r)$, have the common range, [1,0], for values of r from 1 to ∞. These special functions allow one to represent the solutions to many boundary value transport problems in analytical forms.

As one can see, the calculational procedure for the bracket integrals presents the main difficulty in the application of Lees's method. The aforementioned difficulty typically results in the use of various approximate methods which will be considered later in Sections 11.7 and 11.8. Another method for simplifying the analysis is based on a specific form of the moment system which has been generated using some general properties of the Chapman-Enskog solution (see Section 11.10). One method that may be used to alleviate the need for numerical values of the special functions described above has been discussed for cylindrical and spherical geometries in [17,18] and in Appendix B. Since it is impossible to obtain these integrals in analytical forms, numerical values of the special functions, $I_i^{*[j]}(r)$, would be very useful for many boundary value transport problems. Some of these numerical values have been given in Tables B-1 and B-2.

5. BOUNDARY CONDITIONS FOR MOMENT EQUATIONS.

In a general case the microscopic boundary conditions on the body surface can be formulated for reflected molecular distribution function by means of the dispersion operator, A, that may be expressed by the relation:

$$f^{+}(\mathbf{v},\mathbf{r}_S) = \mathrm{A} f^{-}(\mathbf{v},\mathbf{r}_S) \ , \tag{11-23}$$

where $\mathbf{r}_S$ is the position radius vector for a given point on the particle surface. For the arbitrary boundary model the approximating functions given by Eqs. (11-3a), (11-3b) and (11-4) do not coincide with the distribution function of reflected molecules, expressed by Eq. (11-23), for any values of the characteristic parameters that are contained in them. In that case the boundary condition may, on the average, be satisfied only approximately if one uses the integral boundary model which may be written as:

$$\int\limits_{(\mathbf{v}\cdot\mathbf{n})>0} (\mathbf{v}\cdot\mathbf{n}) Q(\mathbf{v}) f^{+} d\mathbf{v} = \int\limits_{(\mathbf{v}\cdot\mathbf{n})>0} (\mathbf{v}\cdot\mathbf{n}) Q(\mathbf{v}) \mathrm{A} f^{-} d\mathbf{v} \ , \tag{11-24}$$

where $\mathbf{n}$ is a unit normal vector to the surface directed into the gas and $Q(\mathbf{v})$ usually represents the molecular properties used in the moment equations. For various $Q(\mathbf{v})$, this relation yields the necessary number of boundary conditions for the moments.

A question arises concerning what molecular properties must be used for a given problem on each of the sections of the boundary. There are some uncertainties connected with this choice. Nevertheless, a general recommendation might be useful to facilitate this choice. Both on inner and outer surfaces it is convenient to use the integral form of the boundary conditions constructed on the integral definition of the accommodation coefficients. For heat and tangential momentum transport problems, the following integral boundary conditions should be used:

$$Q_n(\mathbf{r}_S) = \alpha_T \left[Q_n^-(\mathbf{r}_S) + \left(Q_n^+(\mathbf{r}_S) \right)_{eq} \right], \tag{11-25a}$$

$$P_\tau(\mathbf{r}_S) = \alpha_\tau \left[P_\tau^-(\mathbf{r}_S) + \left(P_\tau^+(\mathbf{r}_S) \right)_{eq} \right]. \tag{11-25b}$$

The second relation is the generalized form of Eq. (6-7) in which $(P_\tau^+(\mathbf{r}_S))_{eq}$ is the tangential momentum that would be carried away by reflected molecules if the tangential momentum accommodation coefficient were equal to unity. Note that, for stationary surfaces, $(P_\tau^+(\mathbf{r}_S))_{eq} = 0$.

Another way of constructing boundary conditions [4] may be used for transport problems for which the approximating distribution function given by Eqs. (11-3a), (11-3b) and (11-4) is adapted to the Maxwellian boundary model (for example, the single-sphere heat transport, drag and torque problems). If the two-sided Maxwellian distribution function is adapted to the boundary model, the boundary conditions may be satisfied exactly by choosing certain characteristic parameters. To simplify this analysis, one can use the general expression for the reflected distribution function (Problem 6.5) which can be written in the form:

$$\begin{aligned} \Phi^+(\mathbf{c},\mathbf{r}_S) &= (1-\alpha_\tau)\Phi^-(\mathbf{c}',\mathbf{r}_S) \\ &+ \alpha_\tau\left(2-c^2\right)\left\{-\alpha_T\tau_w + (1-\alpha_T)\left(\frac{\delta_2}{\pi} - \frac{2\delta_1}{\pi}\right)\right\} - \alpha_\tau\frac{2\delta_1}{\pi}, \end{aligned} \tag{11-26a}$$

where $\Phi^\pm$ is a measure of the perturbation in the distribution function from an absolute Maxwellian, $\mathbf{c}' = \mathbf{c} - 2\mathbf{n}(\mathbf{c}\cdot\mathbf{n})$, τ_w is a correction to the temperature at the surface point, $\mathbf{r}_S$, and δ_1 and δ_2 are given by:

$$\delta_1 = \int_{(\mathbf{c}\cdot\mathbf{n})<0} \Phi^-(\mathbf{c},\mathbf{r}_S)(\mathbf{c}\cdot\mathbf{n})\exp\left(-c^2\right)d\mathbf{c}, \tag{11-26b}$$

$$\delta_2 = \int_{(\mathbf{c}\cdot\mathbf{n})<0} \Phi^-(\mathbf{c},\mathbf{r}_S)c^2(\mathbf{c}\cdot\mathbf{n})\exp(-c^2)d\mathbf{c} \ . \tag{11-26c}$$

Now, equating terms in Eq. (11-26a) with the same molecular velocity factors, one can obtain the boundary conditions necessary to determine the parameters in the two-sided Maxwellian distribution function associated with a given boundary value problem.

6. THERMAL CONDUCTION FROM A HEATED SPHERE.

In this section, the main features of the transition between gas-kinetics and gas-dynamics are analyzed by means of the four-moment approach, utilizing the two-sided Maxwellian distribution function to close the moment system. This approach is illustrated by considering, for arbitrary Knudsen number, the steady heat conduction problem from a heated sphere.

Consider the following brief statement of this problem [18]. Let a sphere of radius, R, be located in an infinite expanse of a gas which, far from the sphere, has a temperature, T_0, and a number density, n_0. The temperature, $T_0 + \Delta T$, of the sphere surface is supposed to be slightly different from that of the surrounding gas $(\Delta T \ll T_0)$. For this assumption the gas is near to the equilibrium state, and, therefore, this problem may be linearized be means of using the small parameters that are proportional to $\Delta T/T_0$.

The problem being considered possesses radial symmetry, and, therefore, the moment equation for transport of any molecular property, $Q(\mathbf{v})$, is:

$$\begin{aligned}
&\frac{1}{r^2}\frac{\partial}{\partial r}\left(r^2\int f v_r Q(\mathbf{v})d\mathbf{v}\right) \\
&-\frac{1}{r}\int\left\{\left[v_\theta^2+v_\phi^2\right]\frac{\partial Q(\mathbf{v})}{\partial v_r}+\left[v_\phi^2\cot(\theta)-v_\theta v_r\right]\frac{\partial Q(\mathbf{v})}{\partial v_\theta}\right. \\
&\qquad\left.-\left[v_\phi v_r+v_\theta v_\phi\cot(\theta)\right]\frac{\partial Q(\mathbf{v})}{\partial v_\phi}\right\} f d\mathbf{v} = Rn\overline{\Delta Q(\mathbf{v})} \ .
\end{aligned} \tag{11-27}$$

Alternately selecting $Q(\mathbf{v})$ to be the collisional invariants, 1, v_r, and v^2, for which $\Delta Q(\mathbf{v}) = 0$, one obtains the ordinary continuity, radial momentum, and energy equations. Since the primary interest is in the radial heat flux, one also takes $Q(\mathbf{v}) = v_r v^2$ which yields a fourth moment equation involving this heat flux [2].

For the four-moment approach the linearized distribution function may be used in the form given by Eq. (11-15). From Eq. (11-27), the following system of moment equations may be derived:

$$\frac{\partial}{\partial r}\left(r^2 n u_r\right)=0 \ , \tag{11-28a}$$

$$\frac{\partial}{\partial r}P_{rr}+\frac{1}{r}\left(2P_{rr}-P_{\theta\theta}-P_{\phi\phi}\right)=0 \ , \tag{11-28b}$$

$$\frac{\partial}{\partial r}\left(r^2 Q_r\right)=0 \ , \tag{11-28c}$$

$$\frac{1}{r^2}\frac{\partial}{\partial r}\int r^2 v^2 v_r^2 f d\mathbf{v}-\frac{1}{r}\int v^2\left(v_\theta^2+v_\phi^2\right) f d\mathbf{v}=Rn\Delta\,\overline{v^2 v_r} \ . \tag{11-28d}$$

All moments of this system can be calculated by means of the method described in Section 11.3. These moments are given by the relations:

$$n(r)=n_0\left(1+\tfrac{1}{2}\nu^+-\tfrac{1}{2}\nu^- x\right) ,$$

$$u_r(r)=\frac{1}{2\sqrt{\pi}r^2}\left(\frac{2kT_0}{m}\right)^{1/2}\left(\nu^-+\tfrac{1}{2}\tau^-\right) ,$$

$$P_{rr}=p_0\left(1+\tfrac{1}{2}\nu^+-\tfrac{1}{2}\nu^- x^3+\tfrac{1}{2}\tau^+-\tfrac{1}{2}\tau^- x^3\right) ,$$

$$p(r)=p_0\left(1+\tfrac{1}{2}\nu^+-\tfrac{1}{2}\nu^- x+\tfrac{1}{2}\tau^+-\tfrac{1}{2}\tau^- x\right) ,$$

$$Q_r(r)=\frac{p_0}{\sqrt{\pi}r^2}\left(\frac{2kT_0}{m}\right)^{1/2}\left(\nu^-+\tfrac{3}{2}\tau^-\right) ,$$

$$\int c^2 c_r^2 f d\mathbf{v} = n_0 \tfrac{5}{4}\left(1 + \tfrac{1}{2}\nu^+ - \tfrac{1}{2}\nu^- x^3 + \tau^+ - \tau^- x^3\right) ,$$

and:

$$\int c^2 \left(c_\theta^2 + c_\phi^2\right) f \, d\mathbf{v} = n_0 \tfrac{5}{2}\left(1 + \tfrac{1}{2}\nu^+ - \tfrac{3}{4}\nu^- x + \tau^+ - \tfrac{3}{2}\tau^- x + \tfrac{1}{4}\nu^- x^3 + \tfrac{1}{2}\tau^- x^3\right) ,$$

where $x = \left(1 - r^{-2}\right)^{1/2}$.

The moment system given by Eqs. (11-28) for the parameters of the distribution function can be expressed as:

$$\frac{d}{dr}\left(\nu^- + \tfrac{1}{2}\tau^-\right) = 0 , \tag{11-29a}$$

$$\frac{d}{dr}\left(\nu^+ + \tau^+\right) = 0 , \tag{11-29b}$$

$$\frac{d}{dr}\left(\nu^- + \tfrac{3}{2}\tau^-\right) = 0 , \tag{11-29c}$$

$$\frac{d}{dr}\left(\nu^+ + 2\tau^+\right) = \tfrac{8}{5}\frac{R}{n_0}\left(\frac{m}{2kT_0}\right)^{1/2} n\Delta\, \overline{c^2 c_r} , \tag{11-29d}$$

where the moment of the collision operator, $n\Delta \overline{c^2 c_r}$, is determined by:

$$n\Delta \overline{c^2 c_r} = -n_0^2 \left\{ \tfrac{1}{2}\left(\nu^- - \tfrac{3}{2}\tau^-\right)\left[\operatorname{sign}\left(c_r - c_r^*\right), c_r c^2\right] + \tfrac{1}{2}\tau^- \left[c^2 \operatorname{sign}\left(c_r - c_r^*\right), c_r c^2\right] \right\} .$$

Now, consider the boundary conditions for these moment equations. Since the distribution function that has been employed is well suited to the Maxwellian boundary model, the ordinary microscopic boundary condition on the sphere surface can be used to yield:

$$\begin{aligned} &\nu_2(1) + \left(c^2 - \tfrac{3}{2}\right)\tau_2(1) \\ &= \left(1 - \alpha_\tau\right)\left[\nu_1(1) + \left(c^2 - \tfrac{3}{2}\right)\tau_1(1)\right] + \alpha_\tau\left[\nu_r + \left(c^2 - \tfrac{3}{2}\right)\tau_r\right] . \end{aligned} \tag{11-30a}$$

This expression contains two unknown quantities, ν_r and τ_r, that may be found from the two additional conditions:

$$N_r = n_0 u_r = 0 \ , \tag{11-30b}$$

$$\alpha_T = \frac{Q_r^+ + Q_r^-}{Q_{rw}^+ + Q_r^-} \ , \tag{11-30c}$$

where the first relationship expresses the fact that the sphere is impenetrable to the gas molecules and the second one is the thermal accommodation condition. At large distances from the sphere the number density and the temperature tend to the equilibrium values that can be written as:

$$\nu^+(\infty) - \nu^-(\infty) = 0 \ , \tag{11-30d}$$

$$\tau^+(\infty) - \tau^-(\infty) = 0 \ . \tag{11-30e}$$

After the integration of Eqs. (11-29a)-(11-29d), one can obtain:

$$\nu^- = -\tfrac{1}{2} A_3 \ , \tag{11-31a}$$

$$\tau^- = A_3 \ , \tag{11-31b}$$

$$\nu^+ = -A_3 \xi(r) + 2A_2 - A_4 \ , \tag{11-31c}$$

$$\tau^+ = A_3 \xi(r) + A_4 - A_2 \ , \tag{11-31d}$$

where A_2, A_3, and A_4 $(A_1 = 0)$ are the arbitrary constants of integration and a function, $\xi(r)$, may be expressed as:

$$\xi(r) = -Kn^{-1} \left[\alpha_1 \int_1^r I_1^{*[1]}(r') dr' + \alpha_2 \int_1^r I_1^{*[2]}(r') dr' \right] . \tag{11-31e}$$

In this expression, the following notations were introduced:

$$\alpha_1 = \frac{8\sqrt{2}}{15\pi}, \quad \alpha_2 = \frac{(8+3\pi)\sqrt{2}}{30\pi},$$

and the functions $I_1^{*[1]}(r)$ and $I_1^{*[2]}(r)$ are determined by Eqs. (11-21) and (11-22).

Using the boundary conditions given by Eqs. (11-30a)-(11-30e) and (11-31a)-(11-31e) and performing some simple transformations, one obtains the following expression for the heat flux through the sphere surface:

$$\frac{Q}{Q_{fm}} = \frac{1}{1+\zeta Kn^{-1}}, \tag{11-32a}$$

where:

$$Q_{fm} = 4\pi R^2 \frac{n_0 k T_0}{\sqrt{\pi}} \left(\frac{2kT_0}{m}\right)^{1/2} \alpha_\tau \alpha_T \frac{\Delta T}{T_0}, \tag{11-32b}$$

is the free-molecular heat flux and the quantity, ζ, is given by:

$$\zeta = \tfrac{1}{2}\alpha_\tau \alpha_T \left[\alpha_1 \int_1^\infty I_1^{*[1]}(r)\,dr + \alpha_2 \int_1^\infty I_1^{*[2]}(r)\,dr \right], \tag{11-33}$$

in which α_τ and α_T are the accommodation coefficients.

This analytical expression gives the exact result for the free-molecular regime when $Kn \to \infty$. For another limiting regime, when $Kn \to 0$, Eq. (11-32a) may be written as:

$$Q = 4\pi R \kappa \Delta T ,$$

where κ is the thermal conductivity coefficient that, for the given approach, can be expressed in the form:

$$\kappa = \frac{2}{\sqrt{\pi}} \left(\frac{2kT_0}{m}\right)^{1/2} n_0 k \lambda \left[\alpha_1 \int_1^\infty I_1^{*[1]}(r)\,dr + \alpha_2 \int_1^\infty I_1^{*[2]}(r)\,dr \right]^{-1} .$$

If $Kn \to 0$, this expression for the thermal conductivity coefficient must tend to that obtained by the same approach for the planar geometry [16]. Using results from [16], one obtains:

$$\alpha_1 \int_1^\infty I_1^{*[1]}(r)\,dr + \alpha_2 \int_1^\infty I_1^{*[2]}(r)\,dr = \alpha_1 + \alpha_2 \ .$$

This relationship may be considered a general property of the special functions. Having employed this property, one can express Eq. (11-33) in the form:

$$\zeta = \tfrac{1}{2}\alpha_\tau \alpha_T (\alpha_1 + \alpha_2) = (0.2508)\alpha_\tau \alpha_T \ . \qquad (11\text{-}34)$$

The results obtained here will be discussed later in detail together with other results that will be derived by the use of the various approximate methods.

7. METHOD OF THE 'SMOOTHED' DISTRIBUTION FUNCTION.

The absence of analytical expressions for the bracket integrals containing discontinuous Lees' distribution functions prevents one from obtaining solutions for boundary value transport problems, with the exception of cases involving the planar geometry and Maxwellian molecules. Some approximate results can be obtained by utilizing a new procedure proposed in [19] that allows one to overcome these difficulties. A continuous distribution function of the Chapman-Enskog type is assumed to be employed in the collision integral instead of the actual discontinuous function. This auxiliary function may be called a 'smoothed' distribution function. The unknown parameters of the 'smoothed' distribution function may be determined by equating its basic moments with those obtained from the actual distribution function.

Employing this method to solve the problem investigated in Sec. 11.6, one assumes that the 'smoothed' distribution function used in the collision operator has the following form:

$$f^* = f^{(0)}\left\{1 + \nu(r) + \left(c^2 - \tfrac{3}{2}\right)\tau(r) + 2c_r u_r^*(r) + A(r)\, c_r S_{3/2}^{(1)}\left(c^2\right)\right\} , \qquad (11\text{-}35)$$

where $\nu(r)$ and $\tau(r)$ are corrections to the number density and the temperature. The quantity, $A(r)$, is determined by:

$$\int c_r c^2 f d\mathbf{c} = \int c_r c^2 f^* d\mathbf{c} \ , \tag{11-36}$$

where, f , is given by Eq. (11-15). All unknown parameters in Eq. (11-35) may be expressed through the basic functions $\nu^{\pm}(r)$ and $\tau^{\pm}(r)$. This 'smoothed' function allows one to obtain analytical results for arbitrary molecular potential interactions. The standard bracket integrals, which are tabulated for many molecular potential models, may be utilized for these analyses. For example, the expression, $n\Delta c^2 c_r$, from the moment system, Eqs. (11-29d), may be written as:

$$\overline{n\Delta c^2 c_r} = n_0^2 A(r)\left[c_r S_{3/2}^{(1)}\left(c^2\right), c_r S_{3/2}^{(1)}\left(c^2\right)\right] .$$

Using this relation in the right-hand-side of Eq. (11-29d), one can obtain (for rigid-sphere molecules) the following final formula for the heat flux:

$$\frac{Q}{Q_{fm}} = \frac{1}{1+\zeta Kn^{-1}} \ , \tag{11-37}$$

where $\zeta = \alpha_\tau \alpha_T \frac{64}{75}\pi^{-1}$.

As one can see, this formula gives the exact expression for the heat flux in a continuum regime when $Kn \to 0$. This method allows one to simplify the calculation of the bracket integrals. Moreover, there are no difficulties in generalizing this analysis for arbitrary models of the intermolecular potential.

8. THE POLYNOMIAL EXPANSION METHOD.

The above mentioned difficulties of calculating the bracket integrals in the analytical form can be overcome by means of the use of a special representation for the function, $\mathrm{sign}\left(c_r - c_r^*\right)$. In this section a method will be presented to calculate the bracket integrals in the analytical form that consists of an expansion of the discontinuous function, $\mathrm{sign}\left(c_r - c_r^*\right)$, in velocity space polynomials. First, consider the problem in a spherical geometry. In Cartesian coordinates, this function can be written as:

$$\mathrm{sign}\left(c_r - c\cos\left(\chi_0\right)\right) = \mathrm{sign}\left(c_r - \sqrt{c_r^2 + c_\theta^2 + c_\phi^2}\left(1 - r^{-2}\right)^{1/2}\right) . \tag{11-38}$$

It is easy to see that this function depends upon the three velocity coordinates and on the radial coordinate of the position of a molecule.

Owing to this notation, one must utilize a three-dimensional expansion of the velocity components in Hermite polynomials. This expansion was first proposed by Ivchenko [16]. But it is necessary to note that the expansion given in [16] is very difficult to be realized for practical calculations as there are many difficulties connected with the use of general schemes for the three-dimensional expansions.

Since the function given by Eq. (11-38) depends only on the sum of the squares of the independent variables, c_θ and c_ϕ, one could hope to find a more simple scheme in which the sum, $c_\theta^2 + c_\phi^2 = \varsigma^2$, is considered as a new independent variable. As $0 \le \varsigma < \infty$, Hermite polynomials cannot be employed for this expansion. Therefore, Sonine polynomials are used instead. A polar coordinate system is introduced for the above velocity components by means of the relations $c_\theta = \varsigma\cos(\beta)$, $c_\phi = \varsigma\sin(\beta)$, and $dc_\theta dc_\phi = \varsigma d\varsigma d\beta$. Integration with respect to these variables is determined by:

$$\int_{-\infty}^{\infty}\int_{-\infty}^{\infty} \exp\left(-\left(c_\theta^2 + c_\phi^2\right)\right) F\left(\varsigma^2\right) dc_\theta dc_\phi = 2\pi \int_0^\infty \exp\left(-\varsigma^2\right) F\left(\varsigma^2\right) \varsigma d\varsigma$$

$$= \pi \int_0^\infty F(x) \exp(-x) dx ,$$

and therefore the integrand contains the weight function, $x^m = 1$. This means that $m = 0$ and the Sonine polynomials, $S_0^{(p)}\left(\varsigma^2\right)$, must used in the expansion.

The orthonormal conditions for this kind of the Sonine polynomials can be expressed as:

$$\int_0^\infty \exp(-x) S_0^{(p)}(x) S_0^{(q)}(x) dx = \delta_{pq} . \tag{11-39}$$

This polynomial can be defined by the expression:

$$S_0^{(n)}(x) = \sum_{p=0}^{n} \frac{(-x)^p n!}{(p!)^2 (n-p)!} , \tag{11-40}$$

where:

$$S_0^{(0)}(x) = 1 , \quad S_0^{(1)}(x) = 1 - x , \quad S_0^{(2)}(x) = 1 - 2x + \tfrac{1}{2}x^2 , \text{ and:}$$

$$S_0^{(3)}(x)=1-3x+\tfrac{3}{2}x^2-\tfrac{1}{6}x^3 \ .$$

Taking into account this analysis, one can use the following two-dimensional expansion:

$$\begin{aligned}&\operatorname{sign}\left(c_r-\sqrt{c_r^2+c_\theta^2+c_\phi^2}\left(1-r^{-2}\right)^{1/2}\right)\\&=\sum_k\sum_n A_{kn}(r)H_k(c_r)S_0^{(n)}\left(c_\theta^2+c_\phi^2\right) .\end{aligned} \tag{11-41}$$

Since this expansion contains only the odd Hermite polynomials, the coefficients, $A_{kn}(r)$, are given by:

$$\begin{aligned}A_{kn}(r)=&\frac{4}{2^k k!\sqrt{\pi}}\int_0^\infty c^2\exp\left(-c^2\right)dc\\&\times\int_{t_0}^1 H_k(c\,t)S_0^{(n)}\left(c^2\left[1-t^2\right]\right)dt \ ,\end{aligned} \tag{11-42}$$

where $t_0=\left(1-r^{-2}\right)^{1/2}$.

Now, consider the features of this expansion for the cylindrical geometry. As one can see from Eq. (11-6), for this case, the quantity, c_r^*, depends only on the two molecular velocity components, c_r and c_θ, and, therefore, the appropriate expansion is defined by:

$$\operatorname{sign}\left(c_r-\sqrt{c_r^2+c_\theta^2}\left(1-r^{-2}\right)^{1/2}\right)=\sum_i\sum_j A_{ij}(r)H_i(c_r)H_j(c_\theta) \ , \tag{11-43}$$

where the coefficients, $A_{ij}(r)$, in this expansion can be expressed as:

$$\begin{aligned}A_{ij}(r)=&\frac{1}{2^{i+j}i!j!\pi}\int_0^\infty c_\varsigma\exp\left(-c_\varsigma^2\right)dc_\varsigma\\&\times\left\{-\int_0^{2\pi}H_i(c_r)H_j(c_\theta)d\theta+2\int_\alpha^{\pi-\alpha}H_i(c_r)H_j(c_\theta)d\theta\right\} ,\end{aligned}$$

where $c_\theta=c_\varsigma\cos(\theta)$ and $c_r=c_\varsigma\sin(\theta)$.

In the planar geometry, one has a simple one-dimensional expansion given by:

$$\operatorname{sign}(c_x)=\sum_{i=0}^{\infty}A_iH_i(c_x)\ , \tag{11-44}$$

where the constant coefficients can be determined by:

$$A_i=\frac{1}{2^i\,i!\sqrt{\pi}}\int_{-\infty}^{\infty}\exp\left(-c_x^2\right)H_i(c_x)\operatorname{sign}(c_x)dc_x\ .$$

This expansion was first proposed by Savkov in [20].

Now, consider some of the features of this method associated with the spherical geometry. If the expansion given by Eq. (11-41) is used in the bracket integrals presented by Eqs. (11-21) and (11-22), one can obtain the following analytical expressions for the special functions, $I_1^{*[i]}(r)$:

$$I_1^{*[1]}(r)=-\tfrac{3}{2}\omega^{-1}\sum_k\sum_n A_{kn}(r)\alpha_{kn}^{[1]}\ , \tag{11-45}$$

$$I_1^{*[2]}(r)=\frac{12}{8+3\pi}\omega^{-1}\sum_k\sum_n A_{kn}(r)\alpha_{kn}^{[2]}\ . \tag{11-46}$$

Here, the constants $\alpha_{kn}^{[1]}$ and $\alpha_{kn}^{[2]}$ are the common bracket integrals:

$$\alpha_{kn}^{[1]}=\left[H_k(c_r)S_0^{(n)}\left(c_\theta^2+c_\phi^2\right),c_rc^2\right]\ , \tag{11-47a}$$

$$\alpha_{kn}^{[2]}=\left[H_k(c_r)c^2S_0^{(n)}\left(c_\theta^2+c_\phi^2\right),c_rc^2\right]\ . \tag{11-47b}$$

Instead of the infinite sums in Eqs. (11-45) and (11-46) one assumes that finite sums can be employed. It is not difficult to determine that the convergence of these series is very rapid. Indeed, if three terms $(n=0\,;k=1,3,5)$ are used, one will find that $I_1^{*[1]}(1)=(1.0102)$ and $I_1^{*[2]}(1)=(1.1198)$. The relative errors of such an approach are about 1% and 12% $(r=1)$ for the first and second functions respectively. When $n=0\ ;\ 1\le k\le 7$, the relative errors are less than 1% $(r=1)$ for both functions. All of the bracket integrals necessary for this analysis can be found in Table 11-1.

For arbitrary r, the other terms for which $n\neq 0$ must be considered. To obtain some appropriate expressions for the special functions, $I_1^{*[i]}(r)$, the

Table 11-1. Expressions for some bracket integrals.

	$\left[c_r^i, c_r c^2\right] = a\sqrt{2\pi}\omega$ †			$\left[c_r^i c^2, c_r c^2\right] = a\sqrt{2\pi}\omega$			
i	3	5	7	1	3	5	7
a	$\frac{4}{5}$	$\frac{30}{7}$	$\frac{287}{12}$	$\frac{4}{3}$	6	$\frac{123}{4}$	$\frac{741}{4}$

† $\omega = \sigma^2 (2kT/m)^{1/2}$

following finite sums, for which $(n=0; k=1,3,5,7)$ and $(k=1; n=1,2,3)$, are employed. Moreover, to obtain the more accurate expressions for $I_1^{*[2]}(r)$, an additional term $(k=3; n=1)$ has been included in the expansion given by Eq. (11-46). For both expansions, the exact values of the bracket integrals are given in Appendix C and the expression $\alpha_{31}^{[2]} = -\frac{92}{5}\sqrt{2\pi}\omega$ has been employed. Having substituted the bracket integrals into Eqs. (11-45) and (11-46), one can obtain:

$$I_1^{*[1]}(r) = \sqrt{2}\left[-\tfrac{53}{105} r^{-6} + \tfrac{53}{84} r^{-4} + \tfrac{163}{280} r^{-2}\right], \tag{11-48}$$

$$I_1^{*[2]}(r) = \frac{12\sqrt{2}}{8+3\pi}\left[-\tfrac{19}{21} r^{-6} + \tfrac{95}{84} r^{-4} + \tfrac{17}{21} r^{-2}\right]. \tag{11-49}$$

These analytical expressions can be readily integrated, after which, one obtains:

$$\int_1^\infty I_1^{*[1]}(r)\,dr = (0.9779) \quad , \quad \int_1^\infty I_1^{*[2]}(r)\,dr = (0.9793) . \tag{11-50}$$

It is necessary to note, however, that the accuracy of the approach proposed here may only be specified by a comparison of these analytical results with those obtained numerically. The deviations of the integrals given in Eq. (11-50) from the corresponding numerical values are approximately 2% for both of these integrals. This very close agreement with the numerical results indicates a rapid convergence of the expansions being used. Therefore, one can see that the approach proposed here is a reasonable approximate method for solving boundary value transport problems.

For the problem of thermal conduction from a heated sphere, the method of polynomial expansion yields:

$$\frac{Q}{Q_{fm}} = \frac{1}{1+\zeta Kn^{-1}} , \tag{11-51a}$$

Table 11-2. Values of the reduced heat flux ratio, Q/Q_{fm} $(\alpha_\tau = \alpha_T = 1)$.

	Q/Q_{fm}		
Kn^{-1}	Exact Moment Solution $(\zeta = 0.2508)$	Smoothed Distribution Function Method $(\zeta = 0.2716)$	Method of Polynomial Expansion $(\zeta = 0.2455)$
0.00	1.0000	1.0000	1.0000
0.25	0.9410	0.9364	0.9422
0.50	0.8886	0.8804	0.8907
0.75	0.8417	0.8308	0.8445
1.00	0.7995	0.7864	0.8029
2.00	0.6660	0.6480	0.6707
3.00	0.5706	0.5510	0.5759
4.00	0.4992	0.4793	0.5045
5.00	0.4437	0.4241	0.4489
6.00	0.3992	0.3803	0.4044
8.00	0.3326	0.3152	0.3374
10.0	0.2851	0.2691	0.2894
12.0	0.2494	0.2348	0.2534
14.0	0.2217	0.2082	0.2254
16.0	0.1995	0.1871	0.2029
18.0	0.1813	0.1698	0.1845
20.0	0.1662	0.1555	0.1692

where, for rigid-sphere molecules:

$$\zeta = (0.2455)\alpha_\tau \alpha_T \ . \qquad (11\text{-}51b)$$

The reduced heat flux given by Eqs. (11-51a) and (11-51b) deviates only slightly from the solution given by Eq. (11-34). Table 11-2 gives values of the reduced heat flux ratio for $(\alpha_\tau = \alpha_T = 1)$ and Fig. 11-2 illustrates these values for the various methods employed here.

9. SOLUTION OF ONE CLASSIC TRANSPORT PROBLEM.

In this section the accuracy of the various methods previously proposed will be analyzed. A comparison is made between the various approximate analyses and the exact moment solution for a classical problem of heat transport between two parallel plates with the temperature difference being $2\Delta T$.

Let the temperature of the lower plate $(\tilde{x} = -L/2)$ be $T = T_0 - \Delta T$ and that of the upper one $(\tilde{x} = L/2)$ be $T = T_0 + \Delta T$, where T_0 is the temperature

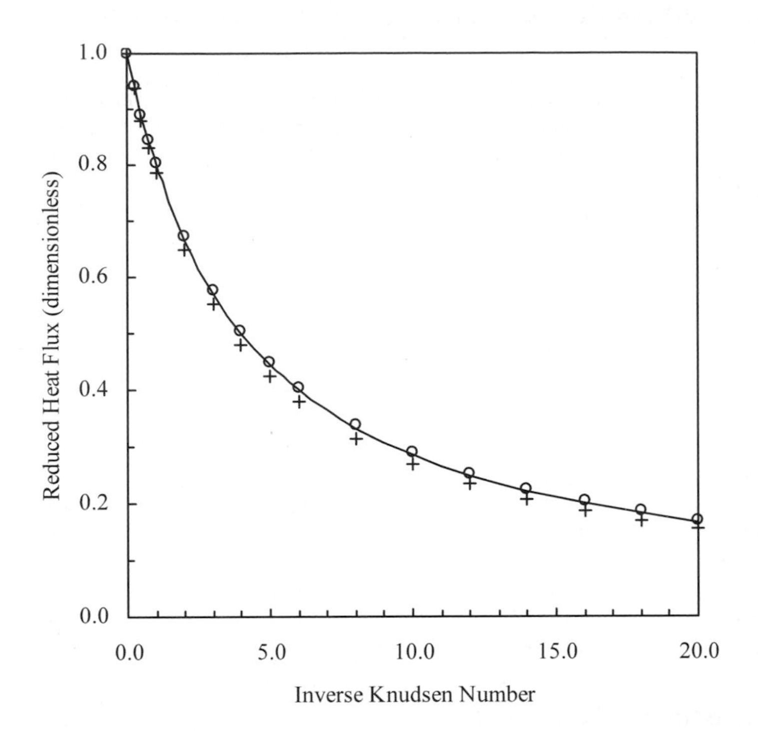

Figure 11-2. A comparison of reduced heat flux values, Q/Q_{fm}, obtained from various approximate methods assuming $(\alpha_\tau = \alpha_T = 1)$ with the Exact Moment Solution given in Table 11-2. The Exact Moment Solution is represented as a solid line while values obtained using the Smoothed Distribution Function method are shown as crosses and values obtained using the Polynomial Expansion Method are shown as open circles.

at $\tilde{x} = 0$. The temperature difference is assumed to be small compared with T_0, and therefore, the linearized theory is used in this problem.

This problem has been previously formulated and studied by Wang-Chang and Uhlenbeck [21], Jackson [22], Gross and Ziering [23], Bassanini, Cercignani and Pagani [24], Loyalka [25], and Ivchenko [16]. Moreover, the heat conduction characteristics and density distribution were experimentally determined by Teagan and Springer [26].

The solution of this planar problem may be based on the Maxwellian integral transport equations which can be directly derived from the Boltzmann equation by means of the relation:

$$\frac{\partial}{\partial \tilde{x}} \int v_x Q(\mathbf{v}) f d\mathbf{v} = \int Q(\mathbf{v}) \frac{\delta f}{\delta t} d\mathbf{v} , \tag{11-52}$$

where $\tilde{x}$ is the normal coordinate to the plate. The corresponding moment equations can be obtained by setting $Q(\mathbf{v})$ equal to 1, v_x, v^2, and $v_x v^2$ in Eq. (11-52). These equations can be written as:

$$\frac{d}{d\tilde{x}}(nu) = 0 \; , \tag{11-53a}$$

$$\frac{d}{d\tilde{x}} P_{xx} = 0 \; , \tag{11-53b}$$

$$\frac{d}{d\tilde{x}} Q_x = 0 \; , \tag{11-53c}$$

$$\frac{d}{d\tilde{x}} \int v_x^2 v^2 f d\mathbf{v} = n\Delta \overline{v_x v^2} \; . \tag{11-53d}$$

In order to close this system of moment equations, the distribution function is chosen to be of the form:

$$f = f^{(0)} \Big\{ 1 + \tfrac{1}{2} \Big[\nu^+(x) + \left(c^2 - \tfrac{3}{2}\right) \tau^+(x) \Big] + \tfrac{1}{2} \Big[\nu^-(x) + \left(c^2 - \tfrac{3}{2}\right) \tau^-(x) \Big] \operatorname{sign}(c_x) \Big\} \; . \tag{11-54}$$

All the moments of the distribution function in Eqs. (11-53a)-(11-53d) may be calculated from the expressions given in Sec. 11.6 by setting $r = 1$. The solution of the moment system then contains four arbitrary constants of integration which can be specified by boundary conditions. For this problem the Maxwellian boundary model is employed in which the tangential accommodation coefficient is assumed to be equal to unity.

Utilizing the values of the bracket integrals which are given in Table 11-1, one obtains the exact analytical solution via a four-moment approach. The heat flux may then be written as [16]:

$$Q = Q_{fm} \left[1 + \frac{\alpha_T}{2 - \alpha_T} \Psi Kn^{-1} \right]^{-1} , \quad \Psi = (0.2508) \; , \tag{11-55}$$

$$Q_{fm} = -\frac{2\alpha_T}{2-\alpha_T} p_0 \left(\frac{2kT_0}{\pi m}\right)^{1/2} \frac{\Delta T}{T_0} ,$$

where Q_{fm} is the free-molecular heat flux and $Kn = \lambda/L$. Molecules are assumed to be rigid spheres. Having obtained the exact moment solution, one can compare this solution with the results of other approaches. Through such comparisons the accuracy of different approaches can be characterized.

If the approximate distribution function given by Eq. (11-35) is used, the 'smoothed' distribution function method gives:

$$Q = Q_{fm}\left[1+\frac{\alpha_T}{2-\alpha_T}\Psi_S Kn^{-1}\right]^{-1} , \quad \Psi_S = (0.2716) . \tag{11-56}$$

The accuracy of the polynomial expansion method depends on the number of terms used in this series. The method proposed here makes it possible to calculate the bracket integrals for arbitrary models of intermolecular potentials to an arbitrary degree of accuracy. Consider the results of this method for the following two finite polynomial sums:

$$\mathrm{sign}(c_x) = \sum_{i=0}^{5} A_i H_i(c_x) , \tag{11-57}$$

$$\mathrm{sign}(c_x) = \sum_{i=0}^{7} A_i H_i(c_x) . \tag{11-58}$$

The first expansion gives $\Psi^{(1)} = (0.2677)$. For the second, this quantity is found to be $\Psi^{(2)} = (0.2523)$. The necessary bracket integrals for this calculation are given in Table 11-1. As one can see, the second approach yields results quite similar to the exact moment solution. Indeed, in the first approach, the maximum difference between the exact results and this approximation is about 7%, while for the second approach, the discrepancy is less than 1%. This indicates that the expansion that has been employed has a rapid convergence to the exact solution. For this reason, there is no significant difference between the exact solution and the second approximation.

It is of interest to compare the results of the analysis described here with those obtained by the half-range moment method. The analysis by Gross and Ziering [23] gives the following expression for the heat flux as obtained via the four-moment approach:

Table 11-3. Values of the reduced heat flux ratio, Q/Q_{fm} $(\alpha_T = 0.826)$.

	Q/Q_{fm}		
Kn^{-1}	Exact Moment Solution $(\Psi = 0.2508)$	Smoothed Distribution Function Method $(\Psi_S = 0.2716)$	Four-Moment Gross-Ziering Method $(\Psi_G = 0.5509)$
0.25	0.9577	0.9544	0.9117
0.50	0.9189	0.9128	0.8377
0.75	0.8831	0.8746	0.7748
1.00	0.8500	0.8396	0.7207
2.00	0.7391	0.7235	0.5633
3.00	0.6538	0.6356	0.4624
4.00	0.5862	0.5668	0.3921
5.00	0.5313	0.5114	0.3404
6.00	0.4857	0.4659	0.3007
8.00	0.4146	0.3954	0.2439
10.0	0.3617	0.3435	0.2051
12.0	0.3208	0.3037	0.1770
14.0	0.2881	0.2721	0.1556
16.0	0.2616	0.2464	0.1389
18.0	0.2394	0.2252	0.1254
20.0	0.2208	0.2074	0.1143

$$Q = Q_{fm}\left[1 + \frac{\alpha_T}{2-\alpha_T}\Psi_G Kn^{-1}\right]^{-1}, \quad \Psi_G = (0.5509) . \tag{11-59}$$

Table 11-3 gives the results obtained by the various analytical methods. The tabulated results show that the ‘exact’ and ‘smoothed’ solutions are very close for all Knudsen numbers. The maximum relative difference with respect to the ‘exact’ solution is about 8% for $Kn \to 0$. But, it is necessary to note, the ‘smoothed’ solution describes a continuum regime much better and gives a correct expression for the heat conductivity coefficient. Analysis of the four-moment Gross-Ziering solution indicates that the thermal conductivity coefficient is approximately one half that predicted by the Chapman-Enskog theory. A failure to include the heat flux moment in the Gross-Ziering technique results in this difference. Thus, it is advisable that this moment be employed in higher-order approaches in order to obtain a better description of the continuum regime.

In Fig. 11-3 a comparison between experimental data [26] and the various analytical results is shown. Both the ‘exact’ and ‘smoothed’ solutions describe the experimental data very well.

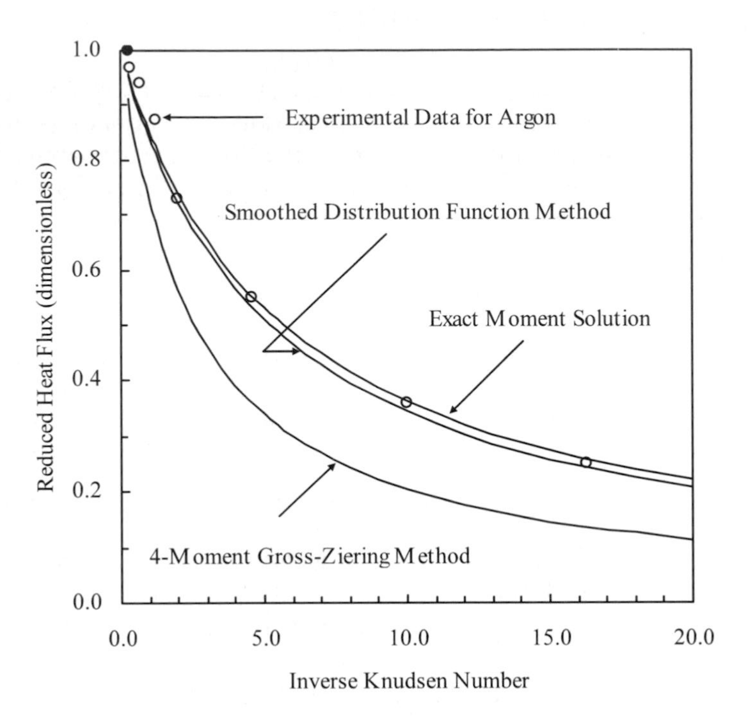

Figure 11-3. A comparison between reduced heat flux data for Argon $(\alpha_T = 0.826)$ and analytical results given in Table 11-3.

10. A SIMPLIFICATION OF MOMENT SYSTEMS FOR CURVILINEAR PROBLEMS.

The Maxwell method (see Section 8.1) using the tangential momentum conservation law yields a very simple scheme to obtain approximate solutions to planar slip-flow problems. The generalization of the Maxwell method proposed by Loyalka (see Section 8.2) is founded on the integral properties of the Chapman-Enskog solutions for viscosity and thermal conductivity which may be expressed by the relations:

$$\left[\Phi(\mathbf{c}), B_{ik}\right] = n^{-2}\int f^{(0)}\Phi(\mathbf{c})\left(c_i c_k - \tfrac{1}{3}\delta_{ik}c^2\right)d\mathbf{v} \ , \tag{11-60}$$

$$\left[\Phi(\mathbf{c}), \mathbf{A}\right] = n^{-2}\int f^{(0)}\Phi(\mathbf{c})\mathbf{v}\left(c^2 - \tfrac{5}{2}\right)d\mathbf{v} \ , \tag{11-61}$$

where $\Phi(\mathbf{c})$ is, in general, a discontinuous function of the molecular velocity.

Loyalka's method is a moment method in which two conservation laws are used to solve planar slip-flow problems. The conservation laws used are exact solutions of the moment system in the two-moment approach. To construct this moment system, one uses the Chapman-Enskog solution as a molecular property of the appropriate transport equation. For planar transport problems, using the conservation laws as exact solutions of the two-moment system results in acceptable accuracy for the slip factors.

For curvilinear geometries, there are no simple general integrals of the moment system such as are found in planar problems. Nevertheless, many difficulties associated with calculation of the bracket integrals may be overcome by using the Chapman-Enskog solutions, $\mathbf{A}$ and B_{ik}, as the molecular property, $Q(\mathbf{v})$, to construct one of the moment equations from Eq. (11-1). Then, the moment equations for the linearized problems contain the bracket integrals expressed by Eqs. (11-60) and (11-61). Therefore, there are no difficulties connected with the calculation procedures. Such bracket integrals may be calculated as ordinary moments of the distribution function. This is a result of considerable importance to the simplification of the analysis. Moreover, it is very important to note that these bracket integrals are independent of the model used for the interaction potential. The only dependence on the intermolecular potential is contained in the left-hand-side of the moment equation that is obtained by use of the Chapman-Enskog solution as a molecular property. Therefore, for any arbitrary intermolecular potential model, standard analytical and numerical data for the Chapman-Enskog coefficients may be used.

Application of this technique to the classical linearized transport problem described in Section 11.6 is considered in this section. The molecular properties, $Q(\mathbf{v}) = 1, v_r, v^2, A_r$, are chosen to construct the moment system. The moment relations given by Eqs. (11-29a)-(11-29c) are not altered, but the fourth relation associated with the molecular property, A_r, becomes:

$$\frac{d\tau^+}{dr} = -\tfrac{8}{5}\frac{R}{a_1}\left\{\tfrac{1}{4}\pi^{-1/2}r^{-2}\left(\nu^- - \tfrac{3}{2}\tau^-\right) - \tfrac{1}{2}\pi^{-1/2}r^{-2}\tau^-\right\}, \qquad (11\text{-}62)$$

where a_1 is the first coefficient of the Chapman-Enskog expansion of A_r. After integrating this equation, one can obtain the following expression for the basic function, $\xi(\mathrm{r})$, which was introduced in Section 11.6:

$$\xi(\mathrm{r}) = \tfrac{8}{5}\frac{R}{a_1}\pi^{-1/2}\left(1 - r^{-1}\right). \qquad (11\text{-}63)$$

It is very important to note that this function contains the first coefficient, a_1, only and, moreover, holds the same form for arbitrary models of the intermolecular potential. Using the analytical solution of the moment system given in Section 11.6 together with Eq. (11-63), one can obtain the reduced thermal flux in the form given by Eqs. (11-32a) and (11-32b). The quantity, ζ, in Eq. (11-32a) may be expressed in the form:

$$\zeta = -\alpha_\tau \alpha_T \tfrac{4}{5} \frac{\lambda}{\sqrt{\pi} a_1} \ . \qquad (11\text{-}64)$$

For rigid-sphere molecules and the first-order Chapman-Enskog solution, $a_1 = a_1^{(1)}$, one can obtain:

$$\zeta = \alpha_\tau \alpha_T \tfrac{64}{75} \pi^{-1} \ . \qquad (11\text{-}65)$$

The method described here makes it possible to overcome the calculational difficulties connected with the bracket integrals and results in analytical solutions of problems involving arbitrary models of the intermolecular potential provided that the summational invariants and the Chapman-Enskog solution are used to construct the moment systems. If one uses a higher-moment approach than that described here, additional bracket integrals must be calculated. This may be done by employing the numerical and analytical techniques described in this chapter.

It is of interest to compare the various analytical results for the reduced heat flux obtained here and previously using the four-moment approach. The maximum relative deviation of the reduced heat flux obtained here (with respect to the values given by Eq. (11-34)) takes place as $Kn \to 0$ and is about 7%. The solution presented here, however, describes the continuum regime much better than that obtained by numerically evaluating the bracket integrals in the four-moment approach because it yields the correct value for the heat conduction coefficient which could not be obtained from the previous solution.

11. THE TORQUE PROBLEM.

Kinetic theory formulations of the torque problem have been given by Halbritter [27] and Smirnov and Chekalov [28]. In both of these cases it was found to be convenient to replace the problem of a rotating body in a gas by a problem involving a stationary body around which the gas is slowly rotating. Generally, the problems are not equivalent as, in rotating reference frames, fictitious inertial forces must be considered. This aspect has been discussed in the literature [29-31]. Such forces are, however, of the second-order in the rotational speed and, for small speeds of rotation, may reasonably be neglected making the two problems equivalent.

A numerical solution of the linearized problem, for rigid-sphere molecules and the Boltzmann form of the collision operator, has been obtained by Loyalka [32]. This numerical analysis may be used to evaluate the accuracy of the analytical method discussed below. Consider a sphere of radius, R, that is slowly rotating with an angular speed, ω, such that $\omega R \ll \left(2kT/m\right)^{1/2}$. In contrast to the work cited above, this development deals directly with the situation where the sphere itself is rotating. As the mean velocity of the gas has only one component, $u_\phi = u_\phi\left(r,\theta\right)$, the two-sided Maxwellian distribution function may be chosen in the form:

$$f_1 = f^{(0)}\left(1 + 2c_\phi G_1\left(r,\theta\right)\right) , \tag{11-66}$$

$$f_2 = f^{(0)}\left(1 + 2c_\phi G_2\left(r,\theta\right)\right) . \tag{11-67}$$

Since only two characteristic mean velocities need to be determined, one may try to employ the two-moment approach in this problem. The problem exhibits azimuthal symmetry in which all quantities depend only on the radial and polar coordinates. For this symmetry condition, the general moment relation given by Eq. (11-1) may be written as:

$$\begin{aligned}
&\frac{1}{r^2}\frac{\partial}{\partial r}\left(r^2 n\overline{v_r Q(\mathbf{v})}\right) + \frac{1}{r\sin(\theta)}\frac{\partial}{\partial\theta}\left(\sin(\theta) n\overline{v_\theta Q(\mathbf{v})}\right) \\
&-\frac{1}{r}\int\left(v_\theta^2 + v_\phi^2\right)\frac{\partial Q(\mathbf{v})}{\partial v_r} f d\mathbf{v} - \frac{1}{r}\int\left(v_\phi^2\cot(\theta) - v_r v_\theta\right)\frac{\partial Q(\mathbf{v})}{\partial v_\theta} f d\mathbf{v} \\
&+\frac{1}{r}\int\left(v_r v_\phi + v_\theta v_\phi \cot(\theta)\right)\frac{\partial Q(\mathbf{v})}{\partial v_\phi} f d\mathbf{v} = Rn\Delta\overline{Q(\mathbf{v})} .
\end{aligned} \tag{11-68}$$

where $r = \tilde{r}/R$.

The moment system in this approach may be obtained from Eq. (11-68) if one employs $Q(\mathbf{v}) = c_\phi$ and $Q(\mathbf{v}) = B_{r\phi} = c_r c_\phi B\left(c^2\right)$ where $B_{r\phi}$ is the Chapman-Enskog solution for viscosity. The function, $B\left(c^2\right)$, is well known to be:

$$B\left(c^2\right) = b_1 \sum_{p=1}^{\infty} b_p^* S_{5/2}^{(p-1)}\left(c^2\right) \quad ; \quad b_1 = \frac{\mu}{nkT} . \tag{11-69}$$

This choice yields:

$$\frac{1}{r^2}\frac{\partial}{\partial r}\left(r^2 n\overline{c_r c_\phi}\right)+\frac{1}{r}n\overline{c_r c_\phi}=0 \ , \tag{11-70}$$

$$\begin{gathered}\frac{1}{r^2}\frac{\partial}{\partial r}\left(r^2 n\overline{c_r^2 c_\phi B\left(c^2\right)}\right)-\frac{1}{r}\left(n\overline{c_\phi c^2 B\left(c^2\right)}\right)\\+\frac{2}{r}\left(n\overline{c_r^2 c_\phi B\left(c^2\right)}\right)=-\left(\frac{m}{2kT}\right)^{1/2} n^2 R\left[\Phi \ , B_{r\phi}\right] \ ,\end{gathered} \tag{11-71}$$

where Φ is the correction to the distribution function. The term involving the partial derivative with respect to θ vanishes as $nv_\theta Q(\mathbf{v})=0$. The moment of the collision integral associated with this system may be easily calculated for arbitrary intermolecular potentials if one takes into account the feature of the Chapman-Enskog solution expressed by Eq. (11-60).

As discussed in Section 11.5, the common Maxwellian boundary model for the reflected distribution function may be used since the distribution function given by Eqs. (11-66) and (11-67) is suited to this model. This boundary model yields:

$$\begin{gathered}f_2\left(\mathbf{c},1,\theta\right)=\alpha_\tau f^{(0)}\left[1+\left(\frac{2m}{kT}\right)^{1/2} c_\phi \omega R\sin\left(\theta\right)\right]\\+\left(1-\alpha_\tau\right)f_1\left(-c_r,c_\theta,c_\phi,1,\theta\right) \ ,\end{gathered} \tag{11-72}$$

$$\lim_{r\to\infty}\left(n\overline{c}_\phi\right)=0 \ . \tag{11-73}$$

The boundary condition expressed by Eq. (11-72) suggests that the distribution function is of the form:

$$f_i=f^{(0)}\left(1+2c_\phi G_i\left(r\right)\sin\left(\theta\right)\right) \ . \tag{11-74}$$

which allows one to separate the radial and polar variables. Using Eq. (11-74) and employing the standard technique (see Section 11.3), one can obtain:

$$n\overline{c_r c_\phi}=n\sin\left(\theta\right)\left[\frac{1}{2\sqrt{\pi}r^4}G^-\left(r\right)\right] \ , \tag{11-75}$$

$$\overline{nc_r^2 c_\phi B(c^2)} = n\sin(\theta)\left[\tfrac{1}{4}G^+(r) + \tfrac{1}{8}(3x^5 - 5x^3)G^-(r)\right]b_1 \ , \tag{11-76}$$

$$\overline{nc^2 c_\phi B(c^2)} = n\sin(\theta)\left[\tfrac{5}{4}G^+(r) + \tfrac{5}{8}(x^3 - 3x)G^-(r)\right]b_1 \ , \tag{11-77}$$

where $x = (1 - r^{-2})^{1/2}$ and $G^{\pm}(r) = G_2(r) \pm G_1(r)$.

After substituting these relations into Eqs. (11-70) and (11-71), the moment system takes the form:

$$\frac{d}{dr}G^-(r) - \frac{1}{r}G^-(r) = 0 \ , \tag{11-78}$$

$$\frac{d\Psi}{dr} - \frac{1}{r}\Psi = \left(\frac{4\zeta}{r^3} - 6\sqrt{\pi}\frac{x}{r^2}\right)\frac{1}{2\sqrt{\pi}r}G^-(r) \ , \tag{11-79}$$

where the following notations have been introduced:

$$\zeta = -\frac{R}{b_1}\left(\frac{m}{2kT}\right)^{1/2} \ , \tag{11-80}$$

$$\Psi = G^+(r) - x^3 G^-(r) \ . \tag{11-81}$$

A general analytical solution of this system is found to be:

$$G^-(r) = 2\sqrt{\pi}rA_1 \ , \tag{11-82}$$

$$G^+(r) = r\left[A_2 + A_1\tfrac{4}{3}\zeta(1 - r^{-3})\right] , \tag{11-83}$$

where A_1 and A_2 are arbitrary constants of integration. The boundary conditions expressed in Eqs. (11-72) and (11-73) yield the following for these constants:

$$A_1 = \tfrac{1}{4}\frac{\alpha_\tau \omega R}{\sqrt{\pi}}\left(\frac{2m}{kT}\right)^{1/2}\left(1 - \tfrac{1}{3}\frac{\alpha_\tau \zeta}{\sqrt{\pi}}\right)^{-1} , \tag{11-84}$$

$$A_2 = -A_1\left(\tfrac{4}{3}\zeta - 2\sqrt{\pi}\right) . \tag{11-85}$$

The torque on the surface element of a sphere, $dS = R^2 \sin(\theta) d\theta d\beta$, is given by $dK_i = \varepsilon_{ikl} R r_k \, dF_l \big|_{r=1}$ where:

$$dF_l = -dS \sum_{\pm} \int_{\pm} m v_r v_l f^{\pm}(\mathbf{v},1,\theta) d\mathbf{v} \ ,$$

ε_{ikl} is the anti-symmetric unit tensor defined by:

$$\varepsilon_{ikl} = \begin{cases} 1 & ; \quad ikl = 123,\ 231,\ 312 \, , \\ -1 & ; \quad ikl = 321,\ 132,\ 213 \, , \\ 0 & ; \quad i = k,\ i = l,\ k = l \, . \end{cases}$$

For the particular conditions formulated above, one can obtain:

$$dK_2 = \varepsilon_{213} R dF_\phi = R dS \sum_{\pm} \int_{\pm} m v_r v_\phi f^{\pm}(\mathbf{v},1,\theta) d\mathbf{v} \ ,$$

and:

$$dK_x = -dK_2 \sin(\theta) \ ,$$

where x indicates the polar axis in the direction of $\boldsymbol{\omega}$. The net torque on the sphere is given by:

$$\begin{aligned} K_x &= -2\pi R^3 \int_0^{\pi} \sin^2(\theta) \left[\sum_{\pm} \int_{\pm} m v_r v_\phi f^{\pm}(\mathbf{v},1,\theta) d\mathbf{v} \right] d\theta \\ &= -\tfrac{8}{3} \sqrt{\pi} R^3 n k T G^{-}(1) \ . \end{aligned} \tag{11-86}$$

Substituting A_1 given by Eq. (11-84) into Eq. (11-86), one can obtain:

$$K_x = K_x^* \left(1 + \alpha_\tau K n^{-1} \tfrac{8}{15} \pi^{-1} \right)^{-1} \ , \tag{11-87}$$

where:

$$K_x^* = -\tfrac{2}{3} \alpha_\tau \pi \rho \bar{V} R^4 \omega \ , \tag{11-88}$$

and $\bar{V} = (8kT/\pi m)^{1/2}$ is the mean speed of the gas molecules which have been assumed to be rigid spheres. Eq. (11-87) gives the exact expressions of the torque for both the free-molecular and continuum regimes.

Table 11-4. Analytical and numerical values of reduced torque on a rotating sphere.

	K_x/K_x^*			
	α_τ=1.0		α_τ=0.75	
Kn^{-1}	Numerical [32]	Eq. (11-87)	Numerical [32]	Eq. (11-87)
0.00	1.0	1.0	1.0	1.0
0.10	0.9901	0.9815	0.9928	0.9861
0.25	0.9803	0.9551	0.9856	0.9659
0.50	0.9601	0.9140	0.9697	0.9341
0.75	0.9427	0.8764	0.9549	0.9043
1.00	0.9206	0.8417	0.9391	0.8764
5.00	0.6080	0.5154	0.6739	0.5864
	α_τ=0.5		α_τ=0.25	
	Numerical [32]	Eq. (11-87)	Numerical [32]	Eq. (11-87)
0.00	1.0	1.0	1.0	1.0
0.10	0.9952	0.9907	0.9976	0.9953
0.25	0.9902	0.9770	0.9948	0.9884
0.50	0.9802	0.9551	0.9896	0.9770
0.75	0.9694	0.9341	0.9844	0.9659
1.00	0.9584	0.9141	0.9788	0.9551
5.00	0.7558	0.6802	0.8600	0.8097

The analytical and numerical values [32] of the torque are given in Table 11-4. Note that, for the purpose of comparison, the analytical values were recalculated using the notations of [32] where the mean free path is given as:

$$\lambda_p = \tfrac{5}{8}\sqrt{\pi}\lambda \ .$$

Comparison of the analytical and numerical results shows that the analytical solution obtained here provides reasonable accuracy in the near free-molecular regime $\left(0 < Kn^{-1} \le 0.25\right)$. In the transition regime $\left(Kn^{-1} \sim 1\right)$ the data differ by up to 8.6 % when $\alpha_\tau = 1$ and by 2.4 % when $\alpha_\tau = 0.25$. For the slip-flow regime, the results differ by 15.2 % and 5.8 %, respectively. This increasing disagreement as one approaches the slip-flow regime is principally due to the lack of accuracy inherent in use of the relatively low-order two-moment approach.

PROBLEMS

11.1. Determine the heat flux in a gas between two concentric spheres the radii and temperatures of which are R_1, T_0 and $R_2, T_0 + \Delta T$ for the inner and outer spheres, respectively. Assume $\Delta T \ll T_0$. The number density of the gas is n_0 at $\tilde{r} = R_1$. Obtain a general solution for all Knudsen number using the four-moment approach.

Solution: The general solution of the moment system given in Eqs. (11-31a)-(11-31d) may be employed. The parameters n_0, T_0 and R_1 are assumed to be the characteristic number density, temperature and length, respectively. This yields $n(1)=n_0$. Two other boundary conditions may be formulated by means of the energy accommodation coefficient defined by:

$$Q_r = \alpha_T\left(Q_{r\,eq}^{\pm} + Q_r^{\mp}\right) \text{ for } r=1 \text{ and } r=\frac{R_2}{R_1} .$$

The equilibrium number densities associated with $Q_{r\,eq}^{\pm}$ may be specified from the impenetrability condition. The procedure outlined above yields:

$$Q = -4\sqrt{\pi}R_1^2 p_0\left(\frac{2kT_0}{m}\right)^{1/2}\frac{\Delta T}{T_0}\frac{\alpha_\tau\alpha_T}{1+z^{-2}\left(1-\alpha_\tau\alpha_T\right)+Kn^{-1}\zeta(z)} ,$$

where:

$$\zeta(z)=\tfrac{1}{2}\alpha_\tau\alpha_T\left\{\alpha_1\int_1^z I_1^{*[1]}(r)\,dr+\alpha_2\int_1^z I_1^{*[2]}(r)\,dr\right\} \text{ and } z=\frac{R_2}{R_1} .$$

11.2. Obtain a solution to Problem 11.1 using the 'smoothed' distribution function method. Molecules are assumed to be rigid spheres.

Solution: The smoothed distribution function method for rigid-sphere molecules yields:

$$\xi(r) = -\tfrac{128}{75}\pi^{-1}Kn^{-1}\left(1-r^{-1}\right) .$$

Using the same boundary conditions as in Problem 11.1, one can obtain:

$$Q = -4\sqrt{\pi}R_1^2 p_0\left(\frac{2kT_0}{m}\right)^{1/2}\frac{\Delta T}{T_0}\frac{\alpha_\tau\alpha_T}{1+z^{-2}\left(1-\alpha_\tau\alpha_T\right)+Kn^{-1}\zeta(z)} ,$$

where:

$$\zeta(z)=\tfrac{64}{75}\pi^{-1}\alpha_\tau\alpha_T\frac{z-1}{z} \text{ and } z=\frac{R_2}{R_1} .$$

11.3. Determine the heat flux in a gas between two coaxial cylinders the radii and temperatures of which are R_1, T_0 and $R_2, T_0+\Delta T$ for the inner and outer cylinders, respectively. Assume $\Delta T \ll T_0$. The number density of the gas is n_0 at $\tilde{r}=R_1$. Obtain a general analytical solution for the four-moment approach by means of the use of the special functions, $I_2^{*[i]}(r)$.

Solution: The moment system for this cylindrical boundary value problem may be written as:

$$\frac{d}{dr}(rnu_r)=0 \ ,$$

$$\frac{d}{dr}P_{rr}+\frac{1}{r}(P_{rr}-P_{\theta\theta})=0 \ ,$$

$$\frac{d}{dr}(rQ_r)=0 \ ,$$

$$\frac{1}{r}\frac{d}{dr}\left(r\int fc_r^2c^2d\mathbf{v}\right)-\frac{1}{r}\int fc_\theta^2c^2d\mathbf{v}=R_1\left(\frac{m}{2kT_0}\right)^{1/2}n\overline{\Delta c_r c^2} \ .$$

For the distribution function given in Eq. (11-5), the moments connected with this system may be calculated by means of the technique described in Section 11.3. For example, the number density, mean velocity, and heat flux are given by:

$$n(r)=n_0\left(1+\tfrac{1}{2}\nu^+-\tfrac{1}{2}\nu^-\frac{2\alpha}{\pi}\right),$$

$$u_r=\tfrac{1}{2}\frac{1}{r\sqrt{\pi}}\left(\frac{2kT_0}{m}\right)^{1/2}\left(\nu^-+\tfrac{1}{2}\tau^-\right),$$

$$Q_r=\frac{p_0}{r\sqrt{\pi}}\left(\frac{2kT_0}{m}\right)^{1/2}\left(\nu^-+\tfrac{3}{2}\tau^-\right),$$

where $\alpha = \arccos\left(r^{-1}\right)$. After some transformations, the moment system takes the same form as that given by Eqs. (11-29a)-(11-29d). Therefore, one can use the general solution described by Eqs. (11-31a)-(11-31d). The function, $\xi(r)$, is given by:

$$\xi(r) = -Kn^{-1}\left\{\alpha_1 \int_1^r I_2^{*[1]}(r')dr' + \alpha_2 \int_1^r I_2^{*[2]}(r')dr'\right\} .$$

The boundary conditions may be written as $n(1) = n_0$ and:

$$Q_r = \alpha_T\left(Q_{r\,eq}^{\pm} + Q_r^{\mp}\right) \text{ for } r = 1 \text{ and } r = z = \frac{R_2}{R_1} .$$

The equilibrium number densities associated with $Q_{r\,eq}^{\pm}$ may be found from the impenetrability condition. This statement gives:

$$Q = -2\sqrt{\pi} R_1 p_0 \left(\frac{2kT_0}{m}\right)^{1/2} \frac{\Delta T}{T_0} \frac{\alpha_\tau \alpha_T}{1 + z^{-1}\left(1 - \alpha_\tau \alpha_T\right) + Kn^{-1}\zeta(z)} ,$$

$$\zeta(z) = \tfrac{1}{2}\alpha_\tau \alpha_T \left\{\alpha_1 \int_1^z I_2^{*[1]}(r)dr + \alpha_2 \int_1^z I_2^{*[2]}(r)dr\right\} ,$$

where Q is the heat flux per unit length of the inner cylinder.

11.4. Using the Chapman-Enskog solution, A_r, as a molecular property in the moment system, determine the heat flux between two coaxial cylinders the radii and temperatures of which are R_1, T_0 and $R_2, T_0 + \Delta T$ for the inner and outer cylinders, respectively. Assume $\Delta T \ll T_0$. The number density of the gas is n_0 at $\tilde{r} = R_1$.

Solution: The moment system for this problem is:

$$\frac{d}{dr}\left(\nu^- + \tfrac{1}{2}\tau^-\right) = 0 ,$$

$$\frac{d}{dr}\left(\nu^+ + \tau^+\right) = 0 ,$$

$$\frac{d}{dr}\left(\nu^- + \tfrac{3}{2}\tau^-\right) = 0 \ ,$$

$$\frac{d}{dr}\tau^+ = \tfrac{8}{5}\frac{R_1 n_0}{a_1}\left(\frac{m}{2kT_0}\right)^{1/2}\left[A_r \, , \Phi\right] \ ,$$

where A_r is the Chapman-Enskog solution for thermal conductivity and Φ is the correction to the distribution function. The basic function, $\xi(r)$, is discussed in Section 11.6 and, in this case, may be written as:

$$\xi(r) = \tfrac{8}{5}\frac{R_1}{a_1}\pi^{-1/2}\ln(r) \ .$$

It is very important to note that this relation is applicable for arbitrary models of the intermolecular potential. For rigid-sphere molecules this expression becomes $\xi(r) = -\tfrac{128}{75}\pi^{-1}Kn^{-1}\ln(r)$. The heat flux per unit length of the inner cylinder is then given by:

$$Q = -2\sqrt{\pi}R_1\left(\frac{2kT_0}{m}\right)^{1/2} p_0 \frac{\Delta T}{T_0}\frac{\alpha_\tau \alpha_T}{1 + z^{-1}\left(1 - \alpha_\tau \alpha_T\right) + Kn^{-1}\zeta(z)} \ ,$$

where:

$$\zeta(z) = \tfrac{64}{75}\pi^{-1}\alpha_\tau \alpha_T \ln(z) \ \text{and} \ z = \frac{R_2}{R_1} \ .$$

11.5. Determine the general expression for the torque on one of two concentric spheres if the inner sphere of radius, R_1, is slowly rotating with an angular speed, ω, such that $\omega R_1 \ll (2kT/m)^{1/2}$ and the outer sphere of radius, R_2, is stationary. Assume that molecules are rigid spheres. From this general expression, derive the torque for the specific limiting case when $h = R_2 - R_1 \ll R_1$.

Solution: To solve this problem, one can use the same distribution function as was obtained in Section 11.11. The corrections, $G^{\pm}(r)$, are given by:

$$G^+(r) = r\left[A_2 + A_1 \tfrac{4}{3}\zeta\left(1 - r^{-3}\right)\right] \ \text{and} \ G^-(r) = 2\sqrt{\pi}A_1 r \ .$$

The constants of integration, A_1 and A_2, may be specified from two boundary conditions at the sphere surfaces. It is convenient to use the integral form of the boundary conditions that are expressed by:

$$P_{r\phi} = \alpha_\tau \left[P_{r\phi}^{\mp} + \left(P_{r\phi}^{\pm} \right)_{eq} \right] \quad ; \quad r = 1, z \ .$$

These relations yield:

$$G^-(1) = \tfrac{1}{2}\alpha_\tau \left[\left(\frac{2m}{kT} \right)^{1/2} \omega R_1 - \left(G^+(1) - G^-(1) \right) \right],$$

$$z^{-4} G^-(z) = \alpha_\tau \left[\tfrac{1}{2} \left(G^+(z) - G^-(z) \right) + z^{-4} G^-(z) \right].$$

The torque on the inner sphere is found to be:

$$K_x = K_x^* \left[1 + \alpha_\tau \tfrac{8}{15} \pi^{-1} Kn^{-1} \frac{\left(z^3 - 1 \right) z}{z^4 + \left(1 - \alpha_\tau \right)} \right]^{-1} \quad \text{where:}$$

$$K_x^* = -\tfrac{2}{3} \alpha_\tau \pi \rho \bar{V} R_1^4 \omega \frac{z^4}{z^4 + \left(1 - \alpha_\tau \right)},$$

$\bar{V} = \left(8kT/\pi m \right)^{1/2}$ is the mean speed of the gas molecules, $Kn^{-1} = R_1/\lambda$, $z = R_2/R_1$, and the x-axis is in the direction of the vector, $\boldsymbol{\omega}$. In the limiting case when $h = R_2 - R_1 \ll R_1$, the torque is given by:

$$K_x = K_{1x}^* \left[1 + \frac{\alpha_\tau}{2 - \alpha_\tau} \tfrac{8}{5} \pi^{-1} \frac{h}{\lambda} \right]^{-1} \quad \text{with:}$$

$$K_{1x}^* = -\frac{\alpha_\tau}{2 - \alpha_\tau} \tfrac{2}{3} \pi \rho \bar{V} R_1^4 \omega \ .$$

11.6. Determine the general expression for the torque on one of two coaxial cylinders if the inner cylinder of radius, R_1, is slowly rotating with an angular speed, ω, such that $\omega R_1 \ll (2kT/m)^{1/2}$ and the outer cylinder of radius, R_2, is stationary. Assume that molecules are rigid spheres. From this general expression, derive the torque for the specific limiting case when $h = R_2 - R_1 \ll R_1$.

Solution: Using $Q(\mathbf{v}) = c_\theta$ and $Q(\mathbf{v}) = c_r c_\theta B(c^2)$ as the molecular properties in Eq. (11-2), one can obtain:

$$\frac{1}{r}\frac{d}{dr}\left(r\overline{nc_r c_\theta}\right) + \frac{1}{r}\left(\overline{nc_r c_\theta}\right) = 0 \ ,$$

$$\frac{1}{r}\frac{d}{dr}\left(r\overline{nc_r^2 c_\theta B(c^2)}\right) - \frac{1}{r}\left(\overline{nc_\theta^3 B(c^2)}\right) + \frac{1}{r}\left(\overline{nc_r^2 c_\theta B(c^2)}\right)$$

$$= R_1\left(\frac{m}{2kT}\right)^{1/2} n\Delta\overline{c_r c_\theta B(c^2)} \ .$$

where $r = \tilde{r}/R_1$. For closing this moment system the two-sided Maxwellian distribution function may be chosen in the form $f_i = f^{(0)}\left(1 + 2c_\theta G_i(r)\right)$. This choice yields:

$$\overline{nc_r c_\theta} = \frac{n}{2\sqrt{\pi} r^3} G^-(r) \ ,$$

$$\overline{nc_r^2 c_\theta B(c^2)} = nb_1\left[\tfrac{1}{4}G^+(r) - \tfrac{1}{2}\pi^{-1}\alpha G^-(r) + \tfrac{1}{2}\pi^{-1}G^-(r)\frac{x}{r}\left(2r^{-2} - 1\right)\right] ,$$

$$\overline{nc_\theta^3 B(c^2)} = nb_1\left[\tfrac{3}{4}G^+(r) - \tfrac{3}{2}\frac{\alpha}{\pi}G^-(r) - \tfrac{1}{2}\pi^{-1}G^-(r)\frac{x}{r}\left(2r^{-2} + 3\right)\right] ,$$

where $\alpha = \arccos\left(r^{-1}\right)$ and $x = \left(1 - r^{-2}\right)^{1/2}$. The solution of the moment system is found to be:

$$G^-(r) = 2\sqrt{\pi} A_1 r \ , \text{ and:}$$

$$G^+(r) = A_2 r + A_1 r\left[2\zeta\left(1 - r^{-2}\right)\right] \ , \text{ where:}$$

$$\zeta = -\frac{R_1}{b_1}\left(\frac{m}{2kT}\right)^{1/2} .$$

The constants, A_1 and A_2, may be specified from the boundary conditions which are given by:

$$P_{r\theta}(1) = \alpha_\tau\left[P_{r\theta}^{-}(1) + \left(P_{r\theta}^{+}\right)_{eq}\right] \text{ and } P_{r\theta}(z) = \alpha_\tau\left[P_{r\theta}^{+}(z) + \left(P_{r\theta}^{-}\right)_{eq}\right].$$

The torque per unit length of the inner cylinder may be expressed as:

$$K_x = K_x^*\left[1 + \alpha_\tau \tfrac{4}{5}\pi^{-1} Kn^{-1}\frac{\left(z^2-1\right)z}{z^3 + \left(1-\alpha_\tau\right)}\right]^{-1} \text{, where:}$$

$$K_x^* = -\tfrac{1}{2}\alpha_\tau \pi\rho\bar{V}R_1^3\omega\frac{z^3}{z^3 + \left(1-\alpha_\tau\right)} ,$$

$\bar{V} = \left(8kT/\pi m\right)^{1/2}$ is the mean speed of the gas molecules, $Kn^{-1} = R_1/\lambda$, $z = R_2/R_1$, and the x-axis is in the direction of the vector, $\boldsymbol{\omega}$. In the limiting case when $h = R_2 - R_1 \ll R_1$, the torque is given by:

$$K_x = K_{1x}^*\left[1 + \frac{\alpha_\tau}{2-\alpha_\tau}\tfrac{8}{5}\pi^{-1}\frac{h}{\lambda}\right]^{-1} ,$$

$$K_{1x}^* = -\frac{\alpha_\tau}{2-\alpha_\tau}\tfrac{1}{2}\pi\rho\bar{V}R_1^3\omega .$$

11.7. Generalize the sphere torque problem described in Section 11.11 for arbitrary models of the intermolecular potential. Derive the reduced torque for the Lennard-Jones (6-12) potential. Consider two different gases (N_2 and CO_2) when $T = 293$ K.

<u>Solution:</u> The quantity, ζ, depending upon the model of the intermolecular potential, may be expressed in the form:

$$\zeta = [\zeta]_{RS}\left(\sigma^*\right)^2 \Omega^{(2,2)å}\left(T^*\right) ,$$

where the first-order Chapman-Enskog approximation for b_1 is utilized, $\sigma^* = \sigma / \sigma_{RS}$, and $T^* = kT/\varepsilon$ is the reduced temperature (σ and ε/k are parameters of the potential function). In terms of the Ω-integral, the torque on the sphere is given by:

$$K_x = K_x^* \left[1 + \alpha_\tau \tfrac{8}{15} \frac{Kn^{-1}}{\pi} \left(\sigma^*\right)^2 \Omega^{(2,2)å}\left(T^*\right) \right]^{-1} ,$$

$$K_x^* = -\tfrac{2}{3} \alpha_\tau \pi \rho \bar{v} R^4 \omega ,$$

where the Knudsen number is calculated for rigid-sphere molecules. For the specific condition of $T = 293$ K, the reduced temperatures and Ω-integrals of the gases under consideration are:

$$T_{N2}^* = \tfrac{293}{91.5} = 3.2022 \quad ; \quad \Omega^{(2,2)å}\left(3.2022\right) = 1.0218 ,$$

$$T_{CO2}^* = \tfrac{293}{190} = 1.5491 \quad ; \quad \Omega^{(2,2)å}\left(1.5491\right) = 1.2988 .$$

The reduced torque for the Lennard-Jones (6-12) potential can be expressed in the form:

$$\frac{K_x}{K_x^*} = \left[1 + \alpha_\tau \Psi Kn^{-1} \right]^{-1} ,$$

where $\Psi_{N2} = (0.1672)$ and $\Psi_{CO2} = (0.1642)$. For comparison, it should be noted that $\Psi_{RS} = (0.1698)$ for rigid-sphere molecules.

11.8. Determine the domains of applicability of the free-molecular and continuum formulas for the heat flux between two coaxial cylinders. Assume that the limiting formulas have to be correct to within about 5%. Consider the case when $\alpha_\tau = \alpha_T = 1$.

Solution: The heat flux expression derived in Problem 11.4 and the limiting expressions derived from it should be combined in the following manner for the free-molecular and continuum regimes:

$$\frac{Q}{Q_{fm}} = \frac{1}{1 + Kn^{-1}\zeta(z)} \quad \text{and} \quad \frac{Q}{Q_c} = \frac{1}{1 + Kn\left[\zeta(z)\right]^{-1}} .$$

For $\zeta(z)=\frac{64}{75}\pi^{-1}\alpha_\tau\alpha_T\ln(z)$, these free-molecular and continuum formulas are valid to within about 5% if the following inequalities are satisfied:

$$Kn\geq(5.43249)\ln(z) \text{ and } Kn\leq(0.01358)\ln(z)$$

where $Kn=\lambda/R_1$ and $z=R_2/R_1$.

11.9. Determine the domains of applicability of the free-molecular and continuum formulas for the heat flux between two concentric spheres. Assume that the limiting formulas have to be correct to within about 5%. Consider the case when $\alpha_\tau=\alpha_T=1$.

Solution: The heat flux expression derived in Problem 11.2 and the limiting expressions derived from it should be combined in the following manner for the free-molecular and continuum regimes:

$$\frac{Q}{Q_{fm}}=\frac{1}{1+Kn^{-1}\zeta(z)} \text{ and } \frac{Q}{Q_c}=\frac{1}{1+Kn\left[\zeta(z)\right]^{-1}}.$$

For $\zeta(z)=\frac{64}{75}\pi^{-1}\alpha_\tau\alpha_T(z-1)/z$, these free-molecular and continuum formulas are valid to within about 5% if the following inequalities are satisfied:

$$Kn\geq(5.43249)\left(1-z^{-1}\right) \text{ and } Kn\leq(0.01358)\left(1-z^{-1}\right)$$

where $Kn=\lambda/R_1$ and $z=R_2/R_1$.

REFERENCES

1. Kogan, M.N., "Recent Developments in the Kinetic Theory of Gases," in *Rarefied Gas Dynamics 6th Symposium* **1**, 1-39, Academic Press Inc., N.Y. (1969).
2. Lees, L., "Kinetic Theory Description of Rarefied Gas Flow," *J. Soc. Indust. Appl. Math.* **13(1)**, 278-311 (1965).
3. Lees, L. and Liu, C.Y., "Kinetic Theory Description of Conductive Heat Transfer from a Fine Wire," *Phys. Fluids* **5(10)**, 1137-1148 (1962).
4. Grad, H., "On the Kinetic Theory of Rarefied Gases," *Comm. Pure and Appl. Math.* **2(4)**, 311-407 (1949).
5. Grad, H., "Principles of the Kinetic Theory of Gases," in *Handbuch der Physik*, vol. 12, ed. Flügge, S. (Springer Verlag, Berlin, 1958).
6. Grad, H., "Asymptotic Theory of the Boltzmann Equation," *Phys. Fluids* **6**, 147 (1963).
7. Kumar, K., "Polynomial Expansions in Kinetic Theory of Gases," *Ann. Phys.* **37(2)**, 113-141 (1966).

8. Muckenfuss, C., "Some Aspects of Shock Structure According the Bimodal Model," *Phys. Fluids* **5(11)**, 1325-1336 (1962).
9. Salwen, H., Grosch, C.E. and Ziering, S., "Extension of the Mott-Smith Method for a One-dimensional Shock Wave," *Phys. Fluids* **7(2)**, 180-189 (1964).
10. Hurlbut, F.C., "Note on Conductive Heat Transfer from a Fine Wire," *Phys. Fluids* **7(6)**,904-906 (1964).
11. Liu, C.Y. and Sigimura, T., "Rarefied Gas Flow Over a Sphere at Low Mach Numbers," in *Rarefied Gas Dynamics 6th Symposium* **1**, 789-794, Academic Press Inc., N.Y. (1969).
12. Phillips, W.F., "Drag on a Small Sphere Moving Through a Gas," *Phys. Fluids* **18(9)**, 1089-1093 (1975).
13. Mott-Smith, H.M., "The Solution of the Boltzmann Equation for a Shock Wave," *Phys. Rev.* **82(6)**, 885-892 (1951).
14. Krook, M., "Continuum Equations in Dynamics of Rarefied Gases," *J. Fluid. Mech.* **6(4)**, 523-541 (1959).
15. Kogan, M.N., *Rarefied Gas Dynamics* (Plenum Press, N.Y., 1969).
16. Ivchenko, I.N., "Generalization of the Lees Method in Boundary Problems of Transfer," *J. Colloid and Interface Sci.* **135(1)**, 16-19 (1990).
17. Ivchenko, I.N., Loyalka, S.K., and Tompson, R.V., "A Solution of the Problem of Heat Transfer Between Two Cylinders for Arbitrary Knudsen Number," *High Temperature* **31(4)**, 776-783 (1993).
18. Ivchenko, I.N., Loyalka, S.K., and Tompson, R.V., "A Method for Solving Linearized Problems of the Transfer Theory for Spherical Geometry at Arbitrary Knudsen Numbers," *Fluid Dynamics* **29(6)**, 888-893 (1994).
19. Ivchenko, I.N., "Evaporation (Condensation) Theory of Spherical Particles with All Knudsen Numbers," *J. Colloid and Interface Sci.* **120(1)**, 1-7 (1987).
20. Savkov, S.A., *The Slip Boundary Conditions of the Non-Uniform Binary Gas Mixture and an Application of Them in Aerosol Dynamics* (M.S. Thesis, Moscow, 1987).
21 Wang-Chang, C.S. and Uhlenbeck, G.E., "Heat Transport Between Parallel Plates," University of Michigan Project M999 (1953).
22. Jackson, E.A., *Boundary Value Problems in Kinetic Theory of Gases* (Ph.D. Dissertation, Syracuse University, 1958).
23. Gross, E.P. and Ziering, S., "Heat Flow Between Parallel Plates," *Phys. Fluids* **2(6)**, 701-712 (1959).
24. Bassanini, P., Cercignani, C., and Pagani, C.D., "Comparison of Kinetic Theory Analyses of Heat Transfer Between Parallel Plates," *Inter. J. Heat Mass Transfer* **10**, 447 (1967).
25. Loyalka, S.K., "Linearized Couette Flow and Heat Transfer Between Two Parallel Plates," in *Rarefied Gas Dynamics 6th Symposium* **1**, 195, Academic Press Inc., N.Y. (1969).
26. Teagan, W.P. and Springer, G.S., "Heat-Transfer and Density-Distribution Measurements Between Parallel Plates in the Transition Regime," *Phys. Fluids* **11(3)**, 497-506 (1968).
27. Halbritter, L., "Torque on a Rotating Ellipsoid in a Rarefied Gas," *Z. Naturforsh.* **29a**, 1717-1722 (1974).
28. Smirnov, L.P. and Chekalov, V.V., "Slow Rotation of a Sphere in a Bounded Rarefied Gas Volume," *Izv. AN SSSR, M. Zh. G.* **(4)**, 117-124 (1978)
29. Chandrasekhar, S., *Hydrodynamic and Hydromagnetic Stability* (Dover, New York, 1961).
30. Landau, L.D. and Lifshitz, E.M., *Mechanics* (Pergamon Press, New York, 1960).
31. Landau, L.D. and Lifshitz, E.M., *Fluid Mechanics* (Addison-Wesley, Reading, Massachusetts, 1960).
32. Loyalka, S.K., "Motion of a Sphere in a Gas: Numerical Solution of the Linearized Boltzmann Equation," *Phys. Fluids A* **4(5)**, 1049-1056 (1992).

Chapter 12

BOUNDARY SLIP PHENOMENA IN A BINARY GAS MIXTURE

1. THE FIRST-ORDER CHAPMAN-ENSKOG APPROXIMATION FOR A BINARY GAS MIXTURE.

The non-uniform state of a binary gas mixture is described by the distribution functions:

$$f_i = f_i^{(0)}(\tilde{\mathbf{r}},t)\left(1+\Psi_i^{(1)}(\mathbf{C}_i,\tilde{\mathbf{r}},t)\right);\ (i=1,2), \tag{12-1}$$

where $\mathbf{C}_i = (m_i/2kT)^{1/2}\,\mathbf{V}_i$ and $f_i^{(0)}(\tilde{\mathbf{r}},t)$ is given by:

$$f_i^{(0)}(\tilde{\mathbf{r}},t) = n_i(\tilde{\mathbf{r}},t)\left(\frac{m_i}{2\pi kT(\tilde{\mathbf{r}},t)}\right)^{3/2}\exp\left(-\frac{m_i\left(\mathbf{v}_i-\mathbf{u}(\tilde{\mathbf{r}},t)\right)^2}{2kT(\tilde{\mathbf{r}},t)}\right). \tag{12-2}$$

The corrections, $\Psi_i^{(1)}(\mathbf{C}_i,\tilde{\mathbf{r}},t)$, may be expressed in terms of the first-order Chapman-Enskog solutions [1], $\mathbf{A}_i$, $\mathbf{D}_i$, and $\mathbf{B}_i$, in the following manner:

$$\Psi_i^{(1)}(\mathbf{C}_i,\tilde{\mathbf{r}},t) = -\mathbf{A}_i^{(1)}\cdot\frac{\partial\ln(T)}{\partial\mathbf{r}} - \mathbf{D}_i^{(1)}\cdot\mathbf{d}_{12} - 2\mathbf{B}_i^{(1)}:\frac{\partial\mathbf{u}}{\partial\tilde{\mathbf{r}}}, \tag{12-3}$$

$$\mathbf{d}_{12} = \nabla x_1 + \frac{n_1 n_2(m_2-m_1)}{n\rho}\nabla\ln(p), \tag{12-4}$$

where n_i is the number density of the i-th constituent, m_i is the molecular mass, $p = p_1 + p_2$ is the hydrostatic pressure, $x_i = n_i / n$, $n = n_1 + n_2$, and $\rho = n_1 m_1 + n_2 m_2$. In Eq. (12-4), an external force field is assumed to be absent.

The first-order Chapman-Enskog solutions for the thermal conductivity, $\mathbf{A}_1^{(1)}$ and $\mathbf{A}_2^{(1)}$, are given by [1]:

$$\mathbf{A}_1^{(1)} = a_1 S_{3/2}^{(1)}\left(\mathbf{C}_1^2\right)\mathbf{C}_1 + \frac{k_T D_{12}}{x_1}\left(\frac{m_1}{2kT}\right)^{1/2}\left(\frac{2m_2}{m}\mathbf{C}_1\right), \tag{12-5}$$

$$\mathbf{A}_2^{(1)} = a_{-1} S_{3/2}^{(1)}\left(\mathbf{C}_2^2\right)\mathbf{C}_2 + \frac{k_T D_{12}}{x_2}\left(\frac{m_2}{2kT}\right)^{1/2}\left(-\frac{2m_1}{m}\mathbf{C}_2\right), \tag{12-6}$$

where a_{-1} and a_1 are the transport coefficients, $k_T = D_T / D_{12}$ is the thermal diffusion ratio, D_{12} is the diffusion coefficient, and $m = x_1 m_1 + x_2 m_2$. In the same manner, the first-order Chapman-Enskog solutions for diffusion, $\mathbf{D}_1^{(1)}$ and $\mathbf{D}_2^{(1)}$, are given by [1]:

$$\mathbf{D}_1^{(1)} = \frac{D_{12}}{x_1}\left(\frac{m_1}{2kT}\right)^{1/2}\left(\frac{2m_2}{m}\mathbf{C}_1\right), \tag{12-7}$$

$$\mathbf{D}_2^{(1)} = \frac{D_{12}}{x_2}\left(\frac{m_2}{2kT}\right)^{1/2}\left(-\frac{2m_1}{m}\mathbf{C}_2\right). \tag{12-8}$$

and the first-order Chapman-Enskog solutions for viscosity, $\mathbf{B}_1^{(1)}$ and $\mathbf{B}_2^{(1)}$, are given in terms of their respective components by [1]:

$$B_{1ij}^{(1)} = b_1\left(C_{1i}C_{1j} - \tfrac{1}{3}C_1^2\delta_{ij}\right), \tag{12-9}$$

$$B_{2ij}^{(1)} = b_{-1}\left(C_{2i}C_{2j} - \tfrac{1}{3}C_2^2\delta_{ij}\right). \tag{12-10}$$

2. THE TRANSPORT COEFFICIENTS FOR A BINARY GAS MIXTURE.

The first-order Chapman-Enskog solutions for the unknown transport coefficients a_{-1}, a_1, d_0, b_{-1}, and b_1, are derived from the following algebraic systems of equations [1]:

$$a_{-1-1}a_{-1} + a_{1-1}a_1 = \alpha_{-1} = -\tfrac{15}{4}\frac{n_2}{n^2}\left(\frac{2kT}{m_2}\right)^{1/2}, \tag{12-11}$$

$$a_{-11}a_{-1} + a_{11}a_1 = \alpha_1 = -\tfrac{15}{4}\frac{n_1}{n^2}\left(\frac{2kT}{m_1}\right)^{1/2}, \tag{12-12}$$

$$a_{00}d_0 = \delta_0 = \tfrac{3}{2}n^{-1}\left(\frac{2kT}{m_0}\right)^{1/2}, \tag{12-13}$$

$$b_{-1-1}b_{-1} + b_{1-1}b_1 = \beta_{-1} = \tfrac{5}{2}\frac{n_2}{n^2}, \tag{12-14}$$

$$b_{-11}b_{-1} + b_{11}b_1 = \beta_1 = \tfrac{5}{2}\frac{n_1}{n^2}, \tag{12-15}$$

The thermal conductivity, diffusion, and viscosity coefficients are then found to be [1]:

$$\kappa = -\tfrac{5}{4}nk\left[x_1\left(\frac{2kT}{m_1}\right)^{1/2}a_1 + x_2\left(\frac{2kT}{m_2}\right)^{1/2}a_{-1}\right], \tag{12-16}$$

$$D_{12} = \tfrac{1}{2}x_1x_2\left(\frac{2kT}{m_0}\right)^{1/2}d_0\,, \tag{12-17}$$

and:

$$\mu = p\left(x_1b_1 + x_2b_{-1}\right);\; p = nkT\,. \tag{12-18}$$

If one introduces the following standard notations of Chapman and Cowling [1]:

$$A = \tfrac{1}{5}\frac{\Omega_{12}^{(2,2)}}{\Omega_{12}^{(1,1)}}, \quad B = 5\Omega_{12}^{(1,2)} - \tfrac{1}{5}\frac{\Omega_{12}^{(1,3)}}{\Omega_{12}^{(1,1)}},$$

$$C = -1 + \tfrac{2}{5}\frac{\Omega_{12}^{(1,2)}}{\Omega_{12}^{(1,1)}}, \text{ and } E = \tfrac{1}{8}\frac{kT}{M_1 M_2 \Omega_{12}^{(1,1)}},$$

then the known algebraic coefficients in Eqs. (12-11)-(12-15), a_{-1-1}, a_{1-1}, a_{-11}, a_{11}, a_{00}, b_{-1-1}, b_{1-1}, b_{-11}, and b_{11}, may be expressed in the following manner:

$$a_{00} = x_1 x_2 \frac{kT}{E}, \tag{12-19}$$

$$a_{11} = x_1^2 \tfrac{5}{2}\frac{kT}{[\mu_1]_1} + 5x_1 x_2 \frac{kT}{E} M_1^{-1} \times\left[\tfrac{1}{4}\left(6M_1^2 + 5M_2^2\right) - M_2^2 B + 2M_1 M_2 A\right], \tag{12-20a}$$

$$a_{-1-1} = x_2^2 \tfrac{5}{2}\frac{kT}{[\mu_2]_1} + 5x_1 x_2 \frac{kT}{E} M_2^{-1} \times\left[\tfrac{1}{4}\left(6M_2^2 + 5M_1^2\right) - M_1^2 B + 2M_1 M_2 A\right], \tag{12-20b}$$

$$a_{1-1} = a_{-11} = -5x_1 x_2 \frac{kT}{E}\left(M_1 M_2\right)^{1/2}\left[\tfrac{11}{4} - B - 2A\right], \tag{12-21}$$

$$b_{11} = x_1^2 \tfrac{5}{2}\frac{kT}{[\mu_1]_1} + 5x_1 x_2 \frac{kT}{E} M_1^{-1}\left(\tfrac{2}{3}M_1 + M_2 A\right), \tag{12-22a}$$

$$b_{-1-1} = x_2^2 \tfrac{5}{2} \frac{kT}{[\mu_2]_1} + 5x_1x_2 \frac{kT}{E} M_2^{-1}\left(\tfrac{2}{3}M_2 + M_1A\right), \tag{12-22b}$$

$$b_{1-1} = b_{-11} = -5x_1x_2 \frac{kT}{E}\left(\tfrac{2}{3} - A\right). \tag{12-23}$$

where $[\mu_1]_1$ is given by Eq. (5-62) with the substitutions $m = m_1$ and $\sigma = \sigma_1$. The coefficients a_{-1-1} and b_{-1-1} have been derived from the coefficients a_{11} and b_{11} by changing the indices associated with the gas constituents; $[\mu_1]_1 \to [\mu_2]_1$, $x_1 \to x_2$, $x_2 \to x_1$, $M_1 \to M_2$, and $M_2 \to M_1$. Then, the transport coefficients may be presented in the form:

$$a_1 = -\tfrac{2}{5}\frac{m_1}{k}\left(\frac{2kT}{m_1}\right)^{1/2} \frac{[\kappa_1]_1}{p[\Delta_a]_1} \times\left\{x_1Q_1 + x_2\left[P_1M_1M_2(11 - 4B - 8A) + 2P_1P_2\right]\right\}, \tag{12-24}$$

$$a_{-1} = -\tfrac{2}{5}\frac{m_2}{k}\left(\frac{2kT}{m_2}\right)^{1/2} \frac{[\kappa_2]_1}{p[\Delta_a]_1} \times\left\{x_2Q_2 + x_1\left[P_2M_1M_2(11 - 4B - 8A) + 2P_1P_2\right]\right\}, \tag{12-25}$$

$$d_0 = \tfrac{3}{2}\left(\frac{2kT}{m_0}\right)^{1/2} \frac{E}{kT}(nx_1x_2)^{-1}, \tag{12-26}$$

$$b_1 = \frac{1}{p[\Delta_b]_1}\left[x_1R_1 + x_2\left(\tfrac{1}{2}E[\mu_2]_1^{-1} + \tfrac{2}{3} - A\right)\right], \tag{12-27}$$

$$b_{-1} = \frac{1}{p[\Delta_b]_1}\left[x_2R_2 + x_1\left(\tfrac{1}{2}E[\mu_1]_1^{-1} + \tfrac{2}{3} - A\right)\right]. \tag{12-28}$$

Here, the following notations have been incorporated:

$$[\Delta_a]_1 = x_1^2 Q_1 + x_2^2 Q_2 + x_1 x_2 Q_{12} , \tag{12-29}$$

$$[\Delta_b]_1 = x_1^2 R_1 [\mu_1]_1^{-1} + x_2^2 R_2 [\mu_2]_1^{-1} + x_1 x_2 R_{12} , \tag{12-30}$$

$$Q_1 = P_1 \left(6M_2^2 + 5M_1^2 - 4M_1^2 B + 8M_1 M_2 A\right), \tag{12-31}$$

$$Q_2 = P_2 \left(6M_1^2 + 5M_2^2 - 4M_2^2 B + 8M_1 M_2 A\right), \tag{12-32}$$

$$Q_{12} = 3\left(M_1 - M_2\right)^2 (5 - 4B) + 4M_1 M_2 A(11 - 4B) + 2P_1 P_2 , \tag{12-33}$$

$$P_1 = M_1 E[\mu_1]_1^{-1}, \;\; P_2 = M_2 E[\mu_2]_1^{-1}, \tag{12-34}$$

$$R_1 = \tfrac{2}{3} + (M_1/M_2) A , \tag{12-35}$$

$$R_2 = \tfrac{2}{3} + (M_2/M_1) A , \tag{12-36}$$

$$R_{12} = \tfrac{1}{2} E[\mu_1]_1^{-1} [\mu_2]_1^{-1} + \tfrac{4}{3} A(EM_1 M_2)^{-1} . \tag{12-37}$$

3. THE SECOND-ORDER CHAPMAN-ENSKOG APPROXIMATION FOR A BINARY GAS MIXTURE.

The second-order Chapman-Enskog solutions for the functions, $\mathbf{A}_i$, $\mathbf{D}_i$, and $\mathbf{B}_i$, can be expressed as:

$$\mathbf{A}_i = a_{\pm 1} \mathbf{C}_i \left[S_{3/2}^{(1)}\left(C_i^2\right) + a_{\pm 2}^* S_{3/2}^{(2)}\left(C_i^2\right) \right] + k_T \mathbf{D}_i , \tag{12-38}$$

$$\mathbf{D}_i = \frac{D_{12}}{x_i}\left(\frac{m_i}{2kT_0}\right)^{1/2} \mathbf{C}_i \left[(-1)^{i-1} 2\frac{m_j}{m} + z_i S_{3/2}^{(1)}\left(C_i^2\right) \right], \tag{12-39}$$

and:

$$\mathbf{B}_i = \mathbf{C}_i \overset{\circ}{\mathbf{C}}_i B_i\left(C_i^2\right) = b_{\pm 1} \mathbf{C}_i \overset{\circ}{\mathbf{C}}_i \left(1 + b_{\pm 2}^* S_{5/2}^{(1)}\left(C_i^2\right)\right), \tag{12-40}$$

where $a_{\pm 2}^* = a_{\pm 2}/a_{\pm 1}$, $b_{\pm 2}^* = b_{\pm 2}/b_{\pm 1}$, $z_i = x_i\left(2kT/m_i\right)^{1/2}\left(d_{\pm 1}/D_{12}\right)$, the upper and lower signs, $(\pm)$, refer to the first $(+)$ and second $(-)$ constituents of the gas mixture, respectively, and $i \neq j$. The transport coefficients of the thermal conductivity, diffusion and viscosity, $a_{\pm 1}$, $a_{\pm 2}$, $d_{\pm 1}$, d_0, $b_{\pm 1}$, and $b_{\pm 2}$ for the second-order Chapman-Enskog approximation can be calculated from the three linear algebraic systems of equations given by:

$$\sum_{p=-2}^{2} a_p a_{pq} = \alpha_q \ ; \ p,q \neq 0 , \tag{12-41}$$

$$\sum_{p=-2}^{2} b_p b_{pq} = \beta_q \ ; \ p,q \neq 0 , \tag{12-42}$$

and:

$$\sum_{p=-1}^{1} d_p a_{pq} = \delta_q , \tag{12-43}$$

where $\alpha_{\pm 1}$, $\beta_{\pm 1}$, and δ_0 are given by Eqs. (12-11)-(12-15) in Section 12-2 while the remaining coefficients, $\alpha_{\pm 2} = \beta_{\pm 2} = \delta_{\pm 1} = 0$.

The coefficient, a_{00}, of the algebraic system for diffusion, Eq. (12-43), is specified by Eq. (12-19) while the other coefficients in this system, $a_{01} = a_{10}$ and $a_{0-1}a_{-10}$, are expressible in the form:

$$a_{01} = -x_1 x_2 \tfrac{5}{2} \frac{kT}{E} C M_2 M_1^{-1/2} , \tag{12-44}$$

and:

$$a_{0-1} = x_1 x_2 \tfrac{5}{2} \frac{kT}{E} C M_1 M_2^{-1/2} . \tag{12-45}$$

All three of the systems given in Eqs. (12-41)-(12-43) involve the coefficients, a_{11}, a_{1-1}, b_{11}, and b_{1-1} which are given by Eqs. (12-20)-(12-23). For $p=1,2$ and $q=\pm 2$, explicit analytical expressions for a_{pq} and b_{pq} may be found if one uses the analytical representations of the appropriate bracket integrals given in [2,3]. These coefficients, a_{pq} and b_{pq}, are then found to be [2,3]:

$$a_{12} = x_1^2 \tfrac{5}{8}\frac{kT}{[\mu_1]_1}\left(7 - 2\frac{\Omega_1^{(2,3)}}{\Omega_1^{(2,2)}}\right)$$
$$+ x_1 x_2 \frac{M_2}{M_1}\frac{kT}{E}\left[\tfrac{35}{16}\left(12M_1^2 + 5M_2^2\right) - \tfrac{21}{8}\left(4M_1^2 + 5M_2^2\right)\frac{\Omega_{12}^{(1,2)}}{\Omega_{12}^{(1,1)}}\right. \tag{12-46a}$$
$$\left. + M_2^2\left(\tfrac{19}{4}\frac{\Omega_{12}^{(1,3)}}{\Omega_{12}^{(1,1)}} - \tfrac{1}{2}\frac{\Omega_{12}^{(1,4)}}{\Omega_{12}^{(1,1)}}\right) + M_1 M_2\left(7\frac{\Omega_{12}^{(2,2)}}{\Omega_{12}^{(1,1)}} - 2\frac{\Omega_{12}^{(2,3)}}{\Omega_{12}^{(1,1)}}\right)\right],$$

$$a_{-1-2} = x_2^2 \tfrac{5}{8}\frac{kT}{[\mu_2]_1}\left(7 - 2\frac{\Omega_2^{(2,3)}}{\Omega_2^{(2,2)}}\right)$$
$$+ x_1 x_2 \frac{M_1}{M_2}\frac{kT}{E}\left[\tfrac{35}{16}\left(12M_2^2 + 5M_1^2\right) - \tfrac{21}{8}\left(4M_2^2 + 5M_1^2\right)\frac{\Omega_{12}^{(1,2)}}{\Omega_{12}^{(1,1)}}\right. \tag{12-46b}$$
$$\left. + M_1^2\left(\tfrac{19}{4}\frac{\Omega_{12}^{(1,3)}}{\Omega_{12}^{(1,1)}} - \tfrac{1}{2}\frac{\Omega_{12}^{(1,4)}}{\Omega_{12}^{(1,1)}}\right) + M_1 M_2\left(7\frac{\Omega_{12}^{(2,2)}}{\Omega_{12}^{(1,1)}} - 2\frac{\Omega_{12}^{(2,3)}}{\Omega_{12}^{(1,1)}}\right)\right],$$

$$a_{1-2} = -x_1 x_2 M_1^{3/2} M_2^{1/2}\frac{kT}{E}\left(\tfrac{595}{16} - \tfrac{189}{8}\frac{\Omega_{12}^{(1,2)}}{\Omega_{12}^{(1,1)}} + \tfrac{19}{4}\frac{\Omega_{12}^{(1,3)}}{\Omega_{12}^{(1,1)}}\right. \tag{12-47a}$$
$$\left. - \tfrac{1}{2}\frac{\Omega_{12}^{(1,4)}}{\Omega_{12}^{(1,1)}} - 7\frac{\Omega_{12}^{(2,2)}}{\Omega_{12}^{(1,1)}} + 2\frac{\Omega_{12}^{(2,3)}}{\Omega_{12}^{(1,1)}}\right),$$

$$a_{-12} = -x_1 x_2 M_2^{3/2} M_1^{1/2}\frac{kT}{E}\left(\tfrac{595}{16} - \tfrac{189}{8}\frac{\Omega_{12}^{(1,2)}}{\Omega_{12}^{(1,1)}} + \tfrac{19}{4}\frac{\Omega_{12}^{(1,3)}}{\Omega_{12}^{(1,1)}}\right. \tag{12-47b}$$
$$\left. - \tfrac{1}{2}\frac{\Omega_{12}^{(1,4)}}{\Omega_{12}^{(1,1)}} - 7\frac{\Omega_{12}^{(2,2)}}{\Omega_{12}^{(1,1)}} + 2\frac{\Omega_{12}^{(2,3)}}{\Omega_{12}^{(1,1)}}\right),$$

$$
\begin{aligned}
a_{22} &= x_1^2 \tfrac{5}{8} \frac{kT}{[\mu_1]_1}\left(\tfrac{77}{4} - 7\frac{\Omega_1^{(2,3)}}{\Omega_1^{(2,2)}} + \frac{\Omega_1^{(2,4)}}{\Omega_1^{(2,2)}}\right) \\
&+ x_1 x_2 \frac{kT}{M_1 E}\Bigg[\tfrac{35}{64}\left(40M_1^4 + 168M_1^2M_2^2 + 35M_2^4\right) \\
&- \tfrac{7}{8}M_2^2\left(84M_1^2 + 35M_2^2\right)\frac{\Omega_{12}^{(1,2)}}{\Omega_{12}^{(1,1)}} \\
&+ \tfrac{1}{8}M_2^2\left(108M_1^2 + 133M_2^2\right)\frac{\Omega_{12}^{(1,3)}}{\Omega_{12}^{(1,1)}} \\
&- \tfrac{7}{2}M_2^4\frac{\Omega_{12}^{(1,4)}}{\Omega_{12}^{(1,1)}} + \tfrac{1}{4}M_2^4\frac{\Omega_{12}^{(1,5)}}{\Omega_{12}^{(1,1)}} + \tfrac{7}{2}M_1M_2\left(4M_1^2 + 7M_2^2\right)\frac{\Omega_{12}^{(2,2)}}{\Omega_{12}^{(1,1)}} \\
&- 14M_1M_2^3\frac{\Omega_{12}^{(2,3)}}{\Omega_{12}^{(1,1)}} + 2M_1M_2^3\frac{\Omega_{12}^{(2,4)}}{\Omega_{12}^{(1,1)}} + 2M_1^2M_2^2\frac{\Omega_{12}^{(3,3)}}{\Omega_{12}^{(1,1)}}\Bigg],
\end{aligned}
\tag{12-48a}
$$

$$
\begin{aligned}
a_{-2-2} &= x_2^2 \tfrac{5}{8} \frac{kT}{[\mu_2]_1}\left(\tfrac{77}{4} - 7\frac{\Omega_2^{(2,3)}}{\Omega_2^{(2,2)}} + \frac{\Omega_2^{(2,4)}}{\Omega_2^{(2,2)}}\right) \\
&+ x_1 x_2 \frac{kT}{M_2 E}\Bigg[\tfrac{35}{64}\left(40M_2^4 + 168M_1^2M_2^2 + 35M_1^4\right) \\
&- \tfrac{7}{8}M_1^2\left(84M_2^2 + 35M_1^2\right)\frac{\Omega_{12}^{(1,2)}}{\Omega_{12}^{(1,1)}} \\
&+ \tfrac{1}{8}M_1^2\left(108M_2^2 + 133M_1^2\right)\frac{\Omega_{12}^{(1,3)}}{\Omega_{12}^{(1,1)}} \\
&- \tfrac{7}{2}M_1^4\frac{\Omega_{12}^{(1,4)}}{\Omega_{12}^{(1,1)}} + \tfrac{1}{4}M_1^4\frac{\Omega_{12}^{(1,5)}}{\Omega_{12}^{(1,1)}} + \tfrac{7}{2}M_1M_2\left(4M_2^2 + 7M_1^2\right)\frac{\Omega_{12}^{(2,2)}}{\Omega_{12}^{(1,1)}} \\
&- 14M_2M_1^3\frac{\Omega_{12}^{(2,3)}}{\Omega_{12}^{(1,1)}} + 2M_2M_1^3\frac{\Omega_{12}^{(2,4)}}{\Omega_{12}^{(1,1)}} + 2M_1^2M_2^2\frac{\Omega_{12}^{(3,3)}}{\Omega_{12}^{(1,1)}}\Bigg],
\end{aligned}
\tag{12-48b}
$$

$$a_{2-2} = -x_1 x_2 M_1^{3/2} M_2^{3/2} \frac{kT}{E} \left(\tfrac{8505}{64} - \tfrac{833}{8} \frac{\Omega_{12}^{(1,2)}}{\Omega_{12}^{(1,1)}} \right.$$
$$+ \tfrac{241}{8} \frac{\Omega_{12}^{(1,3)}}{\Omega_{12}^{(1,1)}} - \tfrac{7}{2} \frac{\Omega_{12}^{(1,4)}}{\Omega_{12}^{(1,1)}} + \tfrac{1}{4} \frac{\Omega_{12}^{(1,5)}}{\Omega_{12}^{(1,1)}} - \tfrac{77}{2} \frac{\Omega_{12}^{(2,2)}}{\Omega_{12}^{(1,1)}} \tag{12-49}$$
$$\left. + 14 \frac{\Omega_{12}^{(2,3)}}{\Omega_{12}^{(1,1)}} - 2 \frac{\Omega_{12}^{(1,1)2,4}}{\Omega_{12}^{(1,1)}} + 2 \frac{\Omega_{12}^{(3,3)}}{\Omega_{12}^{(1,1)}} \right),$$

$$b_{12} = x_1^2 \tfrac{5}{8} \frac{kT}{[\mu_1]_1} \left(7 - 2 \frac{\Omega_1^{(2,3)}}{\Omega_1^{(2,2)}} \right) + x_1 x_2 \tfrac{2}{3} \frac{kT}{E} \frac{M_2}{M_1}$$
$$\times \left(\tfrac{35}{2} M_1 - 7 M_1 \frac{\Omega_{12}^{(1,2)}}{\Omega_{12}^{(1,1)}} + \tfrac{21}{4} M_2 \frac{\Omega_{12}^{(2,2)}}{\Omega_{12}^{(1,1)}} - \tfrac{3}{2} M_2 \frac{\Omega_{12}^{(2,3)}}{\Omega_{12}^{(1,1)}} \right), \tag{12-50a}$$

$$b_{-1-2} = x_2^2 \tfrac{5}{8} \frac{kT}{[\mu_2]_1} \left(7 - 2 \frac{\Omega_2^{(2,3)}}{\Omega_2^{(2,2)}} \right) + x_1 x_2 \tfrac{2}{3} \frac{kT}{E} \frac{M_1}{M_2}$$
$$\times \left(\tfrac{35}{2} M_2 - 7 M_2 \frac{\Omega_{12}^{(1,2)}}{\Omega_{12}^{(1,1)}} + \tfrac{21}{4} M_1 \frac{\Omega_{12}^{(2,2)}}{\Omega_{12}^{(1,1)}} - \tfrac{3}{2} M_1 \frac{\Omega_{12}^{(2,3)}}{\Omega_{12}^{(1,1)}} \right), \tag{12-50b}$$

$$b_{1-2} = x_1 x_2 \tfrac{2}{3} M_2 \frac{kT}{E} \left(-\tfrac{35}{2} + 7 \frac{\Omega_{12}^{(1,2)}}{\Omega_{12}^{(1,1)}} + \tfrac{21}{4} \frac{\Omega_{12}^{(2,2)}}{\Omega_{12}^{(1,1)}} - \tfrac{3}{2} \frac{\Omega_{12}^{(2,3)}}{\Omega_{12}^{(1,1)}} \right), \tag{12-51a}$$

$$b_{-12} = x_1 x_2 \tfrac{2}{3} M_1 \frac{kT}{E} \left(-\tfrac{35}{2} + 7 \frac{\Omega_{12}^{(1,2)}}{\Omega_{12}^{(1,1)}} + \tfrac{21}{4} \frac{\Omega_{12}^{(2,2)}}{\Omega_{12}^{(1,1)}} - \tfrac{3}{2} \frac{\Omega_{12}^{(2,3)}}{\Omega_{12}^{(1,1)}} \right), \tag{12-51b}$$

$$
\begin{aligned}
b_{22} &= x_1^2 \tfrac{5}{8} \frac{kT}{[\mu_1]_1} \left(\tfrac{301}{12} - 7 \frac{\Omega_1^{(2,3)}}{\Omega_1^{(2,2)}} + \frac{\Omega_1^{(2,4)}}{\Omega_1^{(2,2)}} \right) \\
&+ x_1 x_2 \tfrac{2}{3} \frac{kT}{M_1 E} \left[\tfrac{1}{4} M_1 \left(140 M_1^2 + 245 M_2^2 \right) \right. \\
&+ M_1 M_2^2 \left(-49 \frac{\Omega_{12}^{(1,2)}}{\Omega_{12}^{(1,1)}} + 8 \frac{\Omega_{12}^{(1,3)}}{\Omega_{12}^{(1,1)}} \right) \\
&+ \tfrac{1}{8} M_2 \left(154 M_1^2 + 147 M_2^2 \right) \frac{\Omega_{12}^{(2,2)}}{\Omega_{12}^{(1,1)}} \\
&\left. - \tfrac{21}{2} M_2^3 \frac{\Omega_{12}^{(2,3)}}{\Omega_{12}^{(1,1)}} + \tfrac{3}{2} M_2^3 \frac{\Omega_{12}^{(2,4)}}{\Omega_{12}^{(1,1)}} + 3 M_1 M_2^2 \frac{\Omega_{12}^{(3,3)}}{\Omega_{12}^{(1,1)}} \right],
\end{aligned}
\tag{12-52a}
$$

$$
\begin{aligned}
b_{-2-2} &= x_2^2 \tfrac{5}{8} \frac{kT}{[\mu_2]_1} \left(\tfrac{301}{12} - 7 \frac{\Omega_2^{(2,3)}}{\Omega_2^{(2,2)}} + \frac{\Omega_2^{(2,4)}}{\Omega_2^{(2,2)}} \right) \\
&+ x_1 x_2 \tfrac{2}{3} \frac{kT}{M_2 E} \left[\tfrac{1}{4} M_2 \left(140 M_2^2 + 245 M_1^2 \right) \right. \\
&+ M_2 M_1^2 \left(-49 \frac{\Omega_{12}^{(1,2)}}{\Omega_{12}^{(1,1)}} + 8 \frac{\Omega_{12}^{(1,3)}}{\Omega_{12}^{(1,1)}} \right) \\
&+ \tfrac{1}{8} M_1 \left(154 M_2^2 + 147 M_1^2 \right) \frac{\Omega_{12}^{(2,2)}}{\Omega_{12}^{(1,1)}} \\
&\left. - \tfrac{21}{2} M_1^3 \frac{\Omega_{12}^{(2,3)}}{\Omega_{12}^{(1,1)}} + \tfrac{3}{2} M_1^3 \frac{\Omega_{12}^{(2,4)}}{\Omega_{12}^{(1,1)}} + 3 M_2 M_1^2 \frac{\Omega_{12}^{(3,3)}}{\Omega_{12}^{(1,1)}} \right],
\end{aligned}
\tag{12-52b}
$$

$$
\begin{aligned}
b_{2-2} &= -x_1 x_2 \tfrac{2}{3} M_1 M_2 \frac{kT}{E} \left(\tfrac{385}{4} - 49 \frac{\Omega_{12}^{(1,2)}}{\Omega_{12}^{(1,1)}} + 8 \frac{\Omega_{12}^{(1,3)}}{\Omega_{12}^{(1,1)}} \right. \\
&\left. - \tfrac{301}{8} \frac{\Omega_{12}^{(2,2)}}{\Omega_{12}^{(1,1)}} + \tfrac{21}{2} \frac{\Omega_{12}^{(2,3)}}{\Omega_{12}^{(1,1)}} - \tfrac{3}{2} \frac{\Omega_{12}^{(2,4)}}{\Omega_{12}^{(1,1)}} + 3 \frac{\Omega_{12}^{(3,3)}}{\Omega_{12}^{(1,1)}} \right).
\end{aligned}
\tag{12-53}
$$

The coefficients, a_{pq} and b_{pq}, when $p,q<0$, have been specified from the expressions for when $p,q>0$ by changing the indices connected with the gas constituents in the manner shown previously, i.e. $[\mu_1]_1 \to [\mu_2]_1$, $x_1 \to x_2$, $x_2 \to x_1$, $M_1 \to M_2$, and $M_2 \to M_1$. The same procedure applied to the expressions for a_{1-2} and b_{1-2} yield expressions for a_{-12} and b_{-12}. The remaining coefficients, a_{pq} and b_{pq}, can be found from the symmetry conditions, $a_{pq} = a_{qp}$ and $b_{pq} = b_{qp}$.

All of the coefficients described above depend upon the Ω-integrals, many of which have been previously tabulated in [2]. In Appendix F, a full set of these Ω-integrals, sufficient for use with the second-order approximation, have been included. Among these are values for the previously unreported, $\Omega^{(1,4)\star}$, $\Omega^{(1,5)\star}$, and $\Omega^{(3,3)\star}$. All of the Ω-integral values included in Appendix F have been fully recomputed and an annotated *Mathematica®* program for use in further calculations has been provided.

4. ANALYTICAL METHODS OF SOLUTION FOR PLANAR BOUNDARY VALUE PROBLEMS INVOLVING BINARY GAS MIXTURES.

A semi-infinite expanse of a binary gas mixture is considered which is bounded by a planar surface lying in the $y-z$ plane and located at $\tilde{x}=0$. Far from the surface, the gas mixture is maintained at a constant mean-mass velocity gradient normal to the surface, $h=(\partial u/\partial \tilde{x})_\infty$, and constant tangential gradients of the partial concentrations, $d_{12}=(\partial x_1/\partial \tilde{y})_\infty$, and temperature (in the y-direction), $g=(\partial \ln(T)/\partial \tilde{y})_\infty$. The total pressure is assumed to be constant. For these conditions in a binary gas mixture, the mean-mass velocity far from the surface is expressible in the following asymptotic form:

$$u(\tilde{x}) = \tilde{x}h + u_0 , \tag{12-54}$$

where u_0 is a constant velocity that arises owing to the influence of the surface on the gas mixture.

The asymptotic behavior of a binary gas mixture may only be found via solution of the Boltzmann equations with appropriate boundary conditions at the wall. For this geometry, the distribution function for the i-th constituent may be expressed as:

$$f_i^{\pm} = f_i^{(0)}\left[1+\frac{m_i}{kT}\tilde{x}hv_{iy} + \Psi_i(\mathbf{c}_i,\tilde{\mathbf{r}}) + \Phi_i^{\pm}(\mathbf{c}_i,\tilde{x})\right], \tag{12-55}$$

where $\mathbf{c}_i = \left(m_i/2kT\right)^{1/2} \mathbf{v}_i$ is the dimensionless velocity, $\Psi_i\left(\mathbf{c}_i,\tilde{\mathbf{r}}\right)$ is the Chapman-Enskog solution, $\Phi_i^{\pm}\left(\mathbf{c}_i,\tilde{x}\right)$ is a function that allows for the influence of the wall, and the signs, $(+)$ and $(-)$, refer to the reflected and incident molecules, respectively.

The functions, $\Phi_i^{\pm}\left(\mathbf{c}_i,\tilde{x}\right)$, can be found from the following linearized Boltzmann equations:

$$v_{ix} f_i^{(0)} \frac{\partial \Phi_i}{\partial \tilde{x}} = -n_i^2 I_i\left(\Phi_i\right) - n_i n_j I_{ij}\left(\Phi_i + \Phi_j\right), \tag{12-56}$$

where $i = 1,2$ with $i \neq j$, and I_i and I_{ij} are the standard notations [1]. These linearized collision integrals are given by

$$\begin{aligned} n_i^2 I_i\left(\Phi_i\right) = \iint f_i^{(0)}\left(\mathbf{v}_i\right) f_i^{(0)}\left(\mathbf{v}\right)\big[\Phi_i\left(\mathbf{v}_i\right) + \Phi_i\left(\mathbf{v}\right) \\ -\Phi_i\left(\mathbf{v}_i'\right) - \Phi_i\left(\mathbf{v}'\right)\big] g_i \alpha_i\left(\theta, g_i\right) d\Omega d\mathbf{v}\,, \end{aligned} \tag{12-57}$$

$$\begin{aligned} n_i n_j I_{ij}\left(\Phi_i + \Phi_j\right) = \iint f_i^{(0)} f_j^{(0)}\big[\Phi_i + \Phi_j \\ -\Phi_i' - \Phi_j'\big] g_{ij} \alpha_{ij}\left(\theta, g_{ij}\right) d\Omega d\mathbf{v}_j\,. \end{aligned} \tag{12-58}$$

Using the Maxwellian boundary model, Eq. (6-4), the boundary conditions for the functions, $\Phi_i^{\pm}\left(\mathbf{c}_i,\tilde{x}\right)$, can be written in the form:

$$\begin{aligned} \Phi_i^{+}\left(\mathbf{c}_i,0\right) = \alpha_{i\tau}\left[A_{iy} g + D_{iy} d_{12}\right] + 2\left(2-\alpha_{i\tau}\right) B_{ixy} h \\ +\left(1-\alpha_{i\tau}\right)\Phi_i^{-}\left(-c_{ix}, c_{iy}, c_{iz}, 0\right). \end{aligned} \tag{12-59}$$

Let the scalar product be introduced by:

$$\left(\varphi_1\left(\mathbf{c}_i,\tilde{x}\right), \varphi_2\left(\mathbf{c}_i,\tilde{x}\right)\right) = \sum_{i=1}^{2} \int \varphi_1 f_i^{(0)} \varphi_2 d\mathbf{v}_i\,. \tag{12-60}$$

Then, using both the commutative property of the standard bracket integrals and the conservation of momentum present during molecular encounters, the scalar product of Eqs. (12-56) with $m_i v_{iy}$ gives:

$$\frac{\partial}{\partial \tilde{x}}\left(m_i v_{ix} v_{iy}, \Phi_i\left(\mathbf{c}_i, \tilde{x}\right)\right)=-n^2\left\{m_i v_{iy}, \Phi_i\right\}=-n^2\left\{\Phi_i, m_i v_{iy}\right\}$$

$$=-n_1^2\left[\Phi_1, m_1 v_{1y}\right]_1-n_1 n_2\left[\Phi_1+\Phi_2, m_1 v_{1y}+m_2 v_{2y}\right]_{12} \tag{12-61}$$

$$-n_2^2\left[\Phi_2, m_2 v_{2y}\right]_2=0 .$$

where the bracket integral notations of Eq. (12-61) are defined in [1] as:

$$n^2\{F, G\}=n_1^2\left[F_1, G_1\right]_1+n_1 n_2\left[F_1+F_2, G_1+G_2\right]_{12}+n_2^2\left[F_2, G_2\right]_2 ,$$

$$\left[F_i, G_i\right]_i=\int G_i I_i\left(F_i\right) d\mathbf{v}_i ,$$

$$\left[F_1+G_2, H_1+K_2\right]_{12}=\int F_1 I_{12}\left(H_1+K_2\right) d\mathbf{v}_1+\int G_2 I_{21}\left(H_1+K_2\right) d\mathbf{v}_2 ,$$

with F, G, F_i, G_i, H_i, and K_i representing arbitrary functions of the molecular velocities of the subscripted constituents, and I_i and I_{ij} having been previously defined in Eqs. (12-57) and (12-58), respectively. In the dimensionless variables, $\mathbf{c}_i$, this moment solution may be written in the form:

$$\left(c_{ix} c_{iy}, \Phi_i\left(\mathbf{c}_i, \tilde{x}\right)\right)=\text{const} . \tag{12-62}$$

If one takes into account the asymptotic behaviors of the functions, $\Phi_i\left(\mathbf{c}_i, \tilde{x}\right) \sim 2\left(m_i / 2kT\right)^{1/2} u_0 c_{iy}$, then this function in the two-moment approach may be chosen in the form:

$$\Phi_i^{\pm}\left(\mathbf{c}_i, \tilde{x}\right)=M_i^{1/2} a^{\pm}(\tilde{x}) c_{iy} . \tag{12-63}$$

This form of the correction to the distribution function means that Eq. (12-62) may be expressed in the form:

$$\left(c_{ix} c_{iy}, \Phi_i\left(\mathbf{c}_i, \tilde{x}\right)\right)=0 , \tag{12-64}$$

which is the exact analytical solution of Eq. (12-56).

The Maxwellian method [4] involves the use of the analytical solution expressed by Eq. (12-64) at the surface where $\tilde{x}=0$, i.e.:

$$\left(c_{ix}c_{iy}, \Phi_i\left(\mathbf{c}_i, 0\right)\right) = 0\,, \tag{12-65}$$

where $\Phi_i\left(\mathbf{c}_i,0\right) = \eta\left(c_{ix}\right)\Phi_i^+\left(\mathbf{c}_i,0\right) + \eta\left(-c_{ix}\right)\Phi_i^-\left(\mathbf{c}_i,0\right)$ and $\eta\left(c_{ix}\right)$ is the Heaviside step function that is defined as:

$$\eta(x) = \begin{cases} 1\,; & x > 0, \\ 0\,; & x < 0. \end{cases}$$

For this method, the boundary conditions are satisfied only in an integral sense and not in an explicit sense. To get some useful results using this method, additional assumptions must be made. The Maxwellian analysis is constructed around the assumption that, in Eq. (12-63), $a^{\pm}(\infty) = a^{-}(0) = a$. Then, the constant, a, may be calculated if one uses the analytical solution given by Eq. (12-65). The Maxwellian method may be classified as a one-moment approach.

Another analytical method, first proposed by Loyalka [5], involves using the two general moment solutions of the Boltzmann equations, Eq. (12-56). The first of these moment solutions, given by Eq. (12-65), is the same as that used in the Maxwellian method. The second moment solution may be derived from Eqs. (12-56) if one uses a property of the Chapman-Enskog solution for viscosity. For the particular planar geometry being considered here, the Chapman-Enskog corrections may be found from the following expressions:

$$f_1^{(0)} c_{1x} c_{1y} = n_1^2 I_1\left(B_{1xy}\right) + n_1 n_2 I_{12}\left(B_{1xy} + B_{2xy}\right), \tag{12-66}$$

$$f_2^{(0)} c_{2x} c_{2y} = n_2^2 I_2\left(B_{2xy}\right) + n_1 n_2 I_{21}\left(B_{1xy} + B_{2xy}\right). \tag{12-67}$$

Now, the scalar product of Eq. (12-56) with B_{ixy}, taken in accordance with Eq. (12-60), yields:

$$\begin{aligned} \frac{\partial}{\partial \tilde{x}}\left(v_{ix} B_{ixy}, \Phi_i\left(\mathbf{c}_i, \tilde{x}\right)\right) &= -n^2\left\{B_{ixy}, \Phi_i\left(\mathbf{c}_i, \tilde{x}\right)\right\} = -n^2\left\{\Phi_i\left(\mathbf{c}_i, \tilde{x}\right), B_{ixy}\right\} \\ &= -\int \Phi_1\left(\mathbf{c}_1, \tilde{x}\right)\left[n_1^2 I_1\left(B_{1xy}\right) + n_1 n_2 I_{12}\left(B_{1xy} + B_{2xy}\right)\right] d\mathbf{v}_1 \\ &\quad - \int \Phi_2\left(\mathbf{c}_2, \tilde{x}\right)\left[n_2^2 I_2\left(B_{2xy}\right) + n_1 n_2 I_{21}\left(B_{1xy} + B_{2xy}\right)\right] d\mathbf{v}_2\,. \end{aligned}$$

Here, the commutative property of the bracket integrals has been used. From this, taking into account Eqs. (12-64), (12-66) and (12-67), one obtains:

$$\frac{\partial}{\partial \tilde{x}}\left(v_{ix}B_{ixy},\Phi_i\left(\mathbf{c}_i,\tilde{x}\right)\right)=-\left(\Phi_i\left(\mathbf{c}_i,\tilde{x}\right),c_{ix}c_{iy}\right)=0\ ,$$

which yields:

$$\left(M_i^{-1/2}c_{ix}^2c_{iy}B_i\left(c_i^2\right),\Phi_i\left(\mathbf{c}_i,\tilde{x}\right)\right)=\text{const}\ . \tag{12-68}$$

The Loyalka method is a generalization of the Maxwellian method and may be classified as a two-moment approach because two-moment solutions to the Boltzmann equations are used in its formulation. Specifically, for the Loyalka method, the two solutions used are the following exact moment solutions:

$$\left(c_{ix}c_{iy},\Phi_i\left(\mathbf{c}_i,\tilde{x}\right)\right)=0\ , \tag{12-69}$$

$$\left(M_i^{-1/2}c_{ix}^2c_{iy}B_i\left(c_i^2\right),\left[\Phi_i\left(\mathbf{c}_i,0\right)-\Phi_i\left(\mathbf{c}_i,\infty\right)\right]\right)=0\ . \tag{12-70}$$

These two analytical solutions give useful results if one takes into account two assumptions concerning the corrections to the distribution functions. Specifically, one must introduce two constants, $a^{\pm}\left(\infty\right)=a$ and $a^{-}\left(0\right)=a_w$, where $a\neq a_w$, which may be calculated from the two analytical solutions given in Eqs. (12-69) and (12-70).

5. THE SLIP COEFFICIENTS FOR A BINARY GAS MIXTURE.

The slip phenomena in gas mixtures are of fundamental significance when it comes to the specification of appropriate boundary conditions for flows in the slip-flow regime. In this regime, the boundary condition for the mean mass velocity component tangential to a body surface may be presented in the form [6-13]:

$$u_\tau\left(\tilde{\mathbf{r}}_S\right)=-c_m\frac{\lambda_\mu}{\mu}P_{n\tau}+\overset{*}{c}_{Tsl}v^*\frac{\partial\ln\left(T\right)}{\partial\tilde{\mathbf{r}}_\tau}+c_{Dsl}D_{12}\left(d_{12}\right)_\tau\ , \tag{12-71}$$

where $\tilde{\mathbf{r}}_S$ is the radius vector of a given point on the body surface, $\mathbf{n}$ indicates a direction normal to the surface and points into the gas at $\tilde{\mathbf{r}}_S$ while $\boldsymbol{\tau}$ indicates a direction tangential to the body surface at the point $\tilde{\mathbf{r}}_S$. Also, c_m, c^*_{Tsl}, and c_{Dsl} are the velocity-slip, thermal-creep, and diffusion-slip coefficients, respectively and the following notations have been used:

$$\lambda_\mu = \tfrac{8}{5}\frac{\mu}{\sqrt{\pi}\,p}\left(\frac{2kT}{m}\right)^{1/2},$$

$$\nu^* = \frac{\mu}{\rho}\frac{\lambda_\kappa}{\lambda_\mu},$$

$$\lambda_\kappa = \tfrac{64}{75}\frac{\kappa}{\sqrt{\pi}\,p}\left(\frac{m}{2kT}\right)^{1/2} T,$$

in which λ_μ and λ_κ are the mean free paths calculated from the viscosity and thermal conductivity coefficients, μ and κ, respectively, and $P_{n\tau}$ is the pressure tensor component. The first term in Eq. (12-71) yields a correction to continuum expressions which is proportional to the Knudsen number, Kn, and which, when $Kn \to 0$, may be neglected. However, the other terms in Eq. (12-71), which are associated with the thermal-creep and diffusion-slip phenomena, may be significant even if $Kn \to 0$ because the influence of these phenomena depends only upon the value of the tangential component of the external non-uniformities in the gas mixture.

In the slip-flow regime, for the bulk flow of the gas mixture, the Navier-Stokes equations can be used to develop the same description as the Boltzmann equations if one introduces the appropriate slip boundary condition given by Eq. (12-71). For this regime, the use of the Navier-Stokes equations together with the slip boundary condition is preferable to the use of the full Boltzmann equations because it allows one to reduce the number of independent variables from ten (for the Boltzmann equations for a binary gas mixture) to four (in the general case with the Navier-Stokes equations). For stationary boundary value problems and for some specific geometries the number of these independent variables may be additionally reduced. It should be noted that the slip conditions themselves must be determined by a solution of the planar boundary value problem involving the Boltzmann equations. These slip coefficients can then be used with suitable accuracy to formulate transport problems for arbitrary geometries.

The slip-flow phenomena are basic to understanding cross-effects in gas mixtures (where they can be the most important factors), the motions of aerosol particles suspended in non-uniform media, and the flow of gas mixtures through capillary tubes, all of which have numerous technological applications. The slip effects are also of importance in studies of many natural phenomena such as aerosol mechanics and flow in porous media, which also play a significant role in numerous other technological applications such as chemical and physical vapor deposition, nano-fabrication, and most low-pressure applications. In some cases, such as physical vapor transport experiments under micro-gravity conditions, for example, these phenomena may become the dominant transport factors due to the production of side-wall gas creep which may suppress natural convection [14-20]. In spite of considerable work in this field [21-63], there is no completely accurate analysis employing realistic models of the intermolecular interactions that would have reliable accuracy in all applications.

The half-range moment method [34,35,51] described in Chapter 8, was previously believed to be able to yield accurate results for slip problems. Unfortunately, this method involves a very complicated analysis (even in low-moment approaches) and the accuracy of the method is not predictable.

The relatively simple Loyalka's method described in this chapter results in a simpler analysis and good accuracy [5]. To calculate the slip coefficients for binary gas mixtures, Loyalka's method is among the most useful. First, it yields relatively simple expressions and, second, it has no difficulties associated with the use of general intermolecular and gas-surface interaction laws. Moreover, the use of the two general conservation laws as exact moment solutions provides a suitable accuracy to the results. Next in this section, Loyalka's method is used to derive analytical expressions for the slip coefficients of a binary gas mixture. This analysis employs the second-order Chapman-Enskog solutions for the transport parameters. One uses the same statement of the planar boundary value problem as was described in Section 12.4. The velocity, u_0, given by Eq. (12-54) is presented in the form:

$$u_0 = c_m \lambda_\mu h + c_{Dsl} D_{12} d_{12} + c_{Tsl} \nu^* g \,. \tag{12-72}$$

Additionally, this asymptotic velocity may be calculated by simple integration, i.e.:

$$\rho u_0 = \left(m_i v_{iy}, M_i^{1/2} a^{\pm}(\infty) c_{iy} \right),$$

where $a^{\pm}(\infty) = a = \text{const}$. This integration yields:

$$u_0 = \tfrac{1}{2}\left(\frac{2kT}{m_0}\right)^{1/2} a\,. \tag{12-73}$$

Within the framework of the Loyalka method, one has two unknown constants, $a^{\pm}(\infty) = a$ and $a^{-}(0) = a_w$, which can be found from the two exact moment solutions given in Eqs. (12-69) and (12-70). Substitution of the boundary condition given by Eq. (12-59) into Eqs. (12-69) and (12-70) yields the analytical expressions for the constants, a and a_w. Then, Eqs. (12-72) and (12-73) yield the following expressions for the slip coefficients:

$$c_m = \tfrac{5}{8}\frac{pM^{1/2}\pi}{\mu}\sum_{i=1}^{2}(2-\alpha_{i\tau})x_i b_{\pm 1}\left[K_1 + \frac{4b_{\pm 1}}{\pi M_i^{1/2}}K_2\omega_{1i}\right], \tag{12-74}$$

$$\begin{aligned} c_{Dsl} = \sum_{i=1}^{2} 2\alpha_{i\tau}\Bigg[& M_i^{1/2}K_1\left((-1)^{i-1}\frac{m_j}{m} - \tfrac{1}{4}z_i\right) \\ & + b_{\pm 1}K_2\left((-1)^{i-1}\frac{m_j}{m} - \tfrac{1}{2}z_i\omega_{2i}\right)\Bigg], \end{aligned} \tag{12-75}$$

$$\begin{aligned} c_{Tsl}^{*} = \frac{3}{a^{*}}\Bigg[& \sum_{i=1}^{2}\alpha_{i\tau}x_i a_{\pm 1}\left(\tfrac{1}{2}K_1\varepsilon_{1i} + b_{\pm 1}M_i^{-1/2}K_2\varepsilon_{2i}\right) \\ & - k_T D_{12}\left(\frac{m_0}{2kT}\right)^{1/2} c_{Dsl}\Bigg], \end{aligned} \tag{12-76}$$

where:

$$M = x_1 M_1 + x_2 M_2,\quad a^{*} = x_1 a_1 M_1^{-1/2} + x_2 a_{-1} M_2^{-1/2},$$

$$\omega_{1i} = 1 - b_{\pm 2}^{*} + \tfrac{17}{4}\left(b_{\pm 2}^{*}\right)^2,\quad \omega_{2i} = 1 - \tfrac{7}{2}b_{\pm 2}^{*},$$

$$\varepsilon_{1i} = 1 + \tfrac{1}{4}a_{\pm 2}^{*}, \text{ and } \varepsilon_{2i} = 1 - \tfrac{7}{2}b_{\pm 2}^{*} + \tfrac{7}{2}a_{\pm 2}^{*}b_{\pm 2}^{*}.$$

Here, also, the following notations have been used:

$$K_1 = \frac{(2-\alpha_{1\tau})x_1 b_1 + (2-\alpha_{2\tau})x_2 b_{-1}}{\alpha_{1\tau}x_1 M_1^{1/2} + \alpha_{2\tau}x_2 M_2^{1/2}} K_2 , \qquad (12\text{-}77)$$

$$K_2 = \tfrac{1}{4}(x_1 b_1 + x_2 b_{-1})^{-1} . \qquad (12\text{-}78)$$

Equations (12-74)-(12-78) are suitable for the computation of results corresponding to arbitrary models of the intermolecular interaction provided that appropriate values of the transport coefficients, $a_{\pm 1}$, $b_{\pm 1}$, etc, are available.

Now, consider the limiting case of a simple gas. The limiting formulas may be derived from Eqs. (12-74)-(12-78) if one makes the substitutions: $x_1 = 1$, $m_1 = m_2 = m$, $\sigma_1 = \sigma_2 = \sigma$, and $\alpha_{1\tau} = \alpha_{2\tau} = \alpha_\tau$. The factors K_1 and K_2 are then given by:

$$K_1 = \tfrac{1}{4}\sqrt{2}(2-\alpha_\tau)\alpha_\tau^{-1} \text{ and } K_2 = \tfrac{1}{4} b_1^{-1} . \qquad (12\text{-}79)$$

The limiting expressions for the slip coefficients are the same as those given by Eqs. (9-62) and (9-63) and the diffusion-slip coefficient is found to be $c_{Dsl} = 0$.

6. DISCUSSION OF THE SLIP COEFFICIENT RESULTS.

The expressions for the slip coefficients that have been obtained in the preceding sections of this chapter are in simple algebraic forms readily usable in computations. Such computations are straightforward provided that values of the coefficients, $a_{\pm 1}$, $b_{\pm 1}$, etc., are available. These values depend upon the intermolecular interaction (potential) model that is used to compute the Ω-integrals. While many of the necessary Ω-integrals are tabulated in the classical texts [2,3], there are several additional Ω-integrals that have been newly computed for inclusion in this text. The various Ω-integrals, both old and new, are given in Appendix F.

Tables 12-1 and 12-2 contain values of the transport coefficients, μ, κ, and D_{12}, and of the slip coefficients, c_m, c_{Dsl}, c_{Tsl}^*, and c_{Tsl}, that have been computed for a selection of binary gas mixtures using different intermolecular potential models. In Table 12-1 the rigid-sphere potential

Table 12-1. Values of the transport coefficients (μ, κ, and D_{12}) and of the slip coefficients (c_m, c_{Dsl}, c^*_{Tsl}, and c_{Tsl}) for a selection of binary gas mixtures obtained using the rigid-sphere potential model and the first- and second-order Chapman-Enskog approximations (p=1 atm, T_0=293 K). The units of the viscosity, μ, are gm cm^{-1} sec^{-1}, the units of the thermal conductivity, κ, are cal cm^{-1} sec^{-1} K^{-1}, and the units of the diffusion coefficient, D_{12}, are cm^2 sec^{-1}. The slip coefficients are all dimensionless. The thermal conductivity coefficients have been calculated under the assumption that the molecules are effectively monatomic. For all of these calculated values, the assumption has been made that the tangential momentum accommodation coefficient for each gas species is equal to unity, i.e. $\alpha_{1\tau}=\alpha_{2\tau}$=1. The column labeled 'Order' refers to the order of the Chapman-Enskog approximation that was used in calculations.

Mixture	Order	μ ($\times 10^4$)	κ ($\times 10^5$)	D_{12}	c_m	c_{Dsl}	c^*_{Tsl}	c_{Tsl}
N_2-H_2	1	1.6874	4.5642	0.6379	1.1189	-0.1775	1.1102	1.1200
x_1=0.99	2	1.7125	4.6736	0.6857	1.0914	-0.1920	0.9786	0.9961
N_2-Ar	1	1.9262	4.1837	0.1659	1.1206	0.1170	1.1120	1.1029
x_1=0.5	2	1.9545	4.2788	0.1690	1.0936	0.1265	0.9825	0.9822
N_2-O_2	1	1.8202	4.5041	0.1752	1.1167	0.0363	1.1235	1.1211
x_1=0.5	2	1.8471	4.6063	0.1783	1.0893	0.0395	0.9924	0.9979
N_2-CO_2	1	1.5254	3.1895	0.1269	1.1253	0.2211	1.0976	1.1106
x_1=0.5	2	1.5493	3.2666	0.1296	1.0982	0.2374	0.9625	0.9821

model was used while in Table 12-2 the Lennard-Jones (6-12) potential model was used. For all mixtures and models the results obtained using the first-order Chapman-Enskog approximation are compared with those obtained using the second-order Chapman-Enskog approximation. In all of the cases shown in Tables 12-1 and 12-2 molecules have been assumed to be diffusely reflected at the relevant surfaces such that $\alpha_{1\tau}=\alpha_{2\tau}=1$. Table 12-3 summarizes values of the parameters that are associated with each of the intermolecular potential models used.

A variety of additional models are also available in the literature for which results have not been computed here but which may also be of interest to the reader [64]. This source contains tabulated values of the transport collision integrals for a variety of modern potential models and provides substantial information about the various intermolecular collision parameters included in these models. Some approximate methods using the various alternative potential models described in this reference, and which may be used to calculate the various transport coefficients for the second- and third-order Chapman-Enskog approximations, may be found in the literature [65-69] but, since these analyses generally are not complete, they tend to be of limited use.

It is clear from Tables 12-1 and 12-2 that there exists a significant dependence of the slip coefficients upon the intermolecular potential (within the context of the Loyalka method) even when the calculations are limited to

Table 12-2. Values of the transport coefficients (μ, κ, and D_{12}) and of the slip coefficients (c_m, c_{Dsl}, c^*_{Tsl}, and c_{Tsl}) for a selection of binary gas mixtures obtained using the Lennard-Jones (6-12) potential model and the first- and second-order Chapman-Enskog approximations (p=1 atm, T_0=293 K). The units of the viscosity, μ, are gm cm^{-1} sec^{-1}, the units of the thermal conductivity, κ, are cal cm^{-1} sec^{-1} K^{-1}, and the units of the diffusion coefficient, D_{12}, are cm^2 sec^{-1}. The thermal conductivity coefficients have been calculated under the assumption that the molecules are effectively monatomic. The slip coefficients are all dimensionless. For all of these calculated values, the assumption has been made that the tangential momentum accommodation coefficient for each gas species is equal to unity, i.e. $\alpha_{1\tau}=\alpha_{2\tau}$=1. The column labeled 'Order' refers to the order of the Chapman-Enskog approximation that was used in calculations. All of the necessary Ω-integrals have been calculated numerically using the appropriate reduced temperatures and without using the typical interpolation procedure.

Mixture	Order	μ ($\times 10^4$)	κ ($\times 10^5$)	D_{12}	c_m	c_{Dsl}	c^*_{Tsl}	c_{Tsl}
N_2-H_2	1	1.7375	4.7246	0.7361	1.1189	-0.1896	1.1072	1.1228
x_1=0.99	2	1.7446	4.7576	0.7564	1.1021	-0.1959	1.0342	1.0518
N_2-Ar	1	1.9872	4.3334	0.1870	1.1208	0.1307	1.1138	1.1091
x_1=0.5	2	1.9932	4.3543	0.1877	1.1064	0.1331	1.0514	1.0489
N_2-O_2	1	1.8825	4.6598	0.1982	1.1168	0.0380	1.1238	1.1216
x_1=0.5	2	1.8893	4.6858	0.1990	1.1009	0.0390	1.0548	1.0549
N_2-CO_2	1	1.6006	3.3829	0.1477	1.1252	0.2589	1.1051	1.1303
x_1=0.5	2	1.6037	3.3940	0.1481	1.1141	0.2621	1.0560	1.0815

the first-order Chapman-Enskog solutions whereas, in a simple gas, such a dependence is observed only for calculations employing the second- or higher-order Chapman-Enskog solutions. As is evident from Eqs. (12-74)-(12-76), the thermal-creep and diffusion-slip coefficients are significantly more sensitive to the order of the Chapman-Enskog approximation than the velocity-slip coefficient. This is undoubtedly due to the dependence of the thermal-creep and diffusion-slip coefficients on cross-effects; a supposition that is confirmed by direct calculation of the slip coefficients for a selection of gas mixtures. For example, with the rigid-sphere model for a mixture of N_2 and O_2, the relative differences between the first-order Chapman-Enskog derived values with respect to the second-order Chapman-Enskog derived values are 13.5 % for c^*_{Tsl} and 8.1 % for c_{Dsl}. For the same mixture and the rigid-sphere model the relative difference observed for c_m is only 2.5 %. The same situation occurs in a simple gas as well where the relative differences observed for c^*_{Tsl} and c_m are 13 % and 2.5 %, respectively. Of course, for a simple gas, one must remember that there is no diffusion-slip effect to observe.

The relative differences of the slip coefficients that one observes when using the generally more realistic Lennard-Jones (6-12) potential model are somewhat less than those observed for the rigid-sphere model but the same

Table 12-3. Relevant parameters for two of the most commonly used intermolecular potential models; the rigid-sphere model [1] and the Lennard-Jones (6-12) potential model [2].

	Rigid-sphere model	Lennard-Jones (6-12) potential model	
Gas	σ (Å)	σ (Å)	ε/k (K)
He	2.193	2.576	10.22
H_2	2.745	2.94	35.65
N_2	3.784	3.72	85.65
Ar	3.659	3.418	124.0
O_2	3.636	3.487	100.5
CO_2	4.643	3.95	201.5

trend with respect to the sensitivity is observed. In a mixture of N_2 and H_2, the relative differences determined for c_{Tsl}^* and c_{Dsl} (with respect to the second-order approximation values) are 6.7 % and 3.2 %, respectively while the relative difference determined for c_m is only 1.5 %; again indicating that the velocity-slip coefficient is not as sensitive to the order of the approximation used.

Further, if one compares the relative differences between the values of these coefficients using the different potential models with the same order approximation for the same mixture of N_2 and CO_2 one finds that c_{Tsl}^* and c_{Dsl} vary from the Lennard-Jones (6-12) model to the rigid-sphere model by 9.1 % and 9.4 %, respectively; clearly showing that the more realistic model likely produces substantially more realistic results. From this discussion it is clear that whenever possible it is desirable to use the higher-order approximations and most realistic potential models as significant improvements in the theory of the slip phenomena may be realized under these circumstances.

One slip phenomenon of interest is the thermal transpiration effect which is observed in capillary tubes. Table 12-4 compares the experimental and theoretical values of the slip factor, $c_{sl} = c_{Tsl} - c_{Dsl}\left(k_T D_{12}/\nu\right)$, which governs this effect [48]. The experimental values given in Table 12-4 are slightly different from those reported in [48] owing to the use of the second-order Chapman-Enskog approximations in determining the viscosity and velocity-slip coefficients which are used to calculate the slip factor for experimentally obtained pressure difference data. The statement of the problem for the thermal transpiration effect and the method solution are given in Problem 12.5. The specific values of the tangential momentum accommodation coefficients, $\alpha_{1\tau}$ and $\alpha_{2\tau}$, shown in Table 12-4 were chosen so as to produce the best possible agreement between the measured and theoretical slip factors for the selected binary gas mixtures. Table 12-4 shows that excellent agreement between theory and experiment is obtained if these values are assumed. This further demonstrates the substantial improvement in the accuracy of theoretically calculated slip phenomena

Table 12-4. The tangential momentum accommodation coefficients, $\alpha_{i\tau}$, in a binary gas mixture chosen for calculation of the slip coefficients and slip factors. These specific $\alpha_{i\tau}$ were chosen so as to produce the best agreement between the measured and theoretical slip factors for the selected binary gas mixtures.

Gas mixture	$\alpha_{1\tau}$	$\alpha_{2\tau}$	$\left[c_{sl}\right]_{\exp}$	$\left[c_{sl}\right]_{th}$	Error (%) †
He:Ar (4:1)	0.8	0.8	1.1863	1.1708	1.3
He:Ar (2:1)	0.8	0.8	1.1874	1.1559	2.7
He:Ar (1:1)	0.8	0.8	1.1043	1.1146	0.9
He:Ar (1:2)	0.8	0.8	1.0903	1.0714	1.7
He:CO_2 (2:1)	0.8	0.8	1.3846	1.3141	5.1
He:CO_2 (1:1)	0.8	0.8	1.2997	1.2720	2.1
Ar:CO_2 (2:1)	0.8	0.8	0.9593	1.0098	5.3
Ar:CO_2 (1:1)	0.8	0.8	0.9573	1.0143	6.0

† Relative error has been calculated with respect to the experimental values of the slip, i.e. $\left|\left[c_{sl}\right]_{\exp}-\left[c_{sl}\right]_{th}\right|/\left[c_{sl}\right]_{\exp}\times 100\%$.

values that can be obtained when using higher-order Chapman-Enskog approximations; in this case the second-order approximation.

The above discussion has shown that there are strong dependencies of c_{Tsl}^{*} and c_{Dsl} on the potential model used and on the order of the Chapman-Enskog approximation used. However, c_m does not show the same degree of sensitivity as the other slip coefficients. There exists a direct way to assess the accuracy of the analysis reported here by comparing the values of the slip coefficients that it predicts with the very accurate numerical values reported by Takata and Aoki in [62]. The relative deviation of the values calculated from Eqs. (12-75) and (12-76) from the values reported in [62] is less than 2 % over a wide range of the molecular mass ratio. From this, one may conclude that the overall degree of accuracy of the slip coefficients reported here is, almost certainly, less than 2 % as well.

PROBLEMS

12.1. Using the first-order Chapman-Enskog approximation, determine the diffusion-slip coefficient for a binary gas mixture by means of the Maxwell method. Use a planar geometry in which the pressure and temperature are assumed to be constant.

Solution: The distribution function for the i-th constituent may be expressed in the form:

$$f_i^{\pm}=f_i^{(0)}\left[1-c_{iy}\frac{D_{12}}{x_i}\left(\frac{m_i}{2kT}\right)^{1/2}\left((-1)^{i-1}2\frac{m_j}{m}d_{12}\right)+M_i^{1/2}a^{\pm}(\tilde{x})c_{iy}\right],$$

where $d_{12} = \left(\partial x_1 / \partial \tilde{y}\right)_{\infty}$, and $\tilde{x}$, and $\tilde{y}$ are the coordinates normal and tangential to the planar surface, respectively. The Maxwell method is constructed on the assumptions that $a^{\pm}(\infty) = a$ and $a^{-}(0) = a^{-}(\infty) = a$. The constant, a, is found from the moment solution given by Eq. (12-65):

$$\left(c_{ix} c_{iy}, \Phi_i(\mathbf{c}_i, 0)\right) = 0 .$$

The boundary conditions (one for each constituent) are given by:

$$\Phi_i^{+}(\mathbf{c}_i, 0) = 2\alpha_{i\tau} d_{12} \left[(-1)^{i-1} \frac{m_j}{m} \frac{D_{12}}{x_i} \left(\frac{m_i}{2kT} \right)^{1/2} \right] c_{iy} + \left(1 - \alpha_{i\tau}\right) M_i^{1/2} a c_{iy} ,$$

where $\alpha_{i\tau}$ is the tangential momentum accommodation coefficient for the i-th constituent. The constant, a, can then be found by simple integration. Thus, Eq. (12-73) yields:

$$c_{Dsl} = \left(\alpha_{1\tau} \sqrt{m_2} - \alpha_{2\tau} \sqrt{m_1} \right) K_0 ,$$

where:

$$K_0 = \frac{\sqrt{m_1 m_2}}{m\left(\alpha_{1\tau} x_1 \sqrt{m_1} + \alpha_{2\tau} x_2 \sqrt{m_2} \right)} .$$

12.2. For a non-uniform binary gas mixture, determine the mean velocities of the constituent gases and the mean mass velocity of the mixture. Consider the specific case in which there is no net number flow (transport) of molecules and in which the pressure and temperature of the gas mixture are assumed to be uniform.

Solution: The distribution function of the i-th constituent is given by:

$$f_i = f_i^{(0)} \left(1 + \frac{m_i}{kT} \mathbf{u} \cdot \mathbf{v}_i - \mathbf{D}_i \cdot \mathbf{d}_{12} \right),$$

where $\mathbf{u}$ is the mean mass velocity of the mixture and in which $\mathbf{D}_i$ and $\mathbf{d}_{12}$ are expressed as:

$$\mathbf{D}_i = \frac{D_{12}}{x_i} \left(\frac{m_i}{2kT_0} \right)^{1/2} \mathbf{C}_i \left[(-1)^{i-1} 2 \frac{m_j}{m} + z_i S_{3/2}^{(1)}\left(C_i^2\right) \right] \text{ and } \mathbf{d}_{12} = n^{-1} \nabla n_1 .$$

The mean mass velocity of the i-th constituent can be calculated by integrating in the following manner:

$$\bar{v}_{i\beta} = n_i^{-1} \int v_{i\beta} f_i d\mathbf{v}_i = n_i^{-1} \int f_i^{(0)} \left(1 + \frac{m_i}{kT} u_\alpha v_{i\alpha} - D_{i\alpha} \left(d_{12} \right)_\alpha \right) v_{i\beta} d\mathbf{v}_i$$

$$= u_\beta - \left(-1\right)^{i-1} \frac{m_j}{m} \frac{D_{12}}{x_i} \left(d_{12} \right)_\beta .$$

This yields:

$$\bar{\mathbf{v}}_1 = \mathbf{u} - \frac{m_2}{m} \frac{D_{12}}{x_1} \mathbf{d}_{12} , \tag{P-1}$$

and:

$$\bar{\mathbf{v}}_2 = \mathbf{u} + \frac{m_1}{m} \frac{D_{12}}{x_2} \mathbf{d}_{12} . \tag{P-2}$$

For the specific case under consideration in which there is no net number flow of molecules, the velocities, $\bar{\mathbf{v}}_1$ and $\bar{\mathbf{v}}_2$, may be found from the equations:

$$\bar{\mathbf{v}}_1 - \bar{\mathbf{v}}_2 = -D_{12} \left(n/n_1 n_2 \right) \nabla n_1 , \tag{P-3}$$

$$n_1 \bar{\mathbf{v}}_1 + n_2 \bar{\mathbf{v}}_2 = 0 , \tag{P-4}$$

which yield:

$$\bar{\mathbf{v}}_1 = -D_{12} n_1^{-1} \nabla n_1 \quad \text{and} \quad \bar{\mathbf{v}}_2 = D_{12} n_2^{-1} \nabla n_1 .$$

From these, the mean mass velocity is then determined to be:

$$\mathbf{u} = \frac{\rho_1 \bar{\mathbf{v}}_1 + \rho_2 \bar{\mathbf{v}}_2}{\rho_1 + \rho_2} = \frac{m_2 - m_1}{\rho} D_{12} \nabla n_1 . \tag{P-5}$$

12.3. For a non-uniform binary gas mixture, determine the number flux vector (current) of the first constituent assuming that the first constituent is diffusing through the second and that there is no net number flow (current) of the second constituent. Assume the pressure and temperature of the gas mixture to be uniform.

Solution: From the statement of the problem, one has that $\overline{\mathbf{v}}_2 = 0$. Since the gas mixture conditions in this problem are similar to those in Problem 12.2, one then has, from Eq. (P-2), that:

$$\mathbf{u} = -\frac{m_1}{m}\frac{D_{12}}{x_2}\mathbf{d}_{12} .$$

Using this in Eq. (P-1) yields:

$$n_1\overline{\mathbf{v}}_1 = -\left(n/n_2\right)D_{12}\nabla n_1 .$$

12.4. For a non-uniform binary gas mixture, determine the mean velocities of the constituent gases. In this more general problem, assume that the partial densities of the constituents, and the pressure and temperature of the gas mixture are all non-uniform.

Solution: The distribution function of the i-th constituent is given by:

$$f_i = f_i^{(0)}\left(1 + \frac{m_i}{kT}\mathbf{u}\cdot\mathbf{v}_i - \mathbf{A}_i\cdot\nabla\ln\left(T\right) - \mathbf{D}_i\cdot\mathbf{d}_{12} - 2\mathbf{B}_i : \frac{\partial\mathbf{u}}{\partial\tilde{\mathbf{r}}}\right), \tag{P-6}$$

where $\mathbf{d}_{12}$ is specified by Eq. (12-4). The same integration as in Problem 12.2 yields:

$$\overline{\mathbf{v}}_1 = \mathbf{u} - \frac{m_2}{m}\frac{D_{12}}{x_1}\left(\mathbf{d}_{12} + k_T\nabla\ln\left(T\right)\right), \tag{P-7}$$

and:

$$\overline{\mathbf{v}}_2 = \mathbf{u} + \frac{m_1}{m}\frac{D_{12}}{x_2}\left(\mathbf{d}_{12} + k_T\nabla\ln\left(T\right)\right), \tag{P-8}$$

where k_T is the thermal diffusion ratio and $\mathbf{u}$ is the mean mass velocity which is unknown until the exact conditions governing the diffusion are specified. Note: Terms containing Sonine polynomials which occur in the functions, $\mathbf{A}_i$ and $\mathbf{D}_i$, should vanish after the integration procedure and integrations over the function $\mathbf{B}_i$ should be identically zero.

12.5. For a non-uniform binary gas mixture, determine the relationship between the mean velocity of the gas mixture and its mean mass velocity.

Solution: Let $\overline{\mathbf{v}}_i$ be the mean velocity of the i-th constituent of the gas mixture in the laboratory frame of reference. This velocity may be expressed in the form:

$$\overline{\mathbf{v}}_i = \mathbf{w} + \mathbf{W}_i = \mathbf{u} + \mathbf{U}_i \ , \qquad \text{(P-9)}$$

where $\mathbf{w}$ and $\mathbf{u}$ are the mean and mean mass velocities of the gas mixture, respectively, and $\mathbf{W}_i$ and $\mathbf{U}_i$ are the diffusion velocities of the i-th constituent relative to two different frames of reference moving with the velocities, $\mathbf{w}$ and $\mathbf{u}$, respectively. If one takes into account Eqs. (P-7) and (P-8), the relative diffusion velocities, $\mathbf{W}_i$ and $\mathbf{U}_i$, may be found from the two following algebraic systems of equations:

$$\mathbf{W}_1 - \mathbf{W}_2 = \overline{\mathbf{v}}_1 - \overline{\mathbf{v}}_2 = -\frac{D_{12}}{x_1 x_2}\left(\mathbf{d}_{12} + k_T \nabla \ln(T)\right),$$

$$n_1 \mathbf{W}_1 + n_2 \mathbf{W}_2 = 0 \ ,$$

$$\mathbf{U}_1 - \mathbf{U}_2 = \overline{\mathbf{v}}_1 - \overline{\mathbf{v}}_2 = -\frac{D_{12}}{x_1 x_2}\left(\mathbf{d}_{12} + k_T \nabla \ln(T)\right),$$

$$\rho_1 \mathbf{U}_1 + \rho_2 \mathbf{U}_2 = 0 \ .$$

Upon solving these systems for $\mathbf{W}_i$ and $\mathbf{U}_i$, the results may be substituted into Eq. (P-9) to yield:

$$\mathbf{u} = \mathbf{w} - \frac{m_1 - m_2}{m} D_{12}\left(\mathbf{d}_{12} + k_T \nabla \ln(T)\right). \qquad \text{(P-10)}$$

12.6. Calculate the diffusion coefficient for a H_2-N_2 binary gas mixture at $T = 293$ K and $p = 1$ atm. Use the first-order Chapman-Enskog approximation. Assume that the molecules act as rigid spheres.

Solution: Using Eqs. (12-13), (12-17) and (12-19) one obtains:

$$D_{12}^{(1)} = \tfrac{3}{16} \frac{kT}{M_1 M_2 n m_0 \Omega_{12}^{(1,1)}} \ .$$

The Ω-integral for rigid-sphere molecules is given by [3]:

$$\Omega_{12}^{(l,r)} = \pi\sigma_{12}^2 \left(\frac{kT}{2\pi m^*}\right)^{1/2} \tfrac{1}{2}(r+1)! \left[1 - \tfrac{1}{2}\frac{1+(-1)^l}{l+1}\right], \tag{P-11}$$

where $m^* = m_1 m_2/(m_1 + m_2)$. Then, one can obtain the following analytical expression for the diffusion coefficient.

$$D_{12}^{(1)} = \tfrac{3}{8}\frac{kT}{p\sigma_{12}^2}\left(\frac{kT}{2\pi m^*}\right)^{1/2}. \tag{P-12}$$

For the specific mixture under consideration, one may use $\sigma_{12} = 3.2645 \times 10^{-8}$ cm from [1] and, from [70], $p = 1.01325 \times 10^6$ dyne cm^{-2}, $m_H = 1.00794$ gm mol^{-1}, $m_N = 14.00674$ gm mol^{-1}, and $k = 1.38066 \times 10^{-16}$ erg K^{-1} and $N_A = 6.02214 \times 10^{23}$ mol^{-1} for Boltzmann's constant and Avogadro's number, respectively. Then, Eq. (P-12) yields $D_{12}^{(1)} = 0.6379$ cm^2 sec^{-1}.

12.7. For the second-order Chapman-Enskog approximation, obtain an analytical expression for the diffusion coefficient that relates it to the expression for the diffusion coefficient obtained in Problem 12.6 using the first-order Chapman-Enskog approximation.

Solution: The transport coefficient, d_0, is calculated from the following algebraic system of equations:

$$d_{-1}a_{-1-1} + d_0 a_{0-1} + d_1 a_{1-1} = 0,$$

$$d_{-1}a_{-10} + d_0 a_{00} + d_1 a_{10} = \tfrac{3}{2}n^{-1}\left(\frac{2kT}{m_0}\right)^{1/2},$$

$$d_{-1}a_{-11} + d_0 a_{01} + d_1 a_{11} = 0,$$

where the coefficients, a_{00}, a_{11}, a_{-1-1}, and $a_{1-1} = a_{-11}$ are given in Section 12.2 while the coefficients, $a_{01} = a_{10}$ and $a_{0-1} = a_{-10}$, are given by Eqs. (12-44) and (12-45), respectively. Then, taking into account Eqs. (12-13) and (12-17), one can represent the diffusion coefficient obtained using the second-order approximation in the form:

$$D_{12}^{(2)} = D_{12}^{(1)}\frac{1}{1-\Delta},$$

where the following notation has been introduced:

$$\Delta = \frac{a_{0-1}\begin{vmatrix} a_{-10} & a_{10} \\ a_{-11} & a_{11} \end{vmatrix} + a_{01}\begin{vmatrix} a_{-1-1} & a_{1-1} \\ a_{-10} & a_{10} \end{vmatrix}}{a_{00}\begin{vmatrix} a_{-1-1} & a_{1-1} \\ a_{-11} & a_{11} \end{vmatrix}} = \frac{a_{01}^2 a_{-1-1} + a_{0-1}^2 a_{11} - 2a_{0-1}a_{10}a_{-11}}{a_{00}\left(a_{-1-1}a_{11} - a_{1-1}^2\right)} .$$

12.8. For the first-order Chapman-Enskog approximation, derive an analytical expression for the viscosity coefficient for a binary gas mixture.

Solution: The viscosity coefficient is given by $\mu = p(x_1 b_1 + x_2 b_{-1})$. The transport coefficients, b_1 and b_{-1}, are specified from the following algebraic system of equations found in Section 12.2 (Eqs. (12-14) and (12-15)):

$$b_{-1-1}b_{-1} + b_{1-1}b_1 = \beta_{-1} = \tfrac{5}{2}\frac{n_2}{n^2}, \text{ and } b_{-11}b_{-1} + b_{11}b_1 = \beta_1 = \tfrac{5}{2}\frac{n_1}{n^2},$$

The solution of these equations yields:

$$[\mu]_1 = \tfrac{5}{2}kT\frac{x_1^2 b_{-1-1} - 2x_1x_2 b_{1-1} + x_2^2 b_{11}}{b_{-1-1}b_{11} - b_{1-1}^2} .$$

12.9. For the first-order Chapman-Enskog approximation, derive an analytical expression for the thermal conductivity coefficient for a binary gas mixture.

Solution: The thermal conductivity coefficient is given by Eq. (12-16):

$$\kappa = -\tfrac{5}{4}nk\left[x_1\left(\frac{2kT}{m_1}\right)^{1/2} a_1 + x_2\left(\frac{2kT}{m_2}\right)^{1/2} a_{-1}\right],$$

The transport coefficients, a_1 and a_{-1}, are specified from the algebraic system of equations found in Section 12.2 (Eqs. (12-11) and (12-12)):

$$a_{-1-1}a_{-1} + a_{1-1}a_1 = \alpha_{-1} = -\tfrac{15}{4}\frac{n_2}{n^2}\left(\frac{2kT}{m_2}\right)^{1/2}, \text{ and:}$$

$$a_{-11}a_{-1} + a_{11}a_1 = \alpha_1 = -\tfrac{15}{4}\frac{n_1}{n^2}\left(\frac{2kT}{m_1}\right)^{1/2},$$

The solution of these equations yields:

$$[\kappa]_1 = \tfrac{75}{8}k^2T\frac{x_1^2m_1^{-1}a_{-1-1} - 2x_1x_2\left(m_1m_2\right)^{-1/2}a_{1-1} + x_2^2m_2^{-1}a_{11}}{a_{-1-1}a_{11} - a_{1-1}^2}.$$

12.10. For the second-order Chapman-Enskog approximation, derive an analytical expression for the viscosity coefficient for a binary gas mixture.

Solution: The viscosity coefficient is given by $\mu = p(x_1b_1 + x_2b_{-1})$. The transport coefficients, b_1 and b_{-1}, are specified from the following algebraic system of equations:

$$b_{-2-2}b_{-2} + b_{-1-2}b_{-1} + b_{1-2}b_1 + b_{2-2}b_2 = 0,$$

$$b_{-2-1}b_{-2} + b_{-1-1}b_{-1} + b_{1-1}b_1 + b_{2-1}b_2 = \tfrac{5}{2}\frac{n_2}{n^2},$$

$$b_{-21}b_{-2} + b_{-11}b_{-1} + b_{11}b_1 + b_{21}b_2 = \tfrac{5}{2}\frac{n_1}{n^2},$$

$$b_{-22}b_{-2} + b_{-12}b_{-1} + b_{12}b_1 + b_{22}b_2 = 0.$$

The solution of this system of equations is given by:

$$b_1 = \frac{\Delta_{b_1}}{[\Delta_b]_2} \quad \text{and} \quad b_{-1} = \frac{\Delta_{b_{-1}}}{[\Delta_b]_2},$$

where the following notations have been introduced:

$$[\Delta_b]_2 = \begin{vmatrix} b_{-2-2} & b_{-1-2} & b_{1-2} & b_{2-2} \\ b_{-2-1} & b_{-1-1} & b_{1-1} & b_{2-1} \\ b_{-21} & b_{-11} & b_{11} & b_{21} \\ b_{-22} & b_{-12} & b_{12} & b_{22} \end{vmatrix},$$

$$\Delta_{b_1} = \tfrac{5}{2}\frac{n_1}{n^2}\Delta'_{b_1} - \tfrac{5}{2}\frac{n_2}{n^2}\Delta''_{b_1} ,$$

$$\Delta_{b_{-1}} = -\tfrac{5}{2}\frac{n_1}{n^2}\Delta'_{b_{-1}} + \tfrac{5}{2}\frac{n_2}{n^2}\Delta''_{b_{-1}} ,$$

$$\Delta'_{b_1} = \begin{vmatrix} b_{-2-2} & b_{-1-2} & b_{2-2} \\ b_{-2-1} & b_{-1-1} & b_{2-1} \\ b_{-22} & b_{-12} & b_{22} \end{vmatrix}, \quad \Delta''_{b_1} = \begin{vmatrix} b_{-2-2} & b_{-1-2} & b_{2-2} \\ b_{-21} & b_{-11} & b_{21} \\ b_{-22} & b_{-12} & b_{22} \end{vmatrix},$$

$$\Delta'_{b_{-1}} = \begin{vmatrix} b_{-2-2} & b_{1-2} & b_{2-2} \\ b_{-2-1} & b_{1-1} & b_{2-1} \\ b_{-22} & b_{12} & b_{22} \end{vmatrix}, \text{ and } \Delta''_{b_{-1}} = \begin{vmatrix} b_{-2-2} & b_{1-2} & b_{2-2} \\ b_{-21} & b_{11} & b_{21} \\ b_{-22} & b_{12} & b_{22} \end{vmatrix}.$$

Using these notations, the viscosity coefficient is found to be:

$$[\mu]_2 = \tfrac{5}{2}kT\frac{x_1^2\Delta'_{b_1} - x_1x_2\left(\Delta''_{b_1} + \Delta'_{b_{-1}}\right) + x_2^2\Delta''_{b_{-1}}}{[\Delta_b]_2} .$$

12.11. For the second-order Chapman-Enskog approximation, derive an analytical expression for the thermal conductivity coefficient for a binary gas mixture.

Solution: The thermal conductivity coefficient is given by Eq. (12-16):

$$\kappa = -\tfrac{5}{4}nk\left[x_1\left(\frac{2kT}{m_1}\right)^{1/2} a_1 + x_2\left(\frac{2kT}{m_2}\right)^{1/2} a_{-1}\right], \tag{12-16}$$

The transport coefficients, a_1 and a_{-1}, are specified from the following algebraic system of equations:

$$a_{-2-2}a_{-2} + a_{-1-2}a_{-1} + a_{1-2}a_1 + a_{2-2}a_2 = 0 ,$$

$$a_{-2-1}a_{-2} + a_{-1-1}a_{-1} + a_{1-1}a_1 + a_{2-1}a_2 = -\tfrac{15}{4}\frac{n_2}{n^2}\left(\frac{2kT}{m_2}\right)^{1/2} ,$$

$$a_{-21}a_{-2} + a_{-11}a_{-1} + a_{11}a_1 + a_{21}a_2 = -\tfrac{15}{4}\frac{n_1}{n^2}\left(\frac{2kT}{m_1}\right)^{1/2},$$

$$a_{-22}a_{-2} + a_{-12}a_{-1} + a_{12}a_1 + a_{22}a_2 = 0\,.$$

The solution of this system of equations is given by:

$$a_1 = \frac{\Delta_{a_1}}{[\Delta_a]_2} \quad \text{and} \quad a_{-1} = \frac{\Delta_{a_{-1}}}{[\Delta_a]_2},$$

where the following notations have been introduced:

$$[\Delta_a]_2 = \begin{vmatrix} a_{-2-2} & a_{-1-2} & a_{1-2} & a_{2-2} \\ a_{-2-1} & a_{-1-1} & a_{1-1} & a_{2-1} \\ a_{-21} & a_{-11} & a_{11} & a_{21} \\ a_{-22} & a_{-12} & a_{12} & a_{22} \end{vmatrix},$$

$$\Delta_{a_1} = -\tfrac{15}{4}\frac{n_1}{n^2}\left(\frac{2kT}{m_1}\right)^{1/2}\Delta'_{a_1} + \tfrac{15}{4}\frac{n_2}{n^2}\left(\frac{2kT}{m_2}\right)^{1/2}\Delta''_{a_1},$$

$$\Delta_{a_{-1}} = \tfrac{15}{4}\frac{n_1}{n^2}\left(\frac{2kT}{m_1}\right)^{1/2}\Delta'_{a_{-1}} - \tfrac{15}{4}\frac{n_2}{n^2}\left(\frac{2kT}{m_2}\right)^{1/2}\Delta''_{a_{-1}},$$

$$\Delta'_{a_1} = \begin{vmatrix} a_{-2-2} & a_{-1-2} & a_{2-2} \\ a_{-2-1} & a_{-1-1} & a_{2-1} \\ a_{-22} & a_{-12} & a_{22} \end{vmatrix}, \quad \Delta''_{a_1} = \begin{vmatrix} a_{-2-2} & a_{-1-2} & a_{2-2} \\ a_{-21} & a_{-11} & a_{21} \\ a_{-22} & a_{-12} & a_{22} \end{vmatrix},$$

$$\Delta'_{a_{-1}} = \begin{vmatrix} a_{-2-2} & a_{1-2} & a_{2-2} \\ a_{-2-1} & a_{1-1} & a_{2-1} \\ a_{-22} & a_{12} & a_{22} \end{vmatrix}, \quad \text{and } \Delta''_{a_{-1}} = \begin{vmatrix} a_{-2-2} & a_{1-2} & a_{2-2} \\ a_{-21} & a_{11} & a_{21} \\ a_{-22} & a_{12} & a_{22} \end{vmatrix}.$$

Table 12-5. A comparison of diffusion-slip coefficient values. Values determined numerically [62] are compared with corresponding values determined using the Maxwell method for selected values of the mass ratio, ω, for the specific case when $\alpha_{1\tau} = \alpha_{2\tau} = 1$ and $x_1 = x_2 = \frac{1}{2}$.

$\omega = m_2/m_1$	2	4	5	10
c_{Dsl}, Eq. (P-13)	0.3235	0.5333	0.5694	0.5974
c_{Dsl}, numerical	0.1637	0.2514	0.2634	0.2619
$\delta(\%)$	97.6	112	116	128

$\delta(\%) = \left|c_{Dsl.num} - c_{Dsl.Max}\right| / c_{Dsl.num} \times 100$.

Using these notations, the thermal conductivity coefficient is found to be:

$$\left[\kappa\right]_2 = \tfrac{75}{8} k^2 T \frac{x_1^2 m_1^{-1} \Delta'_{a_1} - x_1 x_2 \left(m_1 m_2\right)^{-1/2} \left(\Delta''_{a_1} + \Delta'_{a_{-1}}\right) + x_2^2 m_2^{-1} \Delta''_{a_{-1}}}{\left[\Delta_a\right]_2}.$$

12.12. Evaluate the accuracy of the Maxwell method for determining the diffusion-slip coefficient by comparing it with the numerical results [62] for different values of the mass ratio, $\omega = m_2/m_1$, Consider the specific case when $\alpha_{1\tau} = \alpha_{2\tau} = 1$ and $x_1 = x_2 = \frac{1}{2}$.

Solution: For the specific case under consideration, the diffusion-slip coefficient given in Problem 12.1 may be expressed in the form:

$$c_{Dsl} = \frac{4\omega(\omega - 1)}{\left(\omega^2 + 1\right)\left(\omega + 1\right)}. \quad \text{(P-13)}$$

The numerical values and values computed from Eq. (P-13) can then be assembled in tabular form for ease of comparison. This has been done and is given in Table 12-5. As one can see from Table 12-5, with relative errors on the order of 100%, the Maxwell method does not appear, at least for the specific case under consideration in this problem, to be an effective technique for determination of the diffusion-slip coefficient. However, the current case is quite limited and one should not make the mistake of inferring that the accuracy of the Maxwell method overall is represented by the current problem. Rather, one should understand from the current problem only that the Maxwell method can lead to unacceptably inaccurate results in some circumstances.

12.13. Using the first-order Chapman-Enskog approximation and the Maxwell method, determine the slip-flow coefficient for a binary gas mixture. Consider the case in which the temperature and partial pressures of the gas constituents are uniform.

Solution: The distribution function of the i-th constituent in the gas mixture is given by:

$$f_i^{\pm} = f_i^{(0)}\left[1 + \frac{m_i}{kT}\tilde{x}hv_{iy} - 2c_{ix}c_{iy}b_{\pm 1}h + \Phi_i^{\pm}\left(\mathbf{c}_i, \tilde{x}\right)\right],$$

where one should use b_1 if $i = 1$ and b_{-1} if $i = 2$. The Maxwellian boundary model yields:

$$\Phi_i^{+}\left(\mathbf{c}_i, 0\right) = 2\left(2 - \alpha_{i\tau}\right)c_{ix}c_{iy}b_{\pm 1}h + \left(1 - \alpha_{i\tau}\right)\Phi_i^{-}\left(-c_{ix}, c_{iy}, c_{iz}, 0\right).$$

The correction to the distribution function, $\Phi_i\left(\mathbf{c}_i, \tilde{x}\right)$, can be chosen in the form $\Phi_i^{\pm}\left(\mathbf{c}_i, \tilde{x}\right) = M_i^{1/2}a^{\pm}\left(\tilde{x}\right)c_{iy}$. The Maxwell method is constructed on the assumption that $a^{+}\left(\infty\right) = a^{-}\left(0\right) = a$. Then this constant can be calculated from the exact analytical solution to the Boltzmann equations given by:

$$\left(c_{ix}c_{iy}, \Phi_i^{\pm}\left(\mathbf{c}_i, 0\right)\right) = 0.$$

Taking into account Eqs. (12-72) and (12-73), one can then obtain:

$$c_m = \frac{pM^{1/2}}{\mu}\tfrac{5}{8}\pi\frac{\left(2 - \alpha_{1\tau}\right)x_1 b_1 + \left(2 - \alpha_{2\tau}\right)x_2 b_{-1}}{\alpha_{1\tau}x_1 M_1^{1/2} + \alpha_{2\tau}x_2 M_2^{1/2}}.$$

12.14. Two vessels having volumes, V' and V'', are joined by a long capillary where R and L are the capillary radius and length, respectively, $L \gg R$, and $R \gg \lambda$ where λ is the mean free path of the gas molecules. The two vessels are maintained at the same constant temperature, $T' = T'' = T$, and are filled at the beginning $\left(t = 0\right)$ with a binary gas mixture having slightly different mixture fractions in each vessel such that $x_1'\left(0\right) = x_1\left(0\right)$ and $x_2'\left(0\right) = x_2\left(0\right)$ in the first vessel while in the second vessel $x_1''\left(0\right) = x_1\left(0\right) + \delta x_1\left(0\right)$ and $x_2''\left(0\right) = x_2\left(0\right) - \delta x_1\left(0\right)$. When the capillary is opened a pressure difference, $\delta p = p'' - p'$, will develop as a function of time. Determine the time dependence of the relative density difference $\delta x_1\left(t\right) = x_1''\left(t\right) - x_1'\left(t\right)$ and the time dependence of the pressure difference, $\delta p\left(t\right)$. Also, determine the maximum value of the pressure difference that develops.

Solution: The basic equations of balance for the number of molecules of each constituent gas are given by:

$$-V'\frac{dn_i'}{dt}=J_{iz} \text{ and } V''\frac{dn_i''}{dt}=J_{iz},$$

where $J_{iz}=\pi R^2\overline{n}_i\overline{v}_{iz}$ is the net molecular current (referred to as the molecular number flux in some texts) of the i-th constituent through the capillary. If one takes into account Eq. (P-9) from Problem 12.5, the basic equations can be written in the form:

$$-V'\frac{dn_i'}{dt}=\pi R^2\overline{n}_i\left(\overline{w}_z+W_{iz}\right), \tag{P-14}$$

$$V''\frac{dn_i''}{dt}=\pi R^2\overline{n}_i\left(\overline{w}_z+W_{iz}\right), \tag{P-15}$$

where $\overline{w}_z$ is the mean velocity of the gas mixture and W_{iz} is the diffusion velocity of the i-th constituent relative to a frame of reference moving with velocity, $\mathbf{w}$. For the number density difference, $\delta n_i(t)=n_i''(t)-n_i'(t)$, one can obtain:

$$\frac{d}{dt}\delta n_i=\frac{\pi R^2}{V}\overline{n}_i\left(\overline{w}_z+W_{iz}\right), \tag{P-16}$$

where $V=V'V''/(V'+V'')$. The number density difference may then be expressed in the form:

$$\delta n_i=\delta\left(\frac{px_i}{kT}\right)=\left(\frac{\overline{x}_i}{\overline{p}}\delta p+\delta x_i\right)\overline{n}. \tag{P-17}$$

Then, Eq. (P-16) yields:

$$\frac{d}{dt}\left(\frac{\overline{x}_1}{\overline{p}}\delta p+\delta x_1\right)=\frac{\pi R^2}{V}\overline{x}_1\left(\overline{w}_z+W_{1z}\right), \tag{P-18}$$

$$\frac{d}{dt}\left(\frac{\overline{x}_2}{\overline{p}}\delta p-\delta x_1\right)=\frac{\pi R^2}{V}\overline{x}_2\left(\overline{w}_z+W_{2z}\right). \tag{P-19}$$

The average values introduced here are slightly different from those that would exist in the equilibrium state because the pressure difference that develops is much less than the initial pressure. The mean molecular velocity, $\mathbf{w}$, is related to the mean mass velocity, $\mathbf{u}$, by (see Problem 12.5):

$$\mathbf{w} = \mathbf{u} - \left(\frac{m_2 - m_1}{m} \right) D_{12} \mathbf{d}_{12} , \qquad \text{(P-20)}$$

and the diffusion velocities are then defined by:

$$n_1 \mathbf{W}_1 = -n_2 \mathbf{W}_2 = -n D_{12} \mathbf{d}_{12} , \qquad \text{(P-21)}$$

where the vector, $\mathbf{d}_{12}$, is specified in Eq. (12-4). In the particular case under consideration in this problem, the z-component of this vector can be expressed in the form:

$$\left(d_{12} \right)_z = L^{-1} \left(\delta x_1 + \overline{x}_1 \overline{x}_2 \left(\frac{m_2 - m_1}{\overline{m}} \right) \frac{\delta p}{\overline{p}} \right) . \qquad \text{(P-22)}$$

Note that in the slip-flow regime, the mean mass velocity in the capillary may be specified from the boundary value problem of:

$\nabla \cdot \mathbf{u} = 0$ and $\mu \nabla^2 \mathbf{u} = \nabla p$ with:

$$u_z(R) = -c_m \lambda_\mu \frac{\partial u_z}{\partial r} + c_{Dsl} D_{12} \left(d_{12} \right)_z \text{ and } p(r, L) - p(r, 0) = \delta p .$$

The solution of this boundary value problem is found to be:

$$u_z(r) = \frac{1}{4\mu} \left[R^2 \left(1 + 4 c_m Kn \right) - r^2 \right] \frac{\partial p}{\partial z} + c_{Dsl} D_{12} \left(d_{12} \right)_z .$$

from which one can obtain:

$$\overline{u}_z = \frac{2}{R^2} \int_0^R u_z(r) r dr = -\frac{R^2 \left(1 + 4 c_m Kn \right)}{8 \mu L} \delta p + c_{Dsl} D_{12} \left(d_{12} \right)_z . \qquad \text{(P-23)}$$

Now, taking into account Eqs. (P-20)-(P-23), Eqs. (P-18) and (P-19) can be written in the form:

$$\frac{d}{dt}\delta x_1 = -\omega_{1x}\delta x_1 - \omega_{1p}\delta p\ , \tag{P-24}$$

$$\frac{d}{dt}\delta p = -\omega_x \delta x_1 - \omega_p \delta p\ , \tag{P-25}$$

where the following notations have been employed:

$$\omega_{1x} = aD_{12}\,,\quad \omega_{1p} = a\overline{x}_1\overline{x}_2\left(\frac{m_2 - m_1}{\overline{m}}\right)\frac{D_{12}}{\overline{p}}\,,\quad a = \frac{\pi R^2}{VL}$$

$$\omega_x = -a\overline{p}c_{12}D_{12}\,,\quad c_{12} = c_{Dsl} - \left(\frac{m_2 - m_1}{\overline{m}}\right),\ \text{and}$$

$$\omega_p = a\left[\frac{\overline{p}R^2\left(1 + 4c_m Kn\right)}{8\mu} - c_{12}\overline{x}_1\overline{x}_2\left(\frac{m_2 - m_1}{\overline{m}}\right)D_{12}\right].$$

The characteristic equation for this homogeneous linear system of differential equations is the quadratic equation:

$$\begin{vmatrix} -\omega_{1x} - s & -\omega_{1p} \\ -\omega_x & -\omega_p - s \end{vmatrix} = 0\ ,$$

which has the following roots:

$$s_{1,2} = \tfrac{1}{2}\left[-\left(\omega_p + \omega_{1x}\right) \pm \omega_p\sqrt{\left(1 - \frac{\omega_{1x}}{\omega_p}\right)^2 + \frac{4\omega_x\omega_{1p}}{\omega_p^2}}\right].$$

For the different terms in this expression one can make the following determinations regarding the orders of various quantities, specifically:

$$\mu \sim \rho\overline{V}\lambda\,,\quad D_{12} \sim \overline{V}\lambda\,,\quad \overline{V} \sim \left(\frac{2kT}{m}\right)^{1/2},\ \text{and:}$$

$$\frac{\omega_p}{\omega_{1x}} \sim \frac{R^2 nkT}{\rho \bar{V}^2 \lambda^2} \sim \frac{R^2}{\lambda^2} \sim Kn^{-2} \gg 1 .$$

The last of these implies that:

$$\frac{\omega_{1x}}{\omega_p} \ll 1 , \quad \frac{\omega_x}{\omega_p} \ll 1 \text{, and } \frac{\omega_{1p}}{\omega_p} \ll 1 .$$

Now, taking into account these determinations of order, it follows that $s_1 = -\omega_{1x}$ and $s_2 = -\omega_p$. Thus, since both of the roots of the characteristic equation have been shown to be negative, it further follows that each term in the complete solution to the homogeneous system of equations will decay exponentially to zero as $t \to \infty$. The solution of the system is found to be:

$$\delta x_1(t) = \delta x_1(0)\exp(-\omega_{1x} t) \text{ and:}$$

$$\delta p(t) = K\left(\exp(-\omega_{1x} t) - \exp(-\omega_p t)\right) \text{ where:}$$

$$K = \delta x_1(0) \frac{8\mu c_{12} D_{12}}{R^2(1+4c_m Kn)} \left(1 - \frac{\omega_{1x}}{\omega_p}\right)^{-1} .$$

The maximum pressure difference is found to be:

$$(\Delta p)_{\max} = \delta x_1(0) \frac{8\mu c_{12} D_{12}}{R^2(1+4c_m Kn)} \left(1 - \frac{\omega_{1x}}{\omega_p} \ln\left(\frac{\omega_p}{\omega_{1x}}\right)\right).$$

where $c_{12} = c_{Dsl} - (m_2 - m_1)/\bar{m}$.

12.15. Two vessels having volumes, V' and V'', are joined by a long capillary where R and L are the capillary radius and length, respectively, $L \gg R$, and $R \gg \lambda$ (which corresponds to the slip-flow regime) where λ is the mean free path of the gas molecules. The two vessels are maintained at different temperatures, T' and T'' with $T' > T''$. Assume that the two vessels initially $(t=0)$ contain identical gas mixtures, each with the same pressure and mole ratios, i.e. $x_1'(0) = x_1''(0) = q/(1+q)$ and

$x_2'(0) = x_2''(0) = q/(1+q)$, where q is the initial pressure ratio of the gas constituents. When the capillary is opened a pressure difference, $\delta p(t) = p'' - p'$, will develop as a function of time. Determine the steady-state pressure difference that will develop in this system, $\Delta p_{st} = -\delta p(\infty)$.

Solution: For this problem one can use the same basic system of differential equations that was employed in Problem 12.14 in which the mean velocity, $\overline{w}_z$, and the diffusion velocities, W_{iz}, must be modified by the inclusion of additional terms to account for the non-uniformity of the temperature in the system. This basic system of differential equations is given by:

$$\frac{d}{dt}\left(\frac{\overline{x}_1}{\overline{p}}\delta p + \delta x_1\right) = \frac{\pi R^2}{V}\overline{x}_1\left(\overline{w}_z + W_{1z}\right),$$

$$\frac{d}{dt}\left(\frac{\overline{x}_2}{\overline{p}}\delta p - \delta x_1\right) = \frac{\pi R^2}{V}\overline{x}_2\left(\overline{w}_z + W_{2z}\right).$$

where $\delta p(t) = p'' - p'$ and $\delta x_1 = x_1'' - x_1'$. The diffusion velocities, W_{iz}, are given by:

$$\overline{n}_1 W_{1z} = -\overline{n}_2 W_{2z} = -\overline{n} D_{12}\left[\left(d_{12}\right)_z + k_T \frac{\delta T}{\overline{T}L}\right], \text{ where:}$$

$$\left(d_{12}\right)_z = L^{-1}\left(\delta x_1 + \overline{x}_1\overline{x}_2\left(\frac{m_2 - m_1}{\overline{m}}\right)\frac{\delta p}{\overline{p}}\right),$$

and in which k_T is the thermal diffusion ratio. The mean velocity, $\overline{w}_z$, can be expressed in the form:

$$\overline{w}_z = -\frac{R^2\left(1 + 4c_m Kn\right)}{8\mu L}\delta p + c_{Dsl} D_{12}\left(d_{12}\right)_z + c_{Tsl}^* \nu^* \frac{\delta T}{\overline{T}L} - \left(\frac{m_2 - m_1}{\overline{m}}\right) D_{12}\left[\left(d_{12}\right)_z + k_T\frac{\delta T}{\overline{T}L}\right], \quad \text{(P-26)}$$

where c_m, c_{Tsl}^*, and c_{Dsl} are the velocity-slip, thermal-creep, and diffusion-slip coefficients, respectively. Then, the basic system of equations can be written in the form:

$$\frac{d}{dt}\delta x_1 = -\omega_{1x}\delta x_1 - \omega_{1p}\delta p + \omega_{1T}\delta T \ , \tag{P-27}$$

$$\frac{d}{dt}\delta p = -\omega_x \delta x_1 - \omega_p \delta p + \omega_T \delta T \ , \tag{P-28}$$

where the following notations have been employed:

$$\omega_{1x} = aD_{12}\, , \quad \omega_{1p} = a\overline{x}_1\overline{x}_2\left(\frac{m_2 - m_1}{\overline{m}}\right)\frac{D_{12}}{\overline{p}}\, , \quad a = \frac{\pi R^2}{VL}$$

$$\omega_{1T} = -\frac{ak_T D_{12}}{\overline{T}}\, , \quad \omega_x = -a\overline{p}c_{12}D_{12}\, , \quad c_{12} = c_{Dsl} - \left(\frac{m_2 - m_1}{\overline{m}}\right),$$

$$\omega_p = a\left[\frac{\overline{p}R^2\left(1 + 4c_m Kn\right)}{8\mu} - c_{12}\overline{x}_1\overline{x}_2\left(\frac{m_2 - m_1}{\overline{m}}\right)D_{12}\right],$$

$$\omega_T = \frac{a\overline{p}}{\overline{T}}c_{12}^{(T)}\nu^* \text{ , and } c_{12}^{(T)} = c_{Tsl}^* - \left(\frac{m_2 - m_1}{\overline{m}}\right)\frac{k_T D_{12}}{\nu^*}$$

The last terms of Eqs. (P-27) and (P-28) can be regarded as the driving forces which will determine the ultimate steady-state (or stationary state) of this process. The characteristic equation of the homogeneous system has two negative roots as was shown in Problem 12.14. This again indicates that the solutions to the homogeneous system decay exponentially to zero as $t \to \infty$. However, the steady-state solution for the non-homogeneous system, where the contribution of the particular solutions dominates the behavior of the system, can be specified from the following equations:

$$\omega_{1x}\delta x_1(\infty) + \omega_{1p}\delta p(\infty) = \omega_{1T}\delta T \ ,$$

$$\omega_x \delta x_1(\infty) + \omega_p \delta p(\infty) = \omega_T \delta T \ .$$

The steady-state pressure difference, $\Delta p_{st} = -\delta p(\infty)$, is then found to be:

$$\Delta p_{st} = -\delta T \frac{\omega_T}{\omega_p}\left(1 - \frac{\omega_x \omega_{1T}}{\omega_{1x}\omega_T}\right)\left(1 - \frac{\omega_x \omega_{1p}}{\omega_{1x}\omega_p}\right)^{-1}.$$

An evaluation of the two terms found in the expression for ω_p shows that the first term is the dominant term, i.e.:

$$\frac{R^2 \overline{p}}{8\mu c_{12} D_{12} \overline{x}_1 \overline{x}_2}\left(\frac{m_2 - m_1}{\overline{m}}\right)^{-1} \sim \frac{R^2 nkT}{\mu D_{12}} \sim \frac{R^2 nkT}{mn\overline{V}^2 \lambda^2} \sim \frac{R^2}{\lambda^2} \gg 1,$$

where the following estimations of order have been used:

$$\mu \sim mn\overline{V}\lambda, \quad D_{12} \sim \overline{V}\lambda, \text{ and } \overline{V} \sim \left(\frac{2kT}{m}\right)^{1/2}.$$

This means that:

$$\omega_p = a\overline{p}\frac{R^2(1 + 4c_m Kn)}{8\mu}.$$

Further, the order of the ratio, $\omega_x \omega_{1p} / \omega_{1x}\omega_p$, is determined to be:

$$\frac{\omega_x \omega_{1p}}{\omega_{1x}\omega_p} \sim \frac{D_{12}\mu}{\overline{p}R^2} \sim \frac{\overline{V}^2 \lambda^2 nm}{nkTR^2} \sim \frac{\lambda^2}{R^2} \ll 1.$$

Now, neglecting the small terms, one can obtain:

$$\Delta p_{st} = -\frac{\delta T}{\overline{T}} \frac{8\mu v^* c_{sl}}{R^2(1 + 4c_m Kn)},$$

where the following notation has been introduced:

$$c_{sl} = c_{12}^{(T)} - c_{12}\frac{k_T D_{12}}{v^*} = c_{Tsl}^* - c_{Dsl}\frac{k_T D_{12}}{v^*}.$$

REFERENCES

1. Chapman, S. and Cowling, T.G., *The Mathematical Theory of Non-Uniform Gases* (Cambridge University Press, Cambridge, U.K., 1990).
2. Hirschfelder, J.O., Curtiss, C.F., and Bird, R.B., *Molecular Theory of Gases and Liquids* (John Wiley and Sons, New York, 1954).
3. Ferziger, J.H. and Kaper, H.G., *Mathematical Theory of Transport Processes in Gases* (North-Holland, Amsterdam-London, 1972).
4. Maxwell, J.C., in *The Scientific Papers of J.C. Maxwell*, Vol. 1, ed. by Niven, W.D. (Dover, Mineola, NY, 2003).
5. Loyalka, S.K., "Approximate Method in the Kinetic Theory," *Phys. Fluids* **14(11)**, 2291-2294 (1971).
6. Kennard, E.H., *Kinetic Theory of Gases* (McGraw-Hill, New York, 1938).
7. Williams, M.M.R. and Loyalka, S.K., *Aerosol Science, Theory and Practice* (Pergamon Press, New York, 1991).
8. Kogan, M.N., *Rarefied Gas Dynamics* (Plenum Press, NY, 1969).
9. Cercignani, C., *Theory and Application of the Boltzmann Equation* (Scottish Academic Press, Edinburg, UK, 1975).
10. Hidy, G.M. and Brock, J.R., *The Dynamics of Aerocolloidal Systems*, Vol. 1 (Pergamon Press, New York, 1970).
11. Derjaguin, B.V. and Yalamov, Yu.I., "The Theory of Thermophoresis and Diffusiophoresis of Aerosol Particles and their Experimental Testing," in *Topics in Current Aerosol Research*, vol. 3, part 2, edited by Hidy, G.M. (Pergamon Press, Oxford, 1972).
12. Loeb, L.B., *The Kinetic Theory of Gases*, 3rd edition (Dover, New York, 1961).
13. Present, R.D., *Kinetic Theory of Gases* (McGraw-Hill, New York, 1958).
14. Rosner, D.E., "Side-wall Gas 'Creep' and 'Thermal Stress Convection' in Microgravity Experiments on Film Growth by Vapor Transport," *Phys. Fluids A* **1(11)**, 1761-1763 (1989).
15. Gupta, R.N., Scott, C.D., and Moss, J.N., *Slip-Boundary Equations for Multicomponent Non-equilibrium Air Flow* (NASA TP2455, Nov. 1985).
16. Hess, D.W. and Jensen, K.F., *Microelectronic-Processing Chemical Engineering Aspects* (American Chemical Society, Washington DC, 1989).
17. Wolf, S. and Tauber, R.N., *Silicon Processing for the VLSI ERA, Vol. 1. Process Technology* (Lattice Press, Sunset Beach, CA, 1986).
18. Ikegawa, M. and Kobayashi, J., "Deposition Profile Simulation Using the Direct Simulation Monte Carlo Method," *J. Electrochemical Soc.* **136(10)**, 2982-2986 (1989).
19. Coronell, D.G. and Jensen, K.F., "Analysis of Transition Regime Flows in Low-pressure Chemical Vapor Deposition Reactors Using the Direct Monte Carlo Method," *J. Electrochemical Soc.* **139(8)**, 2264-2273 (1992).
20. Coronell, D.G. and Jensen, K.F., "Simulation of Rarefied Gas Transport and Profile Evolution in Nonplanar Substrate Chemical Vapor Deposition," *J. Electrochemical Soc.* **141(9)**, 2545-2551 (1994).
21. Kramers, H.A. and Kistemaker, J., "On the Slip of a Diffusing Gas Mixture Along a Wall," *Physica (Amsterdam)* **10(8)**, 699-713 (1943).
22. Kucherov, R.Ya. and Richenglas, L.E., "Slip and Temperature Jump at the Boundary of a Gaseous Mixture," *Zh. Eks. Teor. Fis. (Russian)* **36(6)**, 1758-1761 (1959).
23. Kucherov, R.Ya., "Diffusion Slip and Convective Diffusion of a Gas in Capillaries," *Zh. Tekh. Fiz. (Russian)* **27(9)**, 2158-2161 (1957).

24. Schmitt, K.H., "Untersuchungen an Schwebstoffteilchen in Diffundierendem Wasserdampf," *Z. Naturforschg.* **16a**, 144-149 (1961).
25. Brock, J.R., "The State of a Binary Gas Mixture Near a Catalytic Surface," *J. Catalysis* **2**, 248-250 (1963).
26. Brock, J.R., "Forces on Aerosols in Gas Mixtures," *J. Colloid Sci.* **18(6)**, 498-501 (1963).
27. Zhdanov, V.M., "The Theory of Slip on the Boundary of a Gaseous Mixture," *Zh. Tekh. Fiz. (Russian)* **37(1)**, 192-197 (1967).
28. Lang, H. and Eger, K., "Gas Diffusion in Narrow Capillaries," *Z. Physik Chem.* **68**, 130-148 (1969).
29. Lang, H. and Müller, W.J.C., "Slip Effects in Mixtures of Monoatomic Gases for General Surface Accommodation," *Z. Naturforschg.* **30a**, 885-867 (1975).
30. Schmitt, K.H. and Waldmann, L., "Untersuchungen an Schwebstoffteilchen in Diffundierenden Gasen," *Z. Naturforschg.* **15a(10)**, 843-851 (1960).
31. Waldmann, L. and Schmitt, K.H., "Über das bei der Gasdiffusion Auftretende Druckgefälle," *Z. Naturforschg.* **16a**, 1343-1354 (1961).
32. Waldmann, L., "On the Motion of Spherical Particles in Nonhomogeneous Gases," in *Rarefied Gas Dynamics*, ed. Talbot, L. (Academic Press, New York, 1961), pp. 323-344.
33. Waldmann, L. and Schmitt, K.H., "Thermophoresis and Diffusiophoresis of Aerosols," Chapter VI in *Aerosol Science*, ed. Davies, C.N. (Academic Press, New York, 1966), pp.137-162.
34. Breton, J.P., "Interdiffusion of Gases Through Porous Media – Effect of Molecular Interactions," *Phys. Fluids* **12(10)**, 2019-2026 (1969).
35. Breton, J.P., "The Diffusion Equation in Discontinuous Systems," *Physica* **50**, 365-379 (1970).
36. Yalamov, Yu.I., Ivchenko, I.N., and Derjaguin, B.V., "Calculation of the Surface Diffusion Slip Velocity of a Gas Mixture," in *Rarefied Gas Dynamics,* Vol. 1, eds. Trilling, L. and Wachman, H.Y. (Academic Press, New York, 1969), pp. 295-300.
37. Ivchenko, I.N. and Yalamov, Yu.I., "Diffusion Slip of a Binary Gas Mixture," *Mekh. Zhid. G. (Russian)* **4**, 22-26 (1971).
38. Loyalka, S.K. and Ferziger, J.H., "Model Dependence of the Slip Coefficient," *Phys. Fluids* **10(8)**, 1833-1839 (1967).
39. Loyalka, S.K. and Ferziger, J.H., "Model Dependence of the Temperature Slip Coefficient," *Phys. Fluids* **11(8)**, 1668-1671 (1968).
40. Loyalka, S.K., "Momentum and Temperature Slip Coefficient with Arbitrary Accommodation at the Surface," *J. Chem. Phys.* **48(12)**, 5432-5436 (1968).
41. Loyalka, S.K., "Velocity Slip Coefficient and the Diffusion Slip Velocity for a Multicomponent Gas Mixture," *Phys. Fluids* **14(12)**, 2529-2604 (1971).
42. Loyalka, S.K., "Kinetic Theory of Thermal Transpiration and Mechanocaloric Effect," *J. Chem. Phys.* **55(9)**, 4497-4503 (1971).
43. Lang, H. and Loyalka, S.K., "Diffusion Slip Velocity. Theory and Experiment," *Z. Naturforschg.* **27a**, 1307-1319 (1972).
44. Loyalka, S.K., "Temperature Jump in a Gas Mixture," *Phys. Fluids* **17(5)**, 897-899 (1974).
45. Loyalka, S.K., "Kinetic Theory of Thermal Transpiration and Mechanocaloric Effect. II," *J. Chem. Phys.* **63(9)**, 4054-4060 (1975).
46. Loyalka, S.K. and Storvick, T.S., "Kinetic Theory of Thermal Transpiration and Mechanocaloric Effect. III," *J. Chem. Phys.* **71(1)**, 339-350 (1979).
47. Ivchenko, I.N., Loyalka, S.K. and Tompson, R.V., "Slip Coefficients for Binary Gas Mixtures," *JVST A* **15(4)**, 2375-2381 (1997).

48. Huang, C.M., Tompson, R.V., Ghosh, T.K., Ivchenko, I.N., and Loyalka, S.K., "Measurements of Thermal Creep in Binary Gas Mixtures," *Phys. Fluids* **11(6)**, 1662-1671 (1999).
49. Ivchenko, I.N., Loyalka, S.K. and Tompson, R.V., "Boundary Slip Phenomena in a Binary Gas Mixture," *ZAMP* **53(1)**, 58-72 (2002).
50. Yalamov, Yu.I., Shkanov, Yu. and Savkov, S.A., "Boundary Conditions of Sliding of a Binary Gas Mixture and Their Use in the Dynamics of Aerosols. 1. Flow of a Gas Mixture Along a Solid Flat Wall," *Inzh. Fizich. Zh. (Russian)* **66(4)**, 421-426 (1994).
51. Savkov, S.A., *Slip Boundary Conditions for Non-Uniform Binary Gas Mixtures and their Application to Aerosol Dynamics* (M.S. Thesis, Moscow, 1987).
52. Volkov, I.V. and Galkin, V.S., "Analysis of the Slip Coefficients and Temperature Jump in a Binary Gas Mixture," *Mekh. Zhid. G. (Russian)* **6**, 152-159 (1990)
53. Siewert, C.E. and Sharipov, F., "Model Equations in Rarefied Gas Dynamics: Viscous-Slip and Thermal-Slip Coefficients," *Phys. Fluids* **14(12)**, 4123-4129 (2002).
54. Siewert, C.E., "Viscous-Slip, Thermal-Slip, and Temperature-Jump Coefficients as Defined by the Linearized Boltzmann Equation and the Cercignani-Lampis Boundary Condition," *Phys. Fluids* **15(6)**, 1696-1701 (2003).
55. Siewert, C.E., "The Linearized Boltzmann Equation: A Concise and Accurate Solution of the Temperature-Jump Problem," *J. Quantitative Spectroscopy and Radiative Transfer* **77(4)**, 417-432 (2003).
56. Takata, A. and Aoki, K., "Two Surface Problems of a Multicomponent Mixture of Vapors and Non-condensable Gases in the Continuum Limit in the Light of Kinetic Theory," *Phys. Fluids* **11(9)**, 2743-2756 (1999).
57. Sone, Y., Takata, A., and Golse, F, "Notes on the Boundary Conditions for Fluid-Dynamics Equations on the Interface of a Gas and Its Condensed Phase," *Phys. Fluids* **13(1)**, 324-334 (2001).
58. Takata, A. and Aoki, K., "The Ghost Effect in the Continuum Limit for a Vapor-Gas Mixture Around Condensed Phases: Asymptotic Analysis of the Boltzmann Equation," *Transport Theory and Stat. Phys.* **30(2&3)**, 205-237 (2001).
59. Aoki, K., Bardos, C., and Takata, A., "Knudsen Layer for Gas Mixtures," *J. Statistical Phys.* **112(3&4)**, 625-655 (2003).
60. Takata, A. and Aoki, K., "The Ghost Effect in the Continuum Limit for a Vapor-Gas Mixture Around Condensed Phases: Asymptotic Analysis of the Boltzmann Equation," *Transport Theory and Stat. Phys.* **31(3)**, 289-290 (2002).
61. Takata, A., Yasuda, S., Kosuge, S., and Aoki, K., "Numerical Analysis of Thermal-Slip and Diffusion-Slip Flows of a Binary Mixture of Hard-Sphere Molecular Gases," *Phys. Fluids* **15(12)**, 3745-3766 (2003).
62. Takata, A. and Aoki, K., "Numerical Analysis of the Shear Flow of a Binary Mixture of Hard-Sphere Gases Over a Plane Wall," *Phys. Fluids* **16(6)**, 1989-2003 (2004).
63. Takata, A., "Kinetic Theory Analysis of the Two-Surface Problem of a Vapor-Vapor Mixture in the Continuum Limit," *Phys. Fluids* **16(7)**, 2182-2198 (2004).
64. Maitland, G.C., Rigby, M., Smith, E.B., and Wakeham, W.A., *Intermolecular Forces: Their Origin and Determination* (Oxford University Press, New York, 1981, reprinted and corrected edition, 1987).
65. Mason, E.A., "Transport Properties of Gases Obeying a Modified Buckingham (exp-six) Potential," *J. Chem. Phys.* **22(2)**, 169-186 (1954).
66. Mason, E.A., "Higher Approximations for the Transport Properties of Binary Gases Mixtures. I. General Formulas," *J. Chem. Phys.* **27(1)**, 75-84 (1957).

67. Mason, E.A., "Higher Approximations for the Transport Properties of Binary Gases Mixtures. II. Applications," *J. Chem. Phys.* **27(3)**, 782-790 (1957).
68. Mason, E.A. and Saxena, S.C., "Approximate Formula for the Thermal Conductivity of Gas Mixtures," *Phys. Fluids* **1(5)**, 361-369 (1958).
69. Mason, E.A. and Monchick, L., "Heat Conductivity of Polyatomic and Polar Gases," *J. Chem. Phys.* **36(6)**, 1622-1639 (1962).
70. *CRC Handbook of Chemistry and Physics*, 71 edition, ed. Lide, D.R. (CRC Press, Boca Raton, 1990).

Appendix A

BRACKET INTEGRALS FOR THE PLANAR GEOMETRY

1. BRACKET INTEGRALS INVOLVING TWO SONINE POLYNOMIALS.

In this section, bracket integrals of the form:

$$a_{rs} = \left[\mathbf{c}S_{3/2}^{(r)}\left(c^2\right), \mathbf{c}S_{3/2}^{(s)}\left(c^2\right) \right],$$

are discussed. Consider, for example, the bracket integral a_{11} that is contained in the first-order Chapman-Enskog solution for thermal conductivity. This integral can be written as:

$$\begin{aligned} a_{11} &= \left[\mathbf{c}S_{3/2}^{(1)}\left(c^2\right), \mathbf{c}S_{3/2}^{(1)}\left(c^2\right) \right] = 3\left[c_x S_{3/2}^{(1)}\left(c^2\right), c_x S_{3/2}^{(1)}\left(c^2\right) \right] \\ &= 3n^{-2} \int c_x S_{3/2}^{(1)}\left(c^2\right) \left\{ \int \alpha(\theta, g) d\Omega \iint g f_1^{(0)} f^{(0)} \left[c_x S_{3/2}^{(1)}\left(c^2\right) \right.\right. \\ &\quad \left.\left. + c_{1x} S_{3/2}^{(1)}\left(c_1^2\right) - c_x' S_{3/2}^{(1)}\left(c'^2\right) - c_{1x}' S_{3/2}^{(1)}\left(c_1'^2\right) \right] d\mathbf{v}_1 \right\} d\mathbf{v} . \end{aligned} \tag{A-1}$$

Using the momentum conservation equation, one obtains:

$$a_{11} = 3\left[c_x c^2 , c_x c^2 \right] . \tag{A-2}$$

The differential scattering cross section for rigid-sphere molecules can be expressed in the form:

$$\alpha(\theta, g)d\Omega = bdbd\varepsilon = b\left|\frac{db}{d\theta}\right|d\theta d\varepsilon = \tfrac{1}{4}\sigma^2 \sin(\theta)d\theta d\varepsilon \ , \tag{A-3}$$

where σ is the diameter of a molecule.

Let the variables of integration be changed from $\mathbf{v}$ and $\mathbf{v}_1$ to $\tilde{\mathbf{G}}$ and $\tilde{\mathbf{g}} = \mathbf{v}_1 - \mathbf{v}$ (the center of mass and relative molecular velocities, respectively) by the relations:

$$\mathbf{v} = \tilde{\mathbf{G}} - \tfrac{1}{2}\tilde{\mathbf{g}} \ , \tag{A-4a}$$

$$\mathbf{v}_1 = \tilde{\mathbf{G}} + \tfrac{1}{2}\tilde{\mathbf{g}} \ , \tag{A-4b}$$

$$\tfrac{1}{2}m\left(v^2 + v_1^2\right) = m\left(\tilde{G}^2 + \tfrac{1}{4}\tilde{g}^2\right) \ , \tag{A-4c}$$

and:

$$d\mathbf{v}d\mathbf{v}_1 = d\tilde{\mathbf{G}}d\tilde{\mathbf{g}} \ , \tag{A-4d}$$

where $\tilde{\mathbf{G}}$ and $\tilde{\mathbf{g}}$ are dimensional variables.

After changing the variables of integration, one can make the following substitutions:

$$\mathbf{G} = \left(\frac{m}{kT}\right)^{1/2} \tilde{\mathbf{G}} \ , \tag{A-5a}$$

$$\mathbf{g} = \tfrac{1}{2}\left(\frac{m}{kT}\right)^{1/2} \tilde{\mathbf{g}} \ , \tag{A-5b}$$

$$\mathbf{c} = \frac{1}{\sqrt{2}}(\mathbf{G} - \mathbf{g}) \ , \tag{A-5c}$$

$$\mathbf{c}_1 = \frac{1}{\sqrt{2}}(\mathbf{G}+\mathbf{g}) \;, \tag{A-5d}$$

$$d\tilde{\mathbf{G}} = \left(\frac{kT}{m}\right)^{3/2} d\mathbf{G} \;, \tag{A-5e}$$

and:

$$d\tilde{\mathbf{g}} = 8\left(\frac{kT}{m}\right)^{3/2} d\mathbf{g} \;. \tag{A-5f}$$

It is convenient to introduce the spherical coordinates (g,α,β) for $\mathbf{g}$. The relationships between these spherical coordinates and the coordinates in Cartesian velocity space, (g_x, g_y, g_z), are given by:

$$g_x = g\cos(\alpha) \;, \quad g_y = g\sin(\alpha)\cos(\beta) \;, \quad g_z = g\sin(\alpha)\sin(\beta) \;.$$

Substituting the dimensionless variables into a_{11}, and neglecting all vanishing integrals, one can obtain:

$$\begin{aligned} a_{11} &= \tfrac{3}{4}\sigma^2\pi^{-3}\left(\frac{kT}{m}\right)^{1/2} \int_0^{\pi}\sin(\theta)\,d\theta \int_0^{2\pi} d\varepsilon \int_{-\infty}^{\infty}\exp\left(-G_x^2\right)dG_x \\ &\times \int_{-\infty}^{\infty}\exp\left(-G_y^2\right)dG_y \int_{-\infty}^{\infty}\exp\left(-G_z^2\right)dG_z \int_0^{\infty} g^3\exp\left(-g^2\right)dg \\ &\times \int_0^{\pi}\sin(\alpha)\,d\alpha \int_0^{2\pi} d\beta \\ &\times\Big\{\left(G_x^4 + G_x^2G_y^2 + G_x^2G_z^2 + G_x^2g^2 + 2G_x^2g_x^2\right)\left(g_x^2 - g_x'^2\right) \\ &\quad +2G_y^2 g_x g_y\left(g_x g_y - g_x' g_y'\right) + 2G_z^2 g_x g_z\left(g_x g_z - g_x' g_z'\right)\Big\} \;, \end{aligned} \tag{A-6}$$

where g_x', g_y', and g_z' are given by:

$$g_x' = g\left[\cos(\alpha)\cos(\theta) - \sin(\alpha)\sin(\theta)\cos(\varepsilon)\right] \;, \tag{A-7a}$$

$$\begin{aligned} g_y' = g\big[&\sin(\alpha)\cos(\beta)\cos(\theta) + \cos(\alpha)\cos(\beta)\sin(\theta)\cos(\varepsilon) \\ &-\sin(\beta)\sin(\theta)\sin(\varepsilon)\big] \;, \end{aligned} \tag{A-7b}$$

$$g'_z = g\left[\sin(\alpha)\sin(\beta)\cos(\theta) + \cos(\alpha)\sin(\beta)\sin(\theta)\cos(\varepsilon)\right. \\ \left. + \cos(\beta)\sin(\theta)\sin(\varepsilon)\right] \tag{A-7c}$$

After integration over all values of $\mathbf{G}$ and neglecting vanishing integrals one can express the bracket integral, a_{11}, in the following form:

$$\begin{aligned} a_{11} = {} & \tfrac{3}{4}\sigma^2\pi^{-3/2}\left(\frac{kT}{m}\right)^{1/2}\int_0^\infty g^3\exp(-g^2)dg\int_0^\pi \sin(\theta)d\theta \\ & \times\int_0^\pi \sin(\alpha)d\alpha\int_0^{2\pi} d\varepsilon\int_0^{2\pi} d\beta\left\{\left(\tfrac{5}{4}g^2 + \tfrac{1}{2}g^4\right)\right. \\ & \times\left[\cos^2(\alpha)\sin^2(\theta) - \sin^2(\alpha)\sin^2(\theta)\cos^2(\varepsilon)\right] \\ & \left. + g^4\cos^2(\alpha)\sin^2(\theta)\right\} \\ = {} & \tfrac{8}{3}\sqrt{\pi}\sigma^2\left(\frac{kT}{m}\right)^{1/2}\int_0^\infty g^7\exp(-g^2)dg = 8\sqrt{\pi}\sigma^2\left(\frac{kT}{m}\right)^{1/2} . \end{aligned} \tag{A-8}$$

Finally, in summary, the bracket integral, a_{11}, is given by:

$$a_{11} = \left[\mathbf{c}S_{3/2}^{(1)}(c^2), \mathbf{c}S_{3/2}^{(1)}(c^2)\right] = 8\sqrt{\pi}\sigma^2\left(\frac{kT}{m}\right)^{1/2} , \tag{A-9}$$

where one must remember that molecules have been assumed to be rigid spheres.

2. BRACKET INTEGRALS CONTAINING SEVERAL COMPONENTS OF MOLECULAR VELOCITY.

In this section, the bracket integral associated with the first-order Chapman-Enskog solution for viscosity:

$$b'_{11} = \left[c_x c_y\, , c_x c_y\right],$$

is considered. Using the dimensionless variables from Section A.1 and omitting vanishing integrals, one can obtain the following expression for this bracket integral:

$$
\begin{aligned}
b'_{11} = \left[c_x c_y\,, c_x c_y\right] = \tfrac{1}{4}\sigma^2 \pi^{-3}\left(\frac{kT}{m}\right)^{1/2} \\
\times \int_0^{\pi} \sin(\theta) d\theta \int_0^{2\pi} d\varepsilon \int_0^{\infty} g^3 \exp\left(-g^2\right) dg \int_0^{\pi} \sin(\alpha) d\alpha \\
\times \int_0^{2\pi} d\beta \int_{-\infty}^{\infty} \exp\left(-G_x^2\right) dG_x \int_{-\infty}^{\infty} \exp\left(-G_y^2\right) dG_y \\
\times \int_{-\infty}^{\infty} \exp\left(-G_z^2\right) dG_z \left\{ g_x g_y \left(g_x g_y - g'_x g'_y \right) \right\} .
\end{aligned}
\qquad \text{(A-10)}
$$

After integration over the components of $\mathbf{G}$, and using the relations for g'_x and g'_y given by Eqs. (A-7a) and (A-7b), one can obtain:

$$
\begin{aligned}
b'_{11} = \tfrac{1}{4}\sigma^2 \pi^{-3/2}\left(\frac{kT}{m}\right)^{1/2} \int_0^{\pi} \sin(\theta) d\theta \int_0^{2\pi} d\varepsilon \\
\times \int_0^{\infty} g^3 \exp\left(-g^2\right) dg \int_0^{\pi} \sin(\alpha) d\alpha \int_0^{2\pi} d\beta g^4 \\
\times \left\{ \sin^2(\alpha)\cos^2(\alpha)\cos^2(\beta)\sin(\theta)\left[1+\cos^2(\varepsilon)\right] \right\} .
\end{aligned}
\qquad \text{(A-11)}
$$

The remaining simple integrations then yield:

$$
b'_{11} = \left[c_x c_y\,, c_x c_y\right] = \tfrac{4}{5}\sigma^2 \sqrt{\pi}\left(\frac{kT}{m}\right)^{1/2} . \qquad \text{(A-12)}
$$

Molecules have been assumed to be rigid spheres. Analytical expressions for other bracket integrals contained in higher-order Chapman-Enskog solutions and a more general calculation scheme for arbitrary intermolecular potentials may be found in [1-3].

3. BRACKET INTEGRALS CONTAINING TWO DISCONTINUOUS FUNCTIONS.

In this section a method of calculating bracket integrals that contain the discontinuous sign functions is considered. As an example, the following bracket integral will be evaluated for the case of rigid-sphere molecules:

$$\begin{aligned}&\left[c_y \operatorname{sign}(c_x), c_y \operatorname{sign}(c_x)\right]=\\&\quad n^{-2}\int c_y \operatorname{sign}(c_x)\Big\{\int\alpha(\theta,g)\,d\Omega\int\tilde{g}f_1^{(0)}f^{(0)}d\mathbf{v}_1\\&\quad\times\Big[c_y \operatorname{sign}(c_x)+c_{1y}\operatorname{sign}(c_{1x})\\&\qquad -c'_y\operatorname{sign}(c'_x)-c'_{1y}\operatorname{sign}(c'_{1x})\Big]\Big\}d\mathbf{v}\;.\end{aligned} \tag{A-13}$$

This integral may be represented as the sum of four terms:

$$I = I^{(1)} + I^{(2)} + I^{(3)} + I^{(4)}\;. \tag{A-14}$$

The first term does not contain the discontinuous functions and, therefore, may be calculated by the standard method described in Section A.1.

Consider the second integral in detail. The integrand does not contain any collision parameters and, therefore, $\int\alpha(\theta,g)\,d\Omega$ can be calculated independently of the other integrations yielding:

$$\int\alpha(\theta,g)\,d\Omega=\pi\sigma^2\;.$$

Now, let the variables of integration be changed from $(\mathbf{v},\mathbf{v}_1)$ to $(\tilde{\mathbf{G}},\tilde{\mathbf{g}}=\mathbf{v}_1-\mathbf{v})$. Then, after introducing the dimensionless variables $\mathbf{G}$ and $\mathbf{g}$, this integral becomes:

$$\begin{aligned}I^{(2)}&=\sigma^2\pi^{-2}\left(\frac{kT}{m}\right)^{1/2}\int_0^\infty g^3\exp(-g^2)\,dg\int_0^\pi\sin(\alpha)\,d\alpha\\&\times\int_0^{2\pi}d\beta\int_{-\infty}^{\infty}\exp(-G_x^2)\,dG_x\int_{-\infty}^{\infty}\exp(-G_y^2)\,dG_y\\&\times\int_{-\infty}^{\infty}\exp(-G_z^2)\,dG_z\Big\{\left(G_y^2-g^2\sin^2(\alpha)\cos^2(\beta)\right)\\&\qquad\times\operatorname{sign}\left(\left[G_x-g\cos(\alpha)\right]\left[G_x+g\cos(\alpha)\right]\right)\Big\}\;.\end{aligned} \tag{A-15}$$

After integration over all values of G_y, G_z, and β this expression becomes:

$$I^{(2)} = \sigma^2 \left(\frac{kT}{m}\right)^{1/2} \int_0^\infty g^3 \exp\left(-g^2\right) dg \int_0^\pi \sin(\alpha) d\alpha$$
$$\times \int_{-\infty}^{\infty} \exp\left(-G_x^2\right) dG_x \left(1 - g^2 \sin^2(\alpha)\right) \tag{A-16}$$
$$\times \operatorname{sign}\left(\left[G_x - g\cos(\alpha)\right]\left[G_x + g\cos(\alpha)\right]\right) .$$

Note that the method of calculation used in these integrations has been previously described in [4,5]. Here, this method has been generalized by the introduction of appropriate recurrence relations.

Using the new variables of integration, $x = \cos(\alpha)$ and $z = G_x/g$, one can reduce Eq. (A-16) to the form:

$$\begin{aligned} I^{(2)} &= \sigma^2 \left(\frac{kT}{m}\right)^{1/2} \int_{-1}^{1} dx \int_{-\infty}^{\infty} dz \int_0^\infty g^4 \exp\left(-g^2\left[1+z^2\right]\right) dg \\ &\quad \times \left[1 - g^2\left(1-x^2\right)\right] \operatorname{sign}\left(\left[z-x\right]\left[z+x\right]\right) \\ &= \sigma^2 \left(\frac{kT}{m}\right)^{1/2} \int_{-1}^{1} dx \int_{-\infty}^{\infty} dz \operatorname{sign}\left(\left[z-x\right]\left[z+x\right]\right) \\ &\quad \times \tfrac{3}{8}\sqrt{\pi}\left[\left(1+z^2\right)^{-5/2} - \tfrac{5}{2}\left(1-x^2\right)\left(1+z^2\right)^{-7/2}\right] . \end{aligned} \tag{A-17}$$

It is convenient to use the following notations:

$$J_{m+1/2}^{(1)} = 2\int_0^1 dx \int_{-\infty}^{\infty} \frac{\operatorname{sign}\left(\left[z-x\right]\left[z+x\right]\right)}{\left(1+z^2\right)^{m+1/2}} dz , \tag{A-18a}$$

$$J_{m+1/2}^{(2)} = 2\int_0^1 \left(1-x^2\right) dx \int_{-\infty}^{\infty} \frac{\operatorname{sign}\left(\left[z-x\right]\left[z+x\right]\right)}{\left(1+z^2\right)^{m+1/2}} dz . \tag{A-18b}$$

After transformation, these integrals may be written as:

$$J_{m+1/2}^{(1)} = 2\int_0^1 dx \left\{ \int_{-\infty}^{\infty} \left(1+z^2\right)^{-(m+1/2)} dz - 2\int_{-x}^{x} \left(1+z^2\right)^{-(m+1/2)} dz \right\}, \tag{A-19a}$$

Table A-1. Values of the integrals, $J^{(1)}_{m+1/2}$ and $J^{(2)}_{m+1/2}$.

m	$J^{(1)}_{m+1/2}$	$J^{(2)}_{m+1/2}$
1	$J^{(1)}_{3/2}=4\left(3-2\sqrt{2}\right)$	$J^{(2)}_{3/2}=\frac{16}{3}\left(3-2\sqrt{2}\right)$
2	$J^{(1)}_{5/2}=\frac{4}{3}\left(4-3\sqrt{2}\right)$	$J^{(2)}_{5/2}=\frac{8}{9}\left(3-2\sqrt{2}\right)$
3	$J^{(1)}_{7/2}=\frac{2}{15}\left(28-23\sqrt{2}\right)$	$J^{(2)}_{7/2}=\frac{8}{45}\left(15-11\sqrt{2}\right)$
4	$J^{(1)}_{9/2}=\frac{13}{35}\left(8-7\sqrt{2}\right)$	$J^{(2)}_{9/2}=\frac{4}{105}\left(58-45\sqrt{2}\right)$

$$J^{(2)}_{m+1/2}=2\int_0^1\left(1-x^2\right)dx \times\left\{\int_{-\infty}^{\infty}\left(1+z^2\right)^{-(m+1/2)}dz-2\int_{-x}^{x}\left(1+z^2\right)^{-(m+1/2)}dz\right\}. \tag{A-19b}$$

Then, using the following relationship [6]:

$$\int\frac{dz}{\left(1+z^2\right)^{m+1/2}}=\frac{z}{(2m-1)\left(1+z^2\right)^{m-1/2}}+\frac{2m-2}{2m-1}\int\frac{dz}{\left(1+z^2\right)^{m-1/2}}, \tag{A-20}$$

recurrence formulas for $m\geq 2$ may be obtained:

$$J^{(1)}_{m+1/2}=-\frac{8}{(2m-1)(2m-3)}\left(1-\frac{1}{2^{m-3/2}}\right)+\frac{2m-2}{2m-1}J^{(1)}_{m-1/2}, \tag{A-21a}$$

$$J^{(2)}_{m+1/2}=\frac{8}{2m-1}\left(\frac{2^{5/2-m}-2}{2m-3}-\frac{2^{5/2-m}-1}{2m-5}\right)+\frac{2m-2}{2m-1}J^{(2)}_{m-1/2}. \tag{A-21b}$$

Table A-1 gives the expressions for these integrals for several values of m. Using the integrals from Table A-1, one obtains:

$$I^{(2)}=\frac{3}{8}\sqrt{\pi}\sigma^2\left(\frac{kT}{m}\right)^{1/2}\left(J^{(1)}_{5/2}-\frac{5}{2}J^{(2)}_{7/2}\right) =-\sqrt{\pi}\sigma^2\left(\frac{kT}{m}\right)^{1/2}\frac{1}{6}\left(3-2\sqrt{2}\right). \tag{A-22}$$

Now, consider the third integral, $I^{(3)}$, which may be written as:

$$
\begin{aligned}
I^{(3)} = & -\tfrac{1}{4}\sigma^2\pi^{-3}\left(\frac{kT}{m}\right)^{1/2}\int_0^\infty g^3\exp\left(-g^2\right)dg\int_0^\pi \sin(\alpha)\,d\alpha \\
& \times\int_0^{2\pi} d\beta\int_{-\infty}^{\infty}\exp\left(-G_x^2\right)dG_x\int_{-\infty}^{\infty}\exp\left(-G_y^2\right)dG_y \\
& \times\int_{-\infty}^{\infty}\exp\left(-G_z^2\right)dG_z\int_0^\pi \sin(\theta)\,d\theta\int_0^{2\pi} d\varepsilon \\
& \times\left[G_y - g\sin(\alpha)\cos(\beta)\right]\times\left[G_y - g\left(\sin(\alpha)\cos(\beta)\cos(\theta)\right.\right. \\
& \quad +\cos(\alpha)\cos(\beta)\sin(\theta)\cos(\varepsilon) \\
& \quad \left.\left.-\sin(\beta)\sin(\theta)\sin(\varepsilon)\right)\right]sign\left(G_x - g\cos(\alpha)\right) \\
& \times\text{sign}\left(G_x - g\left[\cos(\alpha)\cos(\theta)-\sin(\alpha)\sin(\theta)\cos(\varepsilon)\right]\right)\ .
\end{aligned} \tag{A-23}
$$

First, the variable of integration should be changed from θ to $-\theta$. Then, after integration with respect to β, G_y, and G_z, the integrand may be written as:

$$
\begin{aligned}
& \pi^2\left\{1+g^2\left[\left(\sin^2(\alpha)\cos(\theta)-\sin(\alpha)\cos(\alpha)\sin(\theta)\cos(\varepsilon)\right]\right\}\right. \\
& \times\text{sign}\left(G_x - g\cos(\alpha)\right) \\
& \times\text{sign}\left(G_x - g\left[\cos(\alpha)\cos(\theta)+\sin(\alpha)\sin(\theta)\cos(\varepsilon)\right]\right)\ .
\end{aligned}
$$

Here, it is convenient to introduce new collision angles to simplify integration over the collision parameters. Let the variables of integration be changed from (g,θ,ε) to $(g',\alpha',\varepsilon')$. The relationships between these variables are shown in Fig. A-1. From the spherical triangle, the following expressions may be obtained:

$$\cos(\theta)=\cos(\alpha)\cos(\alpha')+\sin(\alpha)\sin(\alpha')\cos(\varepsilon')\ , \tag{A-24a}$$

$$\cos(\alpha')=\cos(\alpha)\cos(\theta)+\sin(\alpha)\sin(\theta)\cos(\varepsilon)\ . \tag{A-24b}$$

Neglecting vanishing integrals, one obtains:

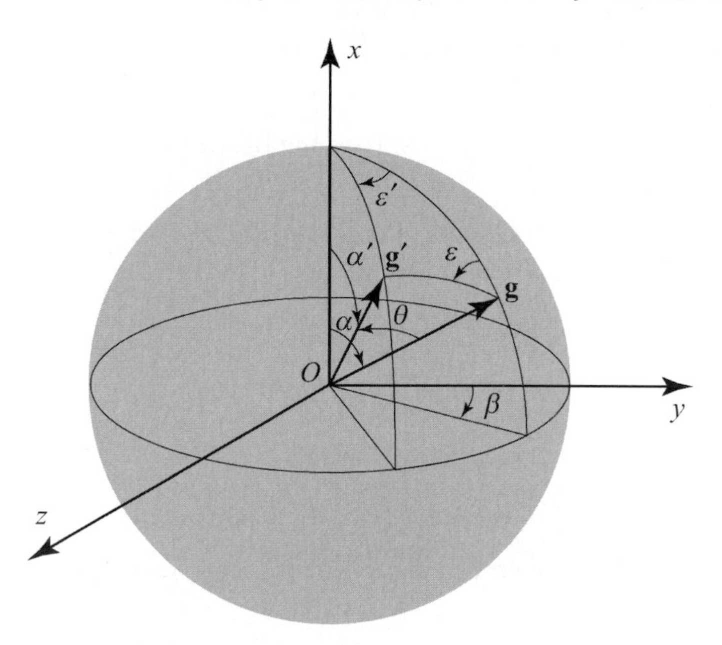

Figure A-1. The relationship between various collision angles.

$$I^{(3)} = -\tfrac{1}{4}\sigma^2\pi^{-1}\left(\frac{kT}{m}\right)^{1/2}\int_0^\infty g^3\exp\left(-g^2\right)dg\int_0^\pi \sin(\alpha)d\alpha$$
$$\times\int_{-\infty}^{\infty}\exp\left(-G_x^2\right)dG_x\int_0^\pi \sin(\alpha')d\alpha'\int_0^\pi d\varepsilon' \tag{A-25}$$
$$\times\text{sign}\left(\left[G_x - g\cos(\alpha)\right]\left[G_x - g\cos(\alpha')\right]\right) .$$

At this point, changing the variables of integration to ($z = G_x/g$, $x = \cos(\alpha)$, $y = \cos(\alpha')$) and then performing the integrations with respect to g and ε', Eq. (A-25) becomes:

$$I^{(3)} = -\tfrac{3}{16}\sqrt{\pi}\sigma^2\left(\frac{kT}{m}\right)^{1/2}\int_{-1}^{1}dx\int_{-1}^{1}dy$$
$$\times\int_{-\infty}^{\infty}dz\left(1+z^2\right)^{-(5/2)}\text{sign}\left([z-x][z-y]\right) . \tag{A-26}$$

For other bracket integrals in planar boundary value transport problems the integrand may contain the following expressions:

$$\left(1+z^2\right)^{-(m+1/2)} \operatorname{sign}\left([z-x][z-y]\right),$$

and:

$$\left(1-x^2\right)\left(1+z^2\right)^{-(m+1/2)} \operatorname{sign}\left([z-x][z-y]\right) .$$

For these terms, let the following notations be introduced:

$$I^{(1)}_{m+1/2} = \int_{-1}^{1} dx \int_{-1}^{1} dy \int_{-\infty}^{\infty} \frac{\operatorname{sign}\left([z-x][z-y]\right)}{\left(1+z^2\right)^{m+1/2}} dz \ , \tag{A-27a}$$

$$I^{(2)}_{m+1/2} = \int_{-1}^{1} \left(1-x^2\right) dx \int_{-1}^{1} dy \int_{-\infty}^{\infty} \frac{\operatorname{sign}\left([z-x][z-y]\right)}{\left(1+z^2\right)^{m+1/2}} dz \ . \tag{A-27b}$$

To facilitate calculations of these integrals it is convenient to obtain recurrence formulas that may be derived by using the relation given by Eq. (A-20).

First, consider the integral, $I^{(1)}_{m+1/2}$, which may be represented in the form:

$$I^{(1)}_{m+1/2} = \int_{-1}^{1} dx \left\{ \int_{-1}^{x} dy \int_{-\infty}^{\infty} \frac{\operatorname{sign}\left([z-x][z-y]\right)}{\left(1+z^2\right)^{m+1/2}} dz + \int_{x}^{1} dy \int_{-\infty}^{\infty} \frac{\operatorname{sign}\left([z-x][z-y]\right)}{\left(1+z^2\right)^{m+1/2}} dz \right\} . \tag{A-28}$$

In the first term $y \le x$ and therefore:

$$\operatorname{sign}\left([z-x][z-y]\right) = \begin{cases} 1 \ ; & -\infty < z < y \ , \quad x < z < \infty \ , \\ -1 \ ; & y < z < x \ . \end{cases}$$

In the second term $y \ge x$ and therefore:

$$\operatorname{sign}\left([z-x][z-y]\right) = \begin{cases} 1 \ ; & -\infty < z < x \ , \quad y < z < \infty \ , \\ -1 \ ; & x < z < y \ . \end{cases}$$

Using these representations for $\text{sign}\left([z-x][z-y]\right)$, the integral $I_{m+1/2}^{(1)}$ may be expressed in the following form:

$$I_{m+1/2}^{(1)} = \int_{-1}^{1} dx \left\{ \int_{-1}^{1} dy \int_{-\infty}^{\infty} \frac{dz}{\left(1+z^2\right)^{m+1/2}} - 2\int_{-1}^{x} dy \int_{y}^{x} \frac{dz}{\left(1+z^2\right)^{m+1/2}} \right. \left. -2\int_{x}^{1} dy \int_{x}^{y} \frac{dz}{\left(1+z^2\right)^{m+1/2}} \right\} . \tag{A-29}$$

Now, by means of Eq. (A-20), the following recurrence formula may be obtained:

$$I_{m+1/2}^{(1)} = \frac{2m-2}{2m-1} I_{m-1/2}^{(1)} + \frac{2^{11/2-m}}{(2m-1)(2m-3)} - \frac{8}{(2m-1)(2m-3)} \int_{-1}^{1} \frac{dx}{\left(1+x^2\right)^{m-3/2}} \quad ; \quad m \geq 2 . \tag{A-30}$$

From this general expression the basic integral, $I_{3/2}^{(1)}$, becomes:

$$I_{3/2}^{(1)} = 8\left(1-\sqrt{2}\right) + 4\int_{-1}^{1} \frac{dx}{\left(1+x^2\right)^{1/2}} . \tag{A-31}$$

In the same way, the following recurrence formula may be obtained for $I_{m+1/2}^{(2)}$:

$$I_{m+1/2}^{(2)} = \frac{2m-2}{2m-1} I_{m-1/2}^{(2)} + \frac{2^{11/2-m}}{3(2m-1)(2m-3)} - \frac{4(2m-2)}{(2m-1)(2m-3)} \int_{-1}^{1} \frac{\left(1-x^2\right)dx}{\left(1+x^2\right)^{m-3/2}} + \frac{4}{(2m-1)} \int_{-1}^{1} \frac{\left(1-x^2\right)dx}{\left(1+x^2\right)^{m-1/2}} \quad ; \quad m \geq 2 , \tag{A-32}$$

with the basic integral:

$$I_{3/2}^{(2)} = \tfrac{16}{3}\left(1-\sqrt{2}\right) + 4\int_{-1}^{1}\frac{\left(1-x^2\right)dx}{\left(1+x^2\right)^{1/2}} \; . \tag{A-33}$$

Expressions for several of these integrals are given in Table A-2. Using these table integrals yields:

$$I^{(3)} = -\tfrac{3}{16}\sqrt{\pi}\sigma^2\left(\frac{kT}{m}\right)^{1/2} I_{5/2}^{(1)} = -\tfrac{1}{2}\sqrt{\pi}\sigma^2\left(\frac{kT}{m}\right)^{1/2}\left(2-\sqrt{2}\right) . \tag{A-34}$$

The integral given by Eq. (A-23) does not alter if the variable of integration, θ, is changed from θ to $\theta+\pi$. This integral then becomes equal to $I^{(4)}$, and:

$$I^{(3)} = I^{(4)} \; .$$

The integral, $I^{(1)}$, after simple integration may be written as:

$$I^{(1)} = \tfrac{7}{3}\sqrt{\pi}\sigma^2\left(\frac{kT}{m}\right)^{1/2} \; . \tag{A-35}$$

Thus, for $\left[c_y\,\mathrm{sign}\left(c_x\right), c_y\,\mathrm{sign}\left(c_x\right)\right]$, the final expression is given by:

$$I = I^{(1)} + I^{(2)} + 2\,I^{(3)} = -\tfrac{1}{6}\sqrt{\pi}\sigma^2\left(\frac{kT}{m}\right)^{1/2}\left(1-8\sqrt{2}\right) . \tag{A-36}$$

It is necessary to note that this bracket integral is different from that used in the Gross-Ziering theory. The relationship between these integrals is:

$$[\Phi\,,\Psi]^* = -[\Phi\,,\Psi]n\pi^{3/2}\left(\frac{m}{2kT}\right)^{1/2} \; , \tag{A-37}$$

where $[\Phi\,,\Psi]^*$ is the Gross-Ziering notation. Using this notation, one obtains:

$$\left[c_y\,\mathrm{sign}\left(c_x\right), c_y\,\mathrm{sign}\left(c_x\right)\right]^* = \tfrac{1}{12}\left(1-8\sqrt{2}\right)\pi\lambda^{-1} \; . \tag{A-38}$$

This expression is identical to that obtained previously in [7].

Table A-2. Values of the integrals, $I^{(1)}_{m+1/2}$ and $I^{(2)}_{m+1/2}$.

m	$I^{(1)}_{m+1/2}$	$I^{(2)}_{m+1/2}$
1	$I^{(1)}_{3/2} = 8\left(1-\sqrt{2}\right) + 4\int_{-1}^{1}\left(1+x^2\right)^{-1/2} dx$	$I^{(2)}_{3/2} = \frac{16}{3}\left(1-\sqrt{2}\right) + 4\int_{-1}^{1}\left(1+x^2\right)^{-1/2}\left(1-x^2\right) dx$
2	$I^{(1)}_{5/2} = \frac{8}{3}\left(2-\sqrt{2}\right)$	$I^{(2)}_{5/2} = \frac{8}{9}\left(4-3\sqrt{2}\right) + \frac{4}{3}\int_{-1}^{1}\left(1+x^2\right)^{-3/2}\left(1-x^2\right) dx$
3	$I^{(1)}_{7/2} = \frac{4}{15}\left(16-9\sqrt{2}\right)$	$I^{(2)}_{7/2} = \frac{4}{45}\left(32-17\sqrt{2}\right)$
4	$I^{(1)}_{9/2} = \frac{2}{105}\left(192-115\sqrt{2}\right)$	$I^{(2)}_{9/2} = \frac{2}{105}\left(128-73\sqrt{2}\right)$

4. BRACKET INTEGRALS CONTAINING ONE DISCONTINUOUS FUNCTION.

In the case of the planar geometry for stationary transport problems, a moment system can be obtained by means of the Maxwellian integral equation given by Eq. (11.52). For these problems it is convenient to employ the discontinuous distribution functions which might satisfy the boundary conditions at the wall surface. In the full velocity space these distribution functions for the linearized transport problems can be presented by the following general expression:

$$f = f^{(0)}\left\{1+\tfrac{1}{2}\left[\Phi_2(\mathbf{c},x)+\Phi_1(\mathbf{c},x)\right] + \tfrac{1}{2}\left[\Phi_2(\mathbf{c},x)-\Phi_1(\mathbf{c},x)\right]\mathrm{sign}(c_x)\right\}, \tag{A-39}$$

where $\Phi_2(\mathbf{c},x)$ and $\Phi_1(\mathbf{c},x)$ are the corrections to the distribution function for which $c_x > 0$ and $c_x < 0$, respectively, and x is a normal coordinate to the wall. If this distribution function is employed for closing the full velocity space moment system, the right-hand-sides of moment equations contain both ordinary bracket integrals and others which can be expressed in the form:

$$\left[\Psi(\mathbf{c})\mathrm{sign}(c_x), Q(\mathbf{c})\right], \tag{A-40}$$

where $\Psi(\mathbf{c})$ and $Q(\mathbf{c})$ are functions of the molecular velocity components.

Consider, as an example, evaluation of the bracket integral defined by:

$$I = \left[\operatorname{sign}(c_x), c^2 c_x \right] . \tag{A-41}$$

This integral occurs in the planar heat transport problem (see Section 11.9) and can be obtained from Eq. (11.16) by substitution of $r = 1$ and $c_r^* = 0$.

Integrating with respect to the variables, θ, ε, and g, and introducing new variables of integration via the following relationships:

$$G_x = xg \ , \quad G_y = yg \ , \quad G_z = zg \ , \quad \cos(\alpha) = t \ ,$$

one can obtain:

$$I = 96\sigma^2 \pi^{-1} \left(\frac{2kT_0}{m} \right)^{1/2} \int_{-1}^{1} dt \int_{t}^{\infty} dx \int_{-\infty}^{\infty} dy \int_{-\infty}^{\infty} F(t,x,y,z) dz \ , \tag{A-42}$$

where:

$$F = x\left(t^2 - \tfrac{1}{3}\right)\left(1 + x^2 + y^2 + z^2\right)^{-5} \ ,$$

and molecules have been assumed to be rigid spheres. In order to facilitate the integration procedure in Eq. (A-42) with respect to variables y and z, it is convenient to introduce polar coordinates by means of the relationships:

$$y = \rho \cos(\psi) \ , \ z = \rho \sin(\psi) \ . \tag{A-43}$$

Then, after integrations with respect to ψ and ρ, this integral becomes:

$$I = 24\sigma^2 \left(\frac{2kT_0}{m} \right)^{1/2} \int_{-1}^{1} \left(t^2 - \tfrac{1}{3}\right) dt \int_{t}^{\infty} \left(1 + x^2\right)^{-4} x dx \ . \tag{A-44}$$

The integrations with respect to x and t offer no difficulty. On integrating over all values of these variables one can obtain:

$$I = -\tfrac{2}{3} \left(\frac{2kT_0}{m} \right)^{1/2} \sigma^2 \ . \tag{A-45}$$

The same value of this integral has been obtained in [8].

REFERENCES

1. Chapman, S. and Cowling, T.G., *The Mathematical Theory of Non-Uniform Gases* (Cambridge University Press, Cambridge, U.K., 1990).
2. Ferziger, J.H. and Kaper, H.G., *Mathematical Theory of Transport Processes in Gases* (North-Holland, Amsterdam-London, 1972).
3. Hirschfelder, J.O., Curtiss, C.F., and Bird, R.B., *Molecular Theory of Gases and Liquids* (John Wiley and Sons, New York, 1954).
4. Rolduguin, V.I., *Application of the Non-Equilibrium Thermodynamics Method in Boundary Problems of the Kinetic Theory of Gases* (M.S. Thesis, Moscow, 1979).
5. Savkov, S.V., *The Slip Boundary Conditions of Non-Uniform Binary Gas Mixture and Application of them in Aerosol Dynamics* (M.S. Thesis, Moscow, 1987).
6. "Mathematical Tables," in *Handbook of Chemistry and Physics*, 71st edition, edited by Lide, D.R. (CRC Press, Boston, 1990), #283, p.A-38.
7. Derjaguin, B.V. and Yalamov, Yu.I., "The Theory of Thermophoresis and Diffusiophoresis of Aerosol Particles and Their Experimental Testing," In *Topics in Current Aerosol Research*, vol. 3, part 2, edited by Hidy, G.M. (Pergamon Press, Oxford, 1972).
8. Ivchenko, I.N., "Generalization of the Lees Method in Boundary Problems of Transfer," *J. Colloid and Interface Sci.* **135(1)**, 16-19 (1990).

Appendix B

BRACKET INTEGRALS FOR CURVILINEAR GEOMETRIES

1. THE SPECIAL FUNCTION OF THE FIRST KIND FOR THE SPHERICAL GEOMETRY.

Here, the special functions introduced in Section 11.4 will be investigated in detail. For spherical transport problems, the function $I_1^{*[1]}(r)$ is determined by Eq. (11-21) and, taking into account Eqs. (11-17) and (11-18), the following expression may be written:

$$I_1^{*[1]}(r) = -72\pi^{-2} \int_{-1}^{1} dt \int_{0}^{2\pi} d\beta \int_{-\infty}^{\infty} dy \int_{-\infty}^{\infty} dz \int_{x_0}^{\infty} F(t, \beta, y, z, x)\, dx \ , \tag{B-1}$$

where the notations introduced in Eq. (11-17) have been used.

It is very difficult to use Eq. (B-1) to obtain numerical values for $I_1^{*[1]}(r)$, however some analytical simplification is readily achievable. The integration with respect to x offers no difficulty if the following recurrence formula is used [1]:

$$\begin{aligned}\int \frac{dx}{\left(a+x^2\right)^{m+1}} &= \frac{1}{2ma} \frac{x}{\left(a+x^2\right)^m} + \frac{2m-1}{2ma} \int \frac{dx}{\left(a+x^2\right)^m} \\ &= \frac{(2m)!}{(m!)^2} \left[\frac{x}{2a} \sum_{r=1}^{m} \frac{r!(r-1)!}{(4a)^{m-r}(2r)!\left(a+x^2\right)^r} + \frac{1}{(4a)^m} \int \frac{dx}{a+x^2} \right] .\end{aligned} \tag{B-2}$$

To integrate with respect to y and z, the variables of integration are changed from y and z to y' and z' which are defined by:

$$y' = y - \left(1 - t^2\right)^{1/2} \cos(\beta) \ , \tag{B-3a}$$

$$z' = z - \left(1 - t^2\right)^{1/2} \sin(\beta) \ . \tag{B-3b}$$

After this substitution it is convenient to introduce the following polar coordinates:

$$y' = \rho \cos(\varphi') \ , \quad z' = \rho \sin(\varphi') \ .$$

Then, the substitution, $\varphi = \varphi' - \beta$, permits one to integrate over all values of β. Having performed this transformation and integration, one obtains the following analytical expression:

$$I_1^{*[1]}(r) = -\frac{144}{\pi} \int_{-1}^{1} dt \int_{0}^{2\pi} d\varphi \int_{0}^{\infty} \rho \left(f_{1,1}^{[1]} \psi_{1,1}^{[1]} - f_{1,2}^{[1]} \psi_{1,2}^{[1]} + f_{1,3}^{[1]} \psi_{1,3}^{[1]} \right) d\rho \ . \tag{B-4}$$

Here, the quantities $f_{1,i}^{[1]}$ and $\psi_{1,i}^{[1]}$ may be written as:

$$f_{1,1}^{[1]} = t^2 - \tfrac{1}{3} \ , \tag{B-5a}$$

$$\begin{aligned} f_{1,2}^{[1]} = &\left[t\left(1 - t^2\right)^{1/2} \cos(\varphi) \rho^2 + t\left(1 - t^2\right) \rho \right] \left(r^2 - 1\right)^{1/2} \\ &+ t^2 \left[\left(1 - t^2\right) + \left(1 - t^2\right)^{1/2} \cos(\varphi) \rho \right] , \end{aligned} \tag{B-5b}$$

$$f_{1,3}^{[1]} = t\left(1 - t^2\right)^{1/2} \cos(\varphi) \rho + t\left(1 - t^2\right) \ , \tag{B-5c}$$

$$\psi_{1,1}^{[1]} = \tfrac{1}{8} \left(a + x_0^2\right)^{-4} \ , \tag{B-5d}$$

Table B-1. Numerical values of the functions, $I_1^{*[i]}(r)$. Note that for $i=1,2$, the apparent values of these functions (evaluated numerically) appear to be the same although the analytical formulations of these functions appear to be different. As of this time these functions have not been proven to be identical.

r	$I^{*[i]}(r)$	r	$I^{*[i]}(r)$	r	$I^{*[i]}(r)$
1.00	9.982E−1	1.50	4.440E−1	3.50	8.143E−2
1.10	8.267E−1	1.75	3.261E−1	4.00	6.052E−2
1.20	6.944E−1	2.00	2.489E−1	5.00	3.982E−2
1.30	5.917E−1	2.50	1.591E−1	6.00	2.791E−2
1.40	5.093E−1	3.00	1.116E−1	10.0	1.049E−2

$$\psi_{1,2}^{[1]} = \tfrac{35}{128} a^{-4}\left(a+x_0^2\right)^{-1} + \tfrac{35}{192} a^{-3}\left(a+x_0^2\right)^{-2} + \tfrac{7}{48} a^{-2}\left(a+x_0^2\right)^{-3} + \tfrac{1}{8} a^{-1}\left(a+x_0^2\right)^{-4}, \tag{B-5e}$$

$$\psi_{1,3}^{[1]} = \tfrac{35}{128} a^{-9/2}\left(\tfrac{1}{2}\pi - \arctan\left(x_0\, a^{-1/2}\right)\right), \tag{B-5f}$$

$$a = \rho^2 + 2\rho\left(1-t^2\right)^{1/2}\cos(\varphi) + \left(2-t^2\right), \tag{B-5g}$$

$$x_0 = t + \rho\left(r^2-1\right)^{1/2}. \tag{B-5h}$$

The numerical values of the functions, $I_1^{*[1]}(r)$ and $I_1^{*[2]}(r)$ (which will be discussed in Section B.2), are given in Table B-1 and the dependence of $I_1^{*[i]}(r)$ on the radial coordinate is shown in Fig. B-1. A numerical analysis shows that the special functions, $I_1^{*[i]}(r)$, are essentially equal to each other for $i=1,2$ with the relative deviations of the numerical values being less than 1% although equality has not been definitively established.

2. THE SPECIAL FUNCTION OF THE SECOND KIND FOR THE SPHERICAL GEOMETRY.

The second special function related to spherical boundary value transport problems can be expressed in the form:

$$I_1^{*[2]}(r) = 1440\pi^{-2}\left(8+3\pi\right)^{-1} \int_{-1}^{1} dt \int_{0}^{2\pi} d\beta \int_{-\infty}^{\infty} dy \int_{-\infty}^{\infty} dz \int_{x_0}^{\infty} F\Phi dx, \tag{B-6}$$

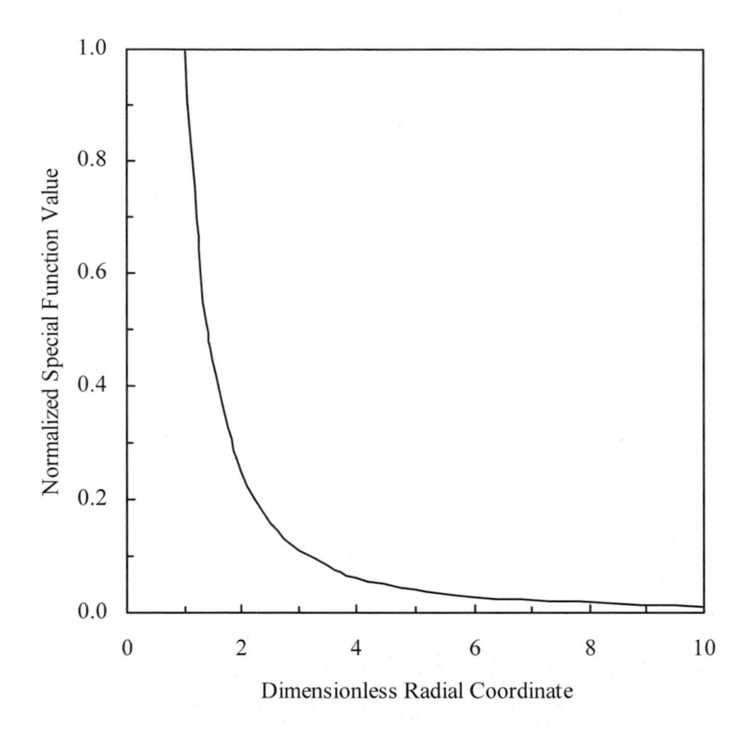

Figure B-1. Dependence of $I_1^{*[i]}(r)$ on the radial coordinate, r.

where the notations of Section 11.4 are used. By means of the transformations employed in Section B.1, this integral can be expressed in the form:

$$I_1^{*[2]}(r) = -20(8+3\pi)^{-1} I_1^{*[1]}(r) + 2880\pi^{-1}(8+3\pi)^{-1} \times \int_{-1}^{1} dt \int_0^{2\pi} d\varphi \int_0^{\infty} \rho d\rho \left\{ f_{1,1}^{[2]}\psi_{1,1}^{[2]} + f_{1,2}^{[2]}\psi_{1,2}^{[2]} + f_{1,3}^{[2]}\psi_{1,3}^{[2]} \right\}, \quad \text{(B-7)}$$

where the quantities $f_{1,i}^{[2]}$ and $\psi_{1,i}^{[2]}$ may be written as:

$$f_{1,1}^{[2]} = -\tfrac{1}{5}\left(1-t^2\right)^{1/2}\left(2t^2-\tfrac{1}{3}\right)\left[\left(1-t^2\right)^{1/2} + \cos(\varphi)\rho\right], \quad \text{(B-8a)}$$

$$f_{1,2}^{[2]} = \tfrac{1}{5}t\left(t^2-\tfrac{1}{3}\right) + \tfrac{9}{5}t\left(1-t^2\right)a^{-1} \times\left[\left(1-t^2\right) + \cos^2(\varphi)\rho^2 + 2\left(1-t^2\right)^{1/2}\cos(\varphi)\rho\right], \quad \text{(B-8b)}$$

$$f_{1,3}^{[2]} = 2t\left(2t^4 - \tfrac{13}{3}t^2 + \tfrac{5}{3}\right) + 4t\left(1-t^2\right)^{1/2}\left(\tfrac{4}{3} - 2t^2\right)\cos(\varphi)\rho$$
$$+ 2t\rho^2\left[\left(1-t^2\right)\cos^2(\varphi) - \left(t^2 - \tfrac{1}{3}\right)\right] , \qquad \text{(B-8c)}$$

$$\psi_{1,1}^{[2]} = \left(a + x_0^2\right)^{-5} , \qquad \text{(B-8d)}$$

$$\psi_{1,2}^{[2]} = x_0\psi_{1,2}^{[1]} - \psi_{1,3}^{[1]} , \qquad \text{(B-8e)}$$

$$\psi_{1,3}^{[2]} = \tfrac{1}{10}x_0 a^{-1}\left(a + x_0^2\right)^{-5} . \qquad \text{(B-8f)}$$

Here, the quantities a and x_0 are the same as those given in Eqs. (B-5g) and (B-5h), respectively. Numerical values of this special function are given in Table B-1.

3. THE SPECIAL FUNCTION OF THE FIRST KIND FOR THE CYLINDRICAL GEOMETRY.

This function is defined by Eqs. (11-18) and (11-21). Since the lower limit of integration with respect to x is independent of the variable, z, it is convenient to perform the first integration over all values of z and then to integrate with respect to x. Having performed these integrations, one can obtain a triple integral which can be numerically evaluated.

To integrate over all values of y, first change the variable of integration to $y' = y - (1-t^2)^{1/2}\cos(\beta)$ and then break this integral into two parts by means of the relationship:

$$\int_{-\infty}^{\infty} dy' = \int_{-\infty}^{0} dy' + \int_{0}^{\infty} dy' .$$

After these very simple transformations, one can obtain:

$$I_2^{*[1]}(r) = -\tfrac{315}{16}\pi^{-1}\int_{-1}^{1} dt \int_{0}^{2\pi} d\beta \int_{0}^{\infty} dy$$
$$\times\left\{ f_{2,1}^{[1]}\psi_{2,1}^{[1]} + f_{2,2}^{[1]}\psi_{2,2}^{[1]} + g_{2,1}^{[1]}\Phi_{2,1}^{[1]} + g_{2,2}^{[1]}\Phi_{2,2}^{[1]} \right\} , \qquad \text{(B-9)}$$

where the following notations are introduced:

$$f_{2,1}^{[1]} = g_{2,1}^{[1]} = \tfrac{1}{7}\left(t^2 - \tfrac{1}{3}\right) , \tag{B-10a}$$

$$f_{2,2}^{[1]} = -t\left(1-t^2\right)^{1/2} \cos(\beta)\left[\left(1-t^2\right)^{1/2} \cos(\beta) + y\right] , \tag{B-10b}$$

$$g_{2,2}^{[1]} = -t\left(1-t^2\right)^{1/2} \cos(\beta)\left[\left(1-t^2\right)^{1/2} \cos(\beta) - y\right] , \tag{B-10c}$$

$$\psi_{2,1}^{[1]} = \left(c + x_0^2\right)^{-7/2} , \tag{B-10d}$$

$$\begin{aligned}\psi_{2,2}^{[1]} = \tfrac{1}{35} x_0 \left(c + x_0^2\right)^{-1/2} \Big[& 16c^{-4} + 8c^{-3}\left(c + x_0^2\right)^{-1} \\ & + 6c^{-2}\left(c + x_0^2\right)^{-2} + 5c^{-1}\left(c + x_0^2\right)^{-3}\Big] .\end{aligned} \tag{B-10e}$$

$$\Phi_{2,1}^{[1]} = \left(d + x_0^2\right)^{-7/2} , \tag{B-10f}$$

$$\begin{aligned}\Phi_{2,2}^{[1]} = \tfrac{1}{35} x_0 \left(d + x_0^2\right)^{-1/2} \Big[& 16d^{-4} + 8d^{-3}\left(d + x_0^2\right)^{-1} \\ & + 6d^{-2}\left(d + x_0^2\right)^{-2} + 5d^{-1}\left(d + x_0^2\right)^{-3}\Big] .\end{aligned} \tag{B-10g}$$

Table B-2. Numerical values of the functions, $I_2^{*[i]}(r)$. Note that again, as in Table B-1, the values of these functions appear to be the same but that analytical equality has not been proven. Here, the values for the different functions have been given separately in order to demonstrate their similarity.

r	$I_2^{*[1]}(r)$	$I_2^{*[2]}(r)$	r	$I_2^{*[1]}(r)$	$I_2^{*[2]}(r)$
1.00	0.9999	1.0000	8.00	0.1249	0.1250
1.10	0.9091	0.9090	9.00	0.1111	0.1110
1.20	0.8333	0.8333	10.00	0.1000	0.0999
1.30	0.7692	0.7691	12.00	0.0833	0.0833
1.40	0.7142	0.7142	15.00	0.0665	0.0667
1.50	0.6666	0.6666	17.00	0.0588	0.0588
1.75	0.5714	0.5714	20.00	0.0500	0.0499
2.00	0.4999	0.5000	25.00	0.0399	0.0400
2.50	0.3999	0.4000	30.00	0.0334	0.0332
3.00	0.3333	0.3333	50.00	0.0200	0.0199
3.50	0.2857	0.2857	60.00	0.0165	0.0167
4.00	0.2499	0.2500	70.00	0.0151	0.0132
5.00	0.2000	0.1999	80.00	0.0123	0.0123
6.00	0.1666	0.1666	90.00	0.0112	0.0110
7.00	0.1426	0.1430	100.00	0.0110	0.0090

In Eqs. (B-10a)-(B-10g) the quantities x_0, c, and d are defined as:

$$x_0 = t + y\left(r^2 - 1\right)^{1/2} , \tag{B-11a}$$

$$c = 1 + \left[y + \left(1 - t^2\right)^{1/2} \cos(\beta) \right]^2 , \tag{B-11b}$$

and:

$$d = 1 + \left[\left(1 - t^2\right)^{1/2} \cos(\beta) - y \right]^2 . \tag{B-11c}$$

Numerical values of the functions, $I_2^{*[1]}(r)$ and $I_2^{*[2]}(r)$, are given in Table B-2 and the dependence of $I_2^{*[1]}(r)$ on the radial coordinate is shown in Fig. B-2. Moreover, the numerical values of $\int_1^z I_2^{*[1]}(r)dr$ for various values of z are given in Table B-3.

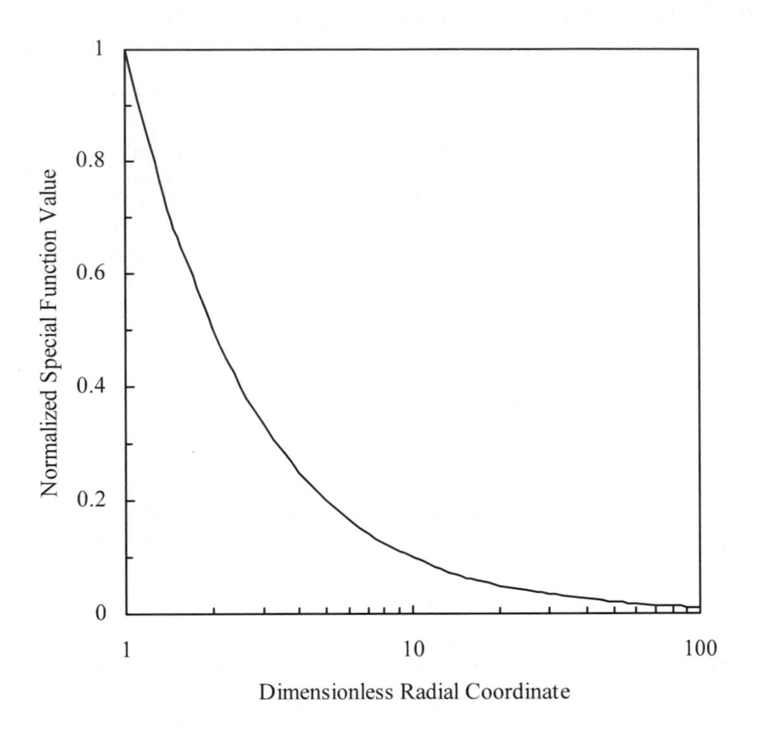

Figure B-2. Dependence of $I_2^{*[1]}(r)$ on the radial coordinate, r.

4. THE SPECIAL FUNCTION OF THE SECOND KIND FOR THE CYLINDRICAL GEOMETRY.

The special function, $I_2^{*[2]}(r)$, is determined by Eqs. (11-19) and (11.22). Performing the same transformations as in Section B.3, one can obtain:

$$I_2^{*[2]}(r) = -20(8+3\pi)^{-1} I_2^{*[1]}(r) - \tfrac{315}{4}\pi^{-1}(8+3\pi)^{-1} \times \int_{-1}^{1} dt \int_0^{2\pi} d\beta \int_0^{\infty} dy \times \left\{ f_{2,1}^{[2]}\psi_{2,1}^{[2]} + f_{2,2}^{[2]}\psi_{2,2}^{[2]} + g_{2,1}^{[2]}\Phi_{2,1}^{[2]} + g_{2,2}^{[2]}\Phi_{2,2}^{[2]} \right\}, \quad \text{(B-12)}$$

where the following functions are employed:

Table B-3. Numerical values of the integrals, $\psi_i(z) = \int_1^{\infty} I_2^{*[i]}(r)\,dr$.

z	ψ_1	ψ_2	z	ψ_1	ψ_2
1.00	0.0000	0.0000	1.75	0.5596	0.5596
1.05	0.0488	0.0488	2.00	0.6931	0.6931
1.10	0.0935	0.0935	3.00	1.0984	1.0983
1.20	0.1823	0.1823	4.00	1.3860	1.3860
1.30	0.2624	0.2624	5.00	1.6090	1.6090
1.40	0.3365	0.3365	7.00	1.9453	1.9454
1.50	0.4055	0.4055	10.00	2.3019	2.3021

$$f_{2,1}^{[2]} = t\left\{\left(t^2 - \tfrac{1}{3}\right) + \left(1 - t^2\right)\left[\sin^2(\beta) + 8c^{-1}\left(\left(1 - t^2\right)^{1/2}\cos(\beta) + y\right)^2 \cos^2(\beta)\right]\right\} , \tag{B-13a}$$

$$f_{2,2}^{[2]} = \left(1 - t^2\right)^{1/2}\left(2t^2 - \tfrac{1}{3}\right)\left(\left(1 - t^2\right)^{1/2}\cos(\beta) + y\right)\cos(\beta) - x_0 c^{-1} t\left(1 - t^2\right)\left(\left(1 - t^2\right)^{1/2}\cos(\beta) + y\right)^2 \cos^2(\beta) + x_0 t\left(t^2 - \tfrac{1}{3}\right) , \tag{B-13b}$$

$$g_{2,1}^{[2]} = t\left\{\left(t^2 - \tfrac{1}{3}\right) + \left(1 - t^2\right)\left[\sin^2(\beta) + 8d^{-1}\left(\left(1 - t^2\right)^{1/2}\cos(\beta) - y\right)^2 \cos^2(\beta)\right]\right\} , \tag{B-13c}$$

$$g_{2,2}^{[2]} = \left(1 - t^2\right)^{1/2}\left(2t^2 - \tfrac{1}{3}\right)\left(\left(1 - t^2\right)^{1/2}\cos(\beta) - y\right)\cos(\beta) - x_0 d^{-1} t\left(1 - t^2\right)\left(\left(1 - t^2\right)^{1/2}\cos(\beta) - y\right)^2 \cos^2(\beta) + x_0 t\left(t^2 - \tfrac{1}{3}\right) , \tag{B-13d}$$

$$\psi_{2,1}^{[2]} = -\psi_{2,2}^{[1]} \ , \tag{B-13e}$$

$$\psi_{2,2}^{[2]} = \left(c + x_0^2\right)^{-1} \psi_{2,1}^{[1]} = \left(c + x_0^2\right)^{-9/2} \ , \tag{B-13f}$$

$$\Phi_{2,1}^{[2]} = -\Phi_{2,2}^{[1]} \ , \tag{B-13g}$$

$$\Phi_{2,2}^{[2]} = \left(d + x_0^2\right)^{-1} \Phi_{2,1}^{[1]} = \left(d + x_0^2\right)^{-9/2} \ . \tag{B-13h}$$

The quantities shown in these formulas are identical to those given in Eqs. (B-10a)-(B-11c). Numerical values of $I_2^{*[2]}(r)$ are given in Table B-2 and numerical values of $\int_1^z I_2^{*[2]}(r)\,dr$ are given in Table B-3.

5. APPROXIMATE EXPRESSIONS FOR THE SPECIAL FUNCTIONS.

Using the general properties of the Boltzmann equation [2], one can obtain very simple approximate expressions for the bracket integrals considered in Sections B.1-B.4. These approximate expressions will be derived in this section by employing the first-order approximation to the Chapman-Enskog solution given in Eq. (5-39). In order to use the first-order approximation of the Chapman-Enskog solution, the bracket integrals must be transformed into the more convenient form:

$$\begin{aligned}\left[c^k \operatorname{sign}\left(c_r - c_r^*\right), c^2 c_r\right] &= -\left[c^k \operatorname{sign}\left(c_r - c_r^*\right), c_r S_{3/2}^{(1)}\left(c^2\right)\right] \\ &= -\int c^k \operatorname{sign}\left(c_r - c_r^*\right) I\left(c_r S_{3/2}^{(1)}\left(c^2\right)\right) d\mathbf{v} \ , \ (k = 0, 2) \ .\end{aligned} \tag{B-14}$$

Now, using Eq. (5-39), these bracket integrals can be approximated by:

$$\begin{aligned}&\left[c^k \operatorname{sign}\left(c_r - c_r^*\right), c^2 c_r\right] \\ &= \frac{1}{n^2 a_1^{(1)}} \int v_r S_{3/2}^{(1)}\left(c^2\right) f^{(0)} c^k \operatorname{sign}\left(c_r - c_r^*\right) d\mathbf{v} \ ,\end{aligned} \tag{B-15}$$

where $a_1^{(1)}$ is given by Eq. (5-40).

It is very important to note that the integrals in the right-hand-side of Eq. (B-15) are ordinary integrals in the velocity space. The values of these integrals depends on the problem geometry. For the spherical geometry, after integration, one can obtain the following expressions:

$$I_1^{*[1]}(r) = \tfrac{4}{5}\sqrt{2}r^{-2} \ , \tag{B-16}$$

$$I_1^{*[2]}(r) = \tfrac{64}{5}\sqrt{2}(8+3\pi)^{-1} r^{-2} \ . \tag{B-17}$$

The appropriate expressions for the cylindrical geometry can be written as:

$$I_2^{*[1]}(r) = \tfrac{4}{5}\sqrt{2}r^{-1} \ , \tag{B-18}$$

$$I_2^{*[2]}(r) = \tfrac{64}{5}\sqrt{2}(8+3\pi)^{-1} r^{-1} \ . \tag{B-19}$$

The accuracy of this analysis can be easily determined by means of a comparison with the appropriate numerical values. The relative errors for $I_i^{*[1]}(r)$ and $I_i^{*[2]}(r)$, for $r=1$, are 13% and 4%, respectively.

REFERENCES

1. "Mathematical Tables," in *Handbook of Chemistry and Physics*, 71st edition, edited by Lide, D.R. (CRC Press, Boston, 1990), #67, p.A-24.
2. Loyalka, S.K., "Approximate Method in the Kinetic Theory," *Phys. Fluids* **14(11)**, 2291-2294 (1971).

Appendix C

BRACKET INTEGRALS FOR POLYNOMIAL EXPANSION METHOD

1. CALCULATION OF THE BRACKET INTEGRALS OF THE FIRST KIND.

To complete the calculations of the polynomial expansion method in the four-moment approach for the spherical geometry, one must know values of the following bracket integrals:

$$\alpha_{kn}^{[1]} = \left[H_k(c_r) S_0^{(n)}\left(c_\theta^2 + c_\phi^2\right), c_r c^2 \right],$$

$$\alpha_{kn}^{[2]} = \left[c^2 H_k(c_r) S_0^{(n)}\left(c_\theta^2 + c_\phi^2\right), c_r c^2 \right],$$

where $H_k(c_r)$ and $S_0^{(n)}\left(c_\theta^2 + c_\phi^2\right)$ are, respectively, Hermite and Sonine polynomials of the molecular velocity components. First, consider the method of calculating the bracket integrals, $\alpha_{kn}^{[1]}$, when molecules are assumed to be rigid spheres.

To obtain general analytical expressions for these integrals, one must use the generating functions for the Hermite and Sonine polynomials, $H(x,t)$ and $S_0(x,s)$, which are given by:

$$H(x,t) = \exp\left(-t^2 + 2xt\right), \quad x = c_r, \tag{C-1}$$

and:

$$S_0(x,s) = \frac{1}{1-s}\exp\left(-\frac{xs}{1-s}\right), \quad x = c_\theta^2 + c_\phi^2 . \tag{C-2}$$

Then, the Hermite and Sonine polynomials may be expressed as:

$$H_k(x) = \frac{\partial^k}{\partial t^k} H(x,t)\bigg|_{t=0} , \tag{C-3}$$

and:

$$S_0^{(n)}(x) = \frac{1}{n!}\frac{\partial^n}{\partial s^n} S_0(x,s)\bigg|_{s=0} . \tag{C-4}$$

Having used Eqs. (C-3) and (C-4), one can obtain:

$$\alpha_{kn}^{[i]} = \frac{1}{n!}\frac{\partial^k}{\partial t^k}\frac{\partial^n}{\partial s^n}\frac{\exp(-t^2)}{1-s}\xi^{[i]}(t,s)\bigg|_{t,s=0} , \tag{C-5}$$

where:

$$\xi^{[i]}(t,s) = \left[c^{2(i-1)}\exp\left(2c_r t - \frac{s}{1-s}\left(c_\theta^2 + c_\phi^2\right)\right), c_r c^2\right] . \tag{C-6}$$

The bracket integral of Eq. (C-6) can be calculated in the same manner as the original one but it has some special properties that can be taken into account to facilitate the calculations. The analysis is simplified if the following new variables of integration are introduced:

$$G_r' = G_r - \frac{t}{\sqrt{2}} , \tag{C-7}$$

$$G_\theta' = G_\theta - \frac{s}{2-s} g\sin(\alpha)\cos(\beta) , \tag{C-8}$$

and:

$$G_\phi' = G_\phi - \frac{s}{2-s} g\sin(\alpha)\sin(\beta) , \tag{C-9}$$

where $\left(G_r, G_\theta, G_\phi\right)$ are the Cartesian coordinates of the dimensionless velocity, $\mathbf{G}$, and $\left(g, \alpha, \beta\right)$ are the spherical coordinates of the dimensionless velocity, $\mathbf{g}$. Then, having performed six integrations with respect to the variables, θ, ε, β, G'_r, G'_θ, and G'_ϕ, and having employed Leibnitz' formula [1] for the k-th derivative with respect to t, one can obtain:

$$\begin{aligned}\alpha_{kn}^{[1]} = \tfrac{16}{3}\pi^{1/2}\sigma^2\left(\frac{2kT}{m}\right)^{1/2}\frac{1}{n!}\frac{\partial^n}{\partial s^n}(2-s)^{-1}\int_0^\pi \sin(\alpha)\,d\alpha \\ \times\int_0^\infty g^3\exp\left(-g^2\left[1+\frac{s}{2-s}\sin^2(\alpha)\right]\right) \\ \times\Psi_k^{[1]}(s,\alpha,g)\,dg\,\Big|_{s=0} ,\end{aligned} \qquad \text{(C-10)}$$

where:

$$\begin{aligned}\Psi_k^{[1]}(s,\alpha,g) = \tfrac{3}{2}\chi^{(k)}(0,g)\frac{s}{2-s}g^3 f_2(\alpha) \\ +\frac{1}{\sqrt{2}}k\chi^{(k-1)}(0,g)g^2 f_1(\alpha) ,\end{aligned}$$

$$\chi(t,g) = \exp\left(-\tfrac{1}{2}t^2 - \sqrt{2}gt\cos(\alpha)\right) ,$$

$$f_1(\alpha) = \tfrac{3}{2}\cos^2(\alpha) - \tfrac{1}{2} ,$$

and:

$$f_2(\alpha) = \sin^2(\alpha)\cos(\alpha) .$$

To calculate the k-th derivative, $\chi^{(k)}(0,g)$, one should use the recurrence relation that is given by:

$$\begin{aligned}\chi^{(k+1)}(0,g) = -\sqrt{2}g\cos(\alpha)\chi^{(k)}(0,g) \\ -k\chi^{(k-1)}(0,g) ; \quad k \geq 0 ,\end{aligned} \qquad \text{(C-11)}$$

Table C-1. Analytical values of the bracket integral, $\alpha_{1n}^{[1]}$, for the polynomial expansion method.

$$\alpha_{1n}^{[1]} = a_n \sigma^2 \sqrt{2\pi} \left(2kT/m\right)^{1/2}$$

n	1	2	3	4
a_n	$-\frac{16}{15}$	$\frac{16}{105}$	$\frac{4}{315}$	$\frac{8}{3465}$

with:

$$\chi^{(0)}\left(0, g\right) = 1 \ . \tag{C-12}$$

For the particular case when the index, $k = 1$, the bracket integrals, $\alpha_{1n}^{[1]}$, may be expressed in the form:

$$\alpha_{1n}^{[1]} = -\tfrac{16}{3}\sqrt{2\pi}\sigma^2 \left(\frac{2kT}{m}\right)^{1/2} \frac{1}{n!} \int_0^1 \varphi_n^{(1)}\left(x\right) dx \ , \tag{C-13}$$

where:

$$\begin{aligned} \varphi_n^{(1)}\left(x\right) = \left(\tfrac{3}{2}x^2 - \tfrac{1}{2}\right)&\left(4\nu_3^{(n)} - 4n\nu_3^{(n-1)} + n\left(n-1\right)\nu_3^{(n-2)}\right) \\ &+9\left(1-x^2\right)x^2\left(4n\nu_4^{(n-1)} - 4n\left(n-1\right)\nu_4^{(n-2)}\right. \\ &\left.+n\left(n-1\right)\left(n-2\right)\nu_4^{(n-3)}\right) \ . \end{aligned}$$

Here, the following notations are introduced:

$$\nu_3^{(m)} = -\frac{3 \cdot 4 \cdot 5 \cdots \left(3+m-1\right)}{2^{3+m}} x^{2m} \ ; \quad m \geq 1 \ , \tag{C-14}$$

$$\nu_4^{(m)} = \frac{4 \cdot 5 \cdot 6 \cdots \left(4+m-1\right)}{2^{4+m}} x^{2m} \ ; \quad m \geq 1 \ . \tag{C-15}$$

Specific values of $\alpha_{1n}^{[1]}$ are given in Table C-1.

2. ANALYTICAL EXPRESSIONS FOR THE BRACKET INTEGRALS OF THE SECOND KIND.

The method described in the preceding section may be used to calculate the integrals, $\alpha_{kn}^{[2]}$, as well. After some fairly complicated algebraic transformations, one can obtain:

$$\alpha_{kn}^{[2]} = \tfrac{8}{3}\sqrt{2\pi}\sigma^2 \left(\frac{kT}{m}\right)^{1/2} \frac{1}{n!}\frac{\partial^n}{\partial s^n}(2-s)^{-1}\int_0^{\pi}\sin(\alpha)\,d\alpha$$
$$\times \int_0^{\infty} g^3 \exp\left(-g^2\left[1+\frac{s}{2-s}\sin^2(\alpha)\right]\right)\Psi_k^{[2]}(s,\alpha,g)\,dg\Bigg|_{s=0}, \qquad \text{(C-16)}$$

where:

$$\Psi_k^{[2]}(s,\alpha,g) = \chi^{(k)}(0,g)\Bigg\{ f_1(\alpha)\left(-g^3\cos(\alpha)\right)$$
$$+ f_2(\alpha)\left[\tfrac{3}{2}g^5\frac{s}{2-s}\left(1+\frac{3s^2-4s}{(2-s)^2}\sin^2(\alpha)\right)+\tfrac{3}{4}g^3\frac{-13s^2+22s-8}{(2-s)^2}\right]\Bigg\}$$
$$+\frac{1}{\sqrt{2}}k\chi^{(k-1)}(0,g)\Bigg\{\left[\tfrac{1}{2}\frac{10-7s}{(2-s)}g^2+g^4\left(1+\frac{3s^2-4s}{(2-s)^2}\sin^2(\alpha)\right)\right]f_1(\alpha)$$
$$-\frac{3s}{(2-s)}g^4\cos(\alpha)f_2(\alpha)\Bigg\}+k(k-1)\chi^{(k-2)}(0,g)$$
$$\times\left\{-g^3\cos(\alpha)f_1(\alpha)+\tfrac{3}{4}\frac{s}{(2-s)}g^3 f_2(\alpha)\right\}$$
$$+\tfrac{1}{2}\frac{1}{\sqrt{2}}k(k-1)(k-2)g^2\chi^{(k-3)}(0,g)f_1(\alpha),$$

$f_1(\alpha)=\tfrac{3}{2}\cos^2(\alpha)-\tfrac{1}{2}$, and $f_2(\alpha)=\sin^2(\alpha)\cos(\alpha)$.

Here, for the specific case of $k=1$, the integrals, $\alpha_{1n}^{[2]}$, may be expressed in the form:

Table C-2. Analytical values of the bracket integral, $\alpha_{1n}^{[2]}$, for the polynomial expansion method.

$\alpha_{1n}^{[2]} = b_n \sigma^2 \sqrt{2\pi} \left(2kT/m\right)^{1/2}$				
n	1	2	3	4
b_n	$-\frac{16}{3}$	$\frac{46}{15}$	$-\frac{38}{105}$	$-\frac{19}{630}$

$$\alpha_{1n}^{[2]} = \tfrac{4}{3}\sqrt{2\pi}\sigma^2 \left(\frac{2kT}{m}\right)^{1/2} \frac{1}{n!}\int_0^1 \varphi_n^{(2)}(x)\,dx \ ,$$

where:

$$\begin{aligned}
\varphi_n^{(2)}(x) = \left(\tfrac{3}{2}x^2 - \tfrac{1}{2}\right)\Big\{ & \left(88 + 96x^2\right)\nu_4^{(n)} + n\left(-168 - 116x^2\right)\nu_4^{(n-1)} \\
& + n(n-1)\left(110 + 36\,x^2\right)\nu_4^{(n-2)} \\
& + n(n-1)(n-2)\left(-24 - x^2\right)\nu_4^{(n-3)}\Big\} \\
+ 3\left(1 - x^2\right)x^2\Big\{ & -96\nu_5^{(n)} + n\left(552 + 48x^2\right)\nu_5^{(n-1)} \\
& + n(n-1)\left(-816 + 12x^2\right)\nu_5^{(n-2)} \\
& + n(n-1)(n-2)\left(474 - 72x^2\right)\nu_5^{(n-3)} \\
& + n(n-1)(n-2)(n-3)\left(-96 + 27x^2\right)\nu_5^{(n-4)}\Big\} \ ,
\end{aligned}$$

and:

$$\nu_5^{(m)} = -\frac{5\cdot 6\cdot 7\cdots(5+m-1)}{2^{5+m}}x^{2m} \ ; \quad m \ge 1 \ .$$

Some values of $\alpha_{1n}^{[2]}$ are given in Table C-2.

REFERENCES

1. Bronshtein, I.N. and Semendyayev, K.A., *A Guidebook to Mathematics* (Springer Verlag, New York, N.Y., 1973).

Appendix D

THE VARIATIONAL PRINCIPLE FOR PLANAR PROBLEMS

1. SOME DEFINITIONS AND PROPERTIES FOR INTEGRAL OPERATORS.

In the Kinetic Theory of Gases, the following form of an integral operator, Q, is frequently used:

$$Q\Phi(\mathbf{c},\mathbf{r}) = \int K(\mathbf{c},\mathbf{c}')\Phi(\mathbf{c}',\mathbf{r})d\mathbf{c}' , \tag{D-1}$$

where the integration extends over all values of the molecular velocity. For planar transport problems, the appropriate operator is defined by:

$$Q\Phi(\mathbf{c},x) = \int K(\mathbf{c},\mathbf{c}')\Phi(\mathbf{c}',x)d\mathbf{c}' . \tag{D-2}$$

To begin with, a fundamental definition for this integral operator is introduced. The adjoint operator, Q^*, in the space defined by the scalar product given by Eqs. (9-16) and (9-17) is specified by:

$$\left[(\Psi , Q\Phi)\right] = \left[\left(\Phi , Q^*\Psi\right)\right] . \tag{D-3}$$

The operator, Q, is termed a self-adjoint operator if:

$$Q^* = Q . \tag{D-4}$$

For a self-adjoint operator, Eq. (D-3) becomes:

$$\left[\left(\Psi , Q\Phi \right) \right] = \left[\left(\Phi , Q\Psi \right) \right] . \tag{D-5}$$

It has been proven in [1,2] that the condition for the operator, Q, to be self-adjoint is that:

$$K\left(\mathbf{c},\mathbf{c}'\right) = K\left(\mathbf{c}',\mathbf{c}\right) . \tag{D-6}$$

Another way of stating Eq. (D-6) is to say that the function, $K\left(\mathbf{c},\mathbf{c}'\right)$, in Eqs. (D-1) and (D-2) is symmetric in the variables, $\mathbf{c}$ and $\mathbf{c}'$. From the symmetry of the kernel, $\mathrm{K}\left(\mathbf{c},\mathbf{c}'\right)$, in Eq. (9-14), one can easily conclude that the operator, H, is self-adjoint. In the formulation of the variational principle, the adjoint operator, $\left(\mathrm{H}L\right)^*$, where L is defined by Eq. (9-13), is employed to facilitate certain transformations. The adjoint operator, $\left(\mathrm{H}L\right)^*$, may be derived from the following relationships:

$$\begin{aligned} \left[\left(\Psi , \mathrm{H}L\Phi \right) \right] &= \left[\left(\mathrm{H}\,\Psi , L\Phi \right) \right] \\ &= \left[\left(\Phi , L^*\mathrm{H}\,\Psi \right) \right] = \left[\left(\Phi , L^*\mathrm{H}^*\Psi \right) \right] , \end{aligned} \tag{D-7}$$

which yield:

$$\left(\mathrm{H}L\right)^* = L^*\mathrm{H}^* = L^*\mathrm{H} . \tag{D-8}$$

2. THE VARIATIONAL PRINCIPLE.

In the linearized planar transport problems studied to date, it has generally been the case that the main quantity of physical interest can be expressed as:

$$I = \left[\left(\Phi^* , p \right) \right] = \left[\left(p^* , \Phi \right) \right] , \tag{D-9}$$

where the following notation has been introduced:

$$\Phi^* = \mathrm{H}\,\Phi . \tag{D-10}$$

and Φ is the correction to the distribution function which satisfies the following integral form of the Boltzmann equation:

$$\Phi(\mathbf{c},x) = L\Phi(\mathbf{c},x) + p(\mathbf{c},x) \ . \tag{D-11}$$

Applying the operator, H, to Eq. (D-11), one obtains:

$$\Phi^*(\mathbf{c},x) = \mathrm{H}L\Phi(\mathbf{c},x) + p^*(\mathbf{c},x) = \bar{L}\Phi^*(\mathbf{c},x) + p^*(\mathbf{c},x) \ , \tag{D-12}$$

where $\bar{L} = \mathrm{H}L\mathrm{H}^{-1}$ and H^{-1} is the inverse of H.

Now, a functional may be constructed of the form:

$$F\left(\tilde{\Phi},\tilde{\Phi}^*\right) = -\left[\left(\tilde{\Phi}^*, p\right)\right] + \left[\left(\tilde{\Phi}^*, \tilde{\Phi} - L\tilde{\Phi} - p\right)\right] , \tag{D-13}$$

where $\tilde{\Phi}$ is a trial function. If this trial function is equal to the real correction to the distribution function, Φ, then the integral, I, is found from:

$$I = -F\left(\Phi,\Phi^*\right) . \tag{D-14}$$

Now, consider a trial function of the form:

$$\tilde{\Phi} = \Phi + \delta\Phi \ . \tag{D-15}$$

For this trial function, the functional of Eq. (D-13) becomes:

$$\begin{aligned} F\left(\Phi + \delta\Phi, \Phi^* + \delta\Phi^*\right) & \\ &= -\left[\left(p^*, \Phi\right)\right] + \left[\left(\delta\Phi, \Phi^* - \bar{L}\Phi^* - p^*\right)\right] \\ &= -\left[\left(p^*, \Phi\right)\right] + O\left[\left(\delta\Phi\right)^2\right] , \end{aligned} \tag{D-16}$$

which implies that:

$$\delta F = O\left[\left(\delta\Phi\right)^2\right] . \tag{D-17}$$

Since the first variation is equal to zero, the functional is stationary when $\tilde{\Phi}(\mathbf{c},x) = \Phi(\mathbf{c},x)$ and $\tilde{\Phi}^*(\mathbf{c},x) = \Phi^*(\mathbf{c},x)$. Now, the basic integral of interest, I, can be expressed in terms of the stationary value of the functional by the following simple expression:

$$I = -F\left(\Phi, \Phi^*\right) = -F_{st}\left(\tilde{\Phi}, \tilde{\Phi}^*\right), \tag{D-18}$$

and, hence, to find an approximate value for I, one need only employ a simple trial function satisfying the stationary condition [3-10]. It should be noted in this context that a non-self-adjoint operator such as the Boltzmann operator cannot lead to an extremum (minimum) but, rather, must yield a saddle point because it has been proven in [4] that the second variation, $\delta^2 F$, may have either sign.

REFERENCES

1. Morse, P.M. and Feshbach, H., *Methods of Theoretical Physics* (McGraw-Hill, NY, 1953).
2. Courant, R. and Hilbert, D., *Methods of Mathematical Physics*, Vol. 1 (Interscience, NY, 1953).
3. Kahan, T., Rideau, G., and Roussopolous, P., *Les Methodés d'Approximation Variationelles Dans la Theorié des Collisions Atomiques et Dans la Physique des Piles Nucléaires* (Gauthier-Villars, Paris, 1956).
4. Pomraning, G.C. and Clark, M.Jr., "The Variational Method Applied to the Monoenergetic Boltzmann Equation. Part I," *Nucl. Sci. Engineering* **16**, 147-154 (1963).
5. Cercignani, C. and Pagani, C.D., "Variational Approach to Boundary-Value Problems in Kinetic Theory," *Phys. Fluids* **9(6)**, 1167-1173 (1966).
6. Loyalka, S.K., "Momentum and Temperature-Slip Coefficients with Arbitrary Accommodation at the Surface," *J. Chem. Phys.* **48(12)**, 5432-5436 (1968).
7. Lang, H., "Ein Variationsprinzip für die Linealisierte, Stationäre Boltzmann-Gleichung der Kinetischen Gastheorie," *Acta Mechanica* **5**, 163-188 (1968).
8. Loyalka, S.K., "Linearized Couette Flow and Heat Transfer Between Two Parallel Plates," in *Rarefied Gas Dynamics*, edited by Trilling, L. and Wachman, H.Y. (Academic Press, New York) Proceedings of the 6th International Symposium on Rarefied Gas Dynamics (1969). pp. 195-203.
9. Loyalka, S.K., "Slip Problems for a Simple Gas," *Z. Naturforsh.* **26a(6)**, 964-972 (1971).
10. Loyalka, S.K., "Slip in the Thermal Creep Flow," *Phys. Fluids* **14(1)**, 21-24 (1971).

Appendix E

SOME DEFINITE INTEGRALS

1. SOME FREQUENTLY ENCOUNTERED INTEGRALS.

One commonly encountered integral is of the form (see Table E-1):

$$i_n = \int_0^\infty x^n \exp\left(-x^2\right) dx \, .$$

For even and odd values of n, respectively, this may be rewritten as:

$$\int_0^\infty x^{2n} \exp\left(-\beta x^2\right) dx = \frac{(2n-1)(2n-3)\cdots 5\cdot 3\cdot 1}{2^{n+1}} \sqrt{\frac{\pi}{\beta^{2n+1}}} \; ; \quad n \geq 1 \, , \tag{E-1}$$

$$\int_0^\infty x^{2n+1} \exp\left(-\beta x^2\right) dx = \tfrac{1}{2} \frac{n!}{\beta^{n+1}} \; ; \quad n \geq 0 \, . \tag{E-2}$$

Table E-1. Values of the commonly encountered i_n integrals.

n	0	1	2	3	4
i_n	$\frac{1}{2}\sqrt{\pi}$	$\frac{1}{2}$	$\frac{1}{4}\sqrt{\pi}$	$\frac{1}{2}$	$\frac{3}{8}\sqrt{\pi}$
n	5	6	7	8	9
i_n	1	$\frac{15}{16}\sqrt{\pi}$	3	$\frac{105}{32}\sqrt{\pi}$	12
n	10	11	12	13	
i_n	$\frac{945}{64}\sqrt{\pi}$	60	$\frac{10395}{128}\sqrt{\pi}$	360	

2. SOME INTEGRALS ENCOUNTERED IN BOUNDARY PROBLEMS.

$$\int_{+} c_x \exp\left(-c^2\right) d\mathbf{c} = \tfrac{1}{2}\pi \ , \tag{E-3}$$

$$\int_{+} c_x c^2 \exp\left(-c^2\right) d\mathbf{c} = \pi \ , \tag{E-4}$$

$$\int_{+} c_x^2 \exp\left(-c^2\right) d\mathbf{c} = \tfrac{1}{4}\pi^{3/2} \ , \tag{E-5}$$

$$\int_{+} c_x c_y^2 \exp\left(-c^2\right) d\mathbf{c} = \tfrac{1}{4}\pi \ , \tag{E-6}$$

$$\int_{+} c_x^2 c_y^2 \exp\left(-c^2\right) d\mathbf{c} = \tfrac{1}{8}\pi^{3/2} \ , \tag{E-7}$$

$$\int_{+} c_x^2 c^2 \exp\left(-c^2\right) d\mathbf{c} = \tfrac{5}{8}\pi^{3/2} \ , \tag{E-8}$$

$$\int_{+} c_x \left(c^2 - \tfrac{3}{2}\right) \exp\left(-c^2\right) d\mathbf{c} = \tfrac{1}{4}\pi \ , \tag{E-9}$$

$$\int_{+} c_x c^2 \left(c^2 - \tfrac{3}{2}\right) \exp\left(-c^2\right) d\mathbf{c} = \tfrac{3}{2}\pi \ , \tag{E-10}$$

$$\int_{+} c_x^2 \left(c^2 - \tfrac{3}{2}\right) \exp\left(-c^2\right) d\mathbf{c} = \tfrac{1}{4}\pi^{3/2} \ , \tag{E-11}$$

$$\int_{+} c_x \left(c^2 - \tfrac{5}{2}\right) \exp\left(-c^2\right) d\mathbf{c} = -\tfrac{1}{4}\pi \ , \tag{E-12}$$

$$\int_{+} c_x c^2 \left(c^2 - \tfrac{5}{2}\right) \exp\left(-c^2\right) d\mathbf{c} = \tfrac{1}{2}\pi \ , \tag{E-13}$$

$$\int_{+} c_x^2 \left(c^2 - \tfrac{5}{2}\right) \exp\left(-c^2\right) d\mathbf{c} = 0 \ , \tag{E-14}$$

$$\int_{+} c_x c_y^2 S_{3/2}^{(1)}\left(c^2\right) \exp\left(-c^2\right) d\mathbf{c} = -\tfrac{1}{8}\pi \ , \tag{E-15}$$

$$\int_{+} c_x^2 c_y^2 S_{3/2}^{(1)}\left(c^2\right) \exp\left(-c^2\right) d\mathbf{c} = -\tfrac{1}{8}\pi^{3/2} \ , \tag{E-16}$$

$$\int_{+} c_x^2 c^2 S_{3/2}^{(1)}\left(c^2\right) \exp\left(-c^2\right) d\mathbf{c} = -\tfrac{5}{8}\pi^{3/2} \ , \tag{E-17}$$

$$\int_{+} c_x^3 S_{3/2}^{(1)}\left(c^2\right) \exp\left(-c^2\right) d\mathbf{c} = -\tfrac{1}{4}\pi \ , \tag{E-18}$$

$$\int_{+} c_x^2 \left[S_{3/2}^{(1)}\left(c^2\right)\right]^2 \exp\left(-c^2\right) d\mathbf{c} = \tfrac{5}{8}\pi^{3/2} \ , \tag{E-19}$$

$$\int_{+} c_x^3 \left[S_{3/2}^{(1)}\left(c^2\right)\right]^2 \exp\left(-c^2\right) d\mathbf{c} = \tfrac{13}{8}\pi \ , \tag{E-20}$$

$$\int_{+} c_x^3 S_{3/2}^{(2)}\left(c^2\right)\exp\left(-c^2\right)d\mathbf{c} = -\tfrac{1}{16}\pi \ , \tag{E-21}$$

$$\int_{+} c_x^3 S_{3/2}^{(1)}\left(c^2\right) S_{3/2}^{(2)}\left(c^2\right)\exp\left(-c^2\right)d\mathbf{c} = -\tfrac{23}{32}\pi \ , \tag{E-22}$$

$$\int_{+} c_x^2 \left(2-c^2\right)\exp\left(-c^2\right)d\mathbf{c} = -\tfrac{1}{8}\pi^{3/2} \ , \tag{E-23}$$

$$\int_{+} c_x c^2 \left(2-c^2\right)\exp\left(-c^2\right)d\mathbf{c} = -\pi \ , \tag{E-24}$$

$$\int_{+} c_x^2 \left(2-c^2\right) S_{3/2}^{(1)}\left(c^2\right)\exp\left(-c^2\right)d\mathbf{c} = \tfrac{5}{8}\pi^{3/2} \ , \tag{E-25}$$

$$\int_{+} c_x^2 \left(2-c^2\right) S_{3/2}^{(2)}\left(c^2\right)\exp\left(-c^2\right)d\mathbf{c} = 0 \ . \tag{E-26}$$

Here, the following notation has been used:

$$\int_{+} d\mathbf{c} = \int_{0}^{\infty} dc_x \int_{-\infty}^{\infty} dc_y \int_{-\infty}^{\infty} dc_z \ .$$

3. SOME INTEGRALS CONNECTED WITH THE SECOND-ORDER CHAPMAN-ENSKOG SOLUTION.

$$\int_{+} c_x \left[S_{3/2}^{(1)}\left(c^2\right) + a_2^{(2)*} S_{3/2}^{(2)}\left(c^2\right)\right]\exp\left(-c^2\right)d\mathbf{c} = \tfrac{1}{4}\pi\left(1+\tfrac{3}{4}a_2^{(2)*}\right) , \tag{E-27}$$

$$\int_{+} c_x \left(c^2 - \tfrac{3}{2}\right)\left[S_{3/2}^{(1)}\left(c^2\right) + a_2^{(2)*} S_{3/2}^{(2)}\left(c^2\right)\right]\exp\left(-c^2\right)d\mathbf{c} = -\tfrac{7}{8}\pi\left(1 + \tfrac{13}{28}a_2^{(2)*}\right) , \tag{E-28}$$

$$\int_{+} c_x c^2 \left[S_{3/2}^{(1)}\left(c^2\right) + a_2^{(2)*} S_{3/2}^{(2)}\left(c^2\right)\right]\exp\left(-c^2\right)d\mathbf{c} = -\tfrac{1}{2}\pi\left(1 + \tfrac{1}{4}a_2^{(2)*}\right) , \tag{E-29}$$

$$\int_{+} c_x c_y^2 \left[S_{3/2}^{(1)}\left(c^2\right) + a_2^{(2)*} S_{3/2}^{(2)}\left(c^2\right)\right]\exp\left(-c^2\right)d\mathbf{c} = -\tfrac{1}{8}\pi\left(1 + \tfrac{1}{4}a_2^{(2)*}\right) , \tag{E-30}$$

$$\int_{+} c_x^2 c^2 \left[S_{3/2}^{(1)}\left(c^2\right) + a_2^{(2)*} S_{3/2}^{(2)}\left(c^2\right)\right]\exp\left(-c^2\right)d\mathbf{c} = -\tfrac{5}{8}\pi^{3/2} , \tag{E-31}$$

$$\int_{+} c_x^3 \left[S_{3/2}^{(1)}\left(c^2\right) + a_2^{(2)*} S_{3/2}^{(2)}\left(c^2\right)\right]^2 \exp\left(-c^2\right)d\mathbf{c} = \tfrac{13}{8}\pi\left(1 - \tfrac{23}{26}a_2^{(2)*} + \tfrac{433}{208}\left[a_2^{(2)*}\right]^2\right) , \tag{E-32}$$

$$\int_{+} c_x^2 c_y^2 \left[S_{3/2}^{(1)}\left(c^2\right) + a_2^{(2)*} S_{3/2}^{(2)}\left(c^2\right)\right]\left[1 + b_2^{(2)*} S_{5/2}^{(1)}\left(c^2\right)\right]\exp\left(-c^2\right)d\mathbf{c} = -\tfrac{1}{8}\pi^{3/2}\left(1 - \tfrac{7}{2}b_2^{(2)*} + \tfrac{7}{2}a_2^{(2)*}b_2^{(2)*}\right) , \tag{E-33}$$

$$\int_{+} c_x^3 c_y^2 \left[1 + b_2^{(2)*} S_{5/2}^{(1)}\left(c^2\right)\right]^2 \exp\left(-c^2\right)d\mathbf{c} = \tfrac{1}{4}\pi\left[1 - b_2^{(2)*} + \tfrac{17}{4}\left(b_2^{(2)*}\right)^2\right] , \tag{E-34}$$

$$\int_{+} c_x^3 c_y^2 \left[1 + b_2^{(2)*} S_{5/2}^{(1)}\left(c^2\right)\right] \exp\left(-c^2\right) d\mathbf{c} = \tfrac{1}{4}\pi\left(1 - \tfrac{1}{2}b_2^{(2)*}\right) . \tag{E-35}$$

4. SOME INTEGRALS CONNECTED WITH NON-LINEAR TRANSPORT PROBLEMS.

$$\int_{u_x}^{\infty} \exp\left(-c_x^2\right) dc_x = \tfrac{1}{2}\pi^{1/2}\left(1 - \operatorname{erf}\left(u_x\right)\right) , \tag{E-36}$$

$$\int_{u_x}^{\infty} c_x \exp\left(-c_x^2\right) dc_x = \tfrac{1}{2}\exp\left(-u_x^2\right) , \tag{E-37}$$

$$\int_{u_x}^{\infty} c_x^2 \exp\left(-c_x^2\right) dc_x = \tfrac{1}{2}\left[u_x \exp\left(-u_x^2\right) + \tfrac{1}{2}\pi^{1/2}\left(1 - \operatorname{erf}\left(u_x\right)\right)\right] , \tag{E-38}$$

$$\int_{u_x}^{\infty} c_x^3 \exp\left(-c_x^2\right) dc_x = \tfrac{1}{2}\exp\left(-u_x^2\right)\left(1 + u_x^2\right) , \tag{E-39}$$

where:

$$\operatorname{erf}\left(x\right) = \frac{2}{\pi^{1/2}} \int_{0}^{x} \exp\left(-t^2\right) dt ,$$

is the error function.

Appendix F

OMEGA-INTEGRALS FOR SECOND-ORDER APPROXIMATION

All of the coefficients associated with the Chapman-Enskog algebraic systems of equations given in Eqs. (12-11)-(12-13) for the first-order Chapman-Enskog approximation and given in Eqs. (12-41)-(12-43) for the second-order Chapman-Enskog approximation are expressible in terms of the Ω-integrals which are dependent upon the specific model of the intermolecular potential that one is using. For a simple gas, these integrals have been described in Section 5.9. For gas mixtures, however, the Ω-integrals must be slightly generalized. In this case, the Ω-integrals are defined by:

$$\Omega_{12}^{(l,r)} = \left[\Omega_{12}^{(l,r)}\right]_{r.s.} \Omega^{(l,r)\mathring{a}} ,$$

where $[\Omega_{12}^{(l,r)}]_{r.s.}$ is the Ω-integral for rigid-sphere molecules which is defined in [1] as:

$$\left[\Omega_{12}^{(l,r)}\right]_{r.s.} = \left(\frac{kT}{2\pi m_{12}^{*}}\right)^{1/2} \frac{(r+1)!}{2}\left[1-\frac{1+(-1)^{l}}{2(l+1)}\right]\pi\sigma_{12}^{2} ,$$

where:

$$m_{12}^{*} = \frac{m_1 m_2}{m_1 + m_2} \quad \text{and} \quad \sigma_{12} = \tfrac{1}{2}(\sigma_1 + \sigma_2) .$$

The functions, $\Omega^{(l,r)\text{å}}$, are known as the reduced Ω-integrals (called the reduced collision integrals in some texts) and are given in terms of the reduced temperature, T^*, by [2]:

$$\Omega^{(l,r)\text{å}}\left(T^*\right)=\left[(r+1)!\left(T^*\right)^{r+2}\right]^{-1}\int_0^{\infty}Q^{(l)\text{å}}\left(E^*\right)\exp\left(-\frac{E^*}{T^*}\right)\left(E^*\right)^{r+1}dE^*,$$

in which the following notations have been used:

$$Q^{(l)\text{å}}\left(E^*\right)=2\left[1-\frac{1+(-1)^l}{2(l+1)}\right]^{-1}\int_0^{\infty}\left(1-\cos^l(\chi)\right)b^*db^*,$$

$$\chi=\pi-2b^*\int_{r_0^*}^{\infty}\left(1-\frac{b^{*2}}{r^{*2}}-\frac{\varphi^*\left(r^*\right)}{E^*}\right)^{-1/2}\frac{dr^*}{r^{*2}},$$

where $r^*=r/\sigma_{12}$ is the reduced intermolecular distance, $b^*=b/\sigma_{12}$ is the reduced impact parameter, $\varphi^*=\varphi/\varepsilon$ is the reduced intermolecular potential energy, $T^*=kT/\varepsilon$ is the reduced temperature, $E^*=\frac{1}{2}m_{12}^*g^2/\varepsilon$ is the reduced relative kinetic energy, ε is the energy parameter of the potential model being used, and r_0^* is determined from the largest, positive, real root of the expression:

$$\left(1-\frac{b^{*2}}{r_0^{*2}}-\frac{\varphi^*\left(r_0^*\right)}{E^*}\right)=0.$$

In Table F-1 below, we have presented values of the $\Omega^{\text{å}}$-integrals computed for selected values of the reduced temperature, T^*, using the Lennard-Jones (6-12) potential model. The values tabulated here are those needed for the first- and second-order Chapman-Enskog approximations. All listed values have been independently recalculated from those tabulated in [3] using a program developed by Maitland. [2] and subsequently revised by us to include increased numerical quadratures in the variables, b^* and E^*.

Additionally, we note that a simple, concise, and very transparent program can be constructed in *Mathematica*® which will also permit good graphical display of the results at several stages in the calculations including plots of the scattering angle, $\chi(b^*, E^*)$, the functions, $Q^{(l)\star}(E^*)$, and the resulting $\Omega^\star$-integrals, $\Omega^{(l,r)\star}(T^*)$. We have constructed this *Mathematica®* program and have included it below as Table F-2 for the convenience of the reader. Given with this program is relevant output which includes a complete alternate set of $\Omega^\star$-integral values calculated using the Lennard-Jones (6-12) potential model which are needed for the first- and second-order Chapman-Enskog approximations. These values are analogous to those given in Table F-1, which were calculated with a modified version of an earlier Fortran program, and are shown computed for the same set of values of the reduced temperature, T^*, that were specified in the earlier Fortran program. We note that while much of the nomenclature in the *Mathematica®* program is self-explanatory, the program has been annotated with a number of comments to assist the reader.

We are suggesting that the reader may use this *Mathematica®* program to explore results for different intermolecular potential models. The reader may also wish to explore ways in which the program may be improved with respect to its accuracy and speed. The major time-consuming parts of this program are the computation of the scattering angle, $\chi(b^*, E^*)$, and, in particular, the determination of $r_0^*(b^*, E^*)$. For potential functions where non-polynomial equations are involved, one might wish to explore the use of the *Mathematica®* command 'FindRoot' (with a suitable starting value $r_0^* \approx b^*$).

Table F-1(a). Values of some $\Omega^{\star}$-integrals computed for selected values of the reduced temperature, T^{*}, using the Lennard-Jones (6-12) potential model. The values tabulated here are the complete set of $\Omega^{\star}$-integrals needed for the first- and second-order Chapman-Enskog approximations. All listed values have been independently recalculated from those tabulated in [3] using a Fortran program developed by [2] and subsequently modified by us to give improved accuracy. Note that the last column of the table on page 541 of [2] appears to be mislabeled. Instead of being labeled $\Omega^{(2,5)\star}$ it should be labeled $\Omega^{(3,3)\star}$.

T^{*}	$\Omega^{(1,1)\star}$	$\Omega^{(1,2)\star}$	$\Omega^{(1,3)\star}$	$\Omega^{(1,4)\star}$	$\Omega^{(1,5)\star}$	$\Omega^{(2,2)\star}$
0.30	2.6505	2.2586	1.9674	1.7426	1.5692	2.8438
0.35	2.4699	2.0816	1.7998	1.5908	1.4363	2.6796
0.40	2.3157	1.9345	1.6659	1.4740	1.3372	2.5337
0.45	2.1826	1.8113	1.5578	1.3828	1.2614	2.4029
0.50	2.0671	1.7077	1.4697	1.3103	1.2021	2.2854
0.55	1.9661	1.6198	1.3971	1.2516	1.1546	2.1798
0.60	1.8775	1.5448	1.3365	1.2034	1.1157	2.0849
0.65	1.7992	1.4803	1.2855	1.1632	1.0834	1.9995
0.70	1.7298	1.4244	1.2419	1.1292	1.0560	1.9226
0.75	1.6679	1.3756	1.2045	1.1001	1.0325	1.8533
0.80	1.6126	1.3328	1.1720	1.0749	1.0121	1.7907
0.85	1.5628	1.2949	1.1435	1.0528	0.9942	1.7341
0.90	1.5178	1.2613	1.1184	1.0333	0.9784	1.6826
0.95	1.4771	1.2312	1.0961	1.0160	0.9642	1.6358
1.00	1.4400	1.2042	1.0761	1.0005	0.9514	1.5932
1.05	1.4062	1.1799	1.0582	0.9865	0.9398	1.5542
1.10	1.3752	1.1578	1.0419	0.9737	0.9292	1.5184
1.15	1.3468	1.1376	1.0271	0.9621	0.9195	1.4856
1.20	1.3206	1.1192	1.0136	0.9515	0.9105	1.4553
1.25	1.2964	1.1024	1.0012	0.9416	0.9022	1.4274
1.30	1.2740	1.0868	0.9897	0.9325	0.8945	1.4015
1.35	1.2531	1.0724	0.9791	0.9240	0.8873	1.3776
1.40	1.2337	1.0591	0.9693	0.9161	0.8805	1.3553
1.45	1.2157	1.0467	0.9601	0.9087	0.8741	1.3346
1.50	1.1988	1.0352	0.9516	0.9018	0.8681	1.3152
1.55	1.1829	1.0244	0.9435	0.8952	0.8624	1.2971
1.60	1.1681	1.0143	0.9360	0.8891	0.8570	1.2802
1.65	1.1541	1.0048	0.9288	0.8832	0.8519	1.2643
1.70	1.1409	0.9959	0.9221	0.8777	0.8470	1.2494
1.75	1.1285	0.9874	0.9158	0.8724	0.8424	1.2353
1.80	1.1167	0.9795	0.9097	0.8674	0.8380	1.2221
1.85	1.1056	0.9719	0.9040	0.8626	0.8337	1.2095
1.90	1.0950	0.9648	0.8985	0.8580	0.8296	1.1977
1.95	1.0850	0.9580	0.8933	0.8536	0.8257	1.1865
2.00	1.0755	0.9515	0.8883	0.8494	0.8219	1.1758
2.10	1.0577	0.9394	0.8790	0.8415	0.8148	1.1561
2.20	1.0416	0.9284	0.8704	0.8341	0.8082	1.1383
2.30	1.0268	0.9183	0.8625	0.8273	0.8019	1.1220
2.40	1.0132	0.9090	0.8551	0.8209	0.7961	1.1071
2.50	1.0007	0.9003	0.8482	0.8149	0.7906	1.0934
2.60	0.9891	0.8923	0.8418	0.8092	0.7854	1.0809

T^*	$\Omega^{(1,1)\star}$	$\Omega^{(1,2)\star}$	$\Omega^{(1,3)\star}$	$\Omega^{(1,4)\star}$	$\Omega^{(1,5)\star}$	$\Omega^{(2,2)\star}$
2.70	0.9783	0.8848	0.8357	0.8039	0.7804	1.0692
2.80	0.9683	0.8777	0.8300	0.7988	0.7757	1.0584
2.90	0.9589	0.8711	0.8245	0.7940	0.7713	1.0483
3.00	0.9501	0.8649	0.8194	0.7894	0.7670	1.0389
3.10	0.9418	0.8590	0.8145	0.7850	0.7629	1.0301
3.20	0.9340	0.8534	0.8099	0.7808	0.7590	1.0218
3.30	0.9267	0.8481	0.8054	0.7768	0.7552	1.0141
3.40	0.9197	0.8430	0.8012	0.7730	0.7516	1.0067
3.50	0.9131	0.8382	0.7971	0.7693	0.7481	0.9998
3.60	0.9069	0.8336	0.7932	0.7657	0.7448	0.9932
3.70	0.9009	0.8292	0.7895	0.7623	0.7416	0.9870
3.80	0.8953	0.8250	0.7859	0.7590	0.7384	0.9810
3.90	0.8898	0.8210	0.7824	0.7558	0.7354	0.9754
4.00	0.8847	0.8171	0.7791	0.7527	0.7325	0.9700
4.10	0.8797	0.8134	0.7758	0.7497	0.7296	0.9649
4.20	0.8749	0.8098	0.7727	0.7468	0.7269	0.9599
4.30	0.8704	0.8063	0.7697	0.7440	0.7242	0.9552
4.40	0.8660	0.8030	0.7667	0.7413	0.7216	0.9507
4.50	0.8618	0.7997	0.7639	0.7386	0.7191	0.9463
4.60	0.8577	0.7966	0.7611	0.7361	0.7166	0.9422
4.70	0.8538	0.7936	0.7584	0.7336	0.7142	0.9381
4.80	0.8500	0.7906	0.7558	0.7311	0.7119	0.9343
4.90	0.8464	0.7878	0.7533	0.7288	0.7096	0.9305
5.00	0.8429	0.7850	0.7508	0.7264	0.7074	0.9269
6.00	0.8129	0.7610	0.7292	0.7060	0.6877	0.8963
7.00	0.7898	0.7419	0.7117	0.6894	0.6716	0.8728
8.00	0.7712	0.7261	0.6970	0.6753	0.6580	0.8539
9.00	0.7556	0.7126	0.6844	0.6632	0.6462	0.8381
10.0	0.7423	0.7008	0.6733	0.6526	0.6359	0.8245
20.0	0.6641	0.6295	0.6052	0.5865	0.5714	0.7437
30.0	0.6235	0.5913	0.5684	0.5507	0.5364	0.7008
40.0	0.5963	0.5655	0.5435	0.5265	0.5127	0.6717
50.0	0.5760	0.5461	0.5248	0.5083	0.4949	0.6498
60.0	0.5599	0.5308	0.5099	0.4938	0.4808	0.6323
70.0	0.5465	0.5181	0.4977	0.4819	0.4691	0.6178
80.0	0.5352	0.5073	0.4872	0.4718	0.4592	0.6055
90.0	0.5254	0.4979	0.4782	0.4630	0.4506	0.5947
100	0.5168	0.4897	0.4702	0.4552	0.4431	0.5852
200	0.4630	0.4383	0.4207	0.4071	0.3962	0.5256
300	0.4339	0.4106	0.3940	0.3812	0.3710	0.4931
400	0.4142	0.3919	0.3761	0.3641	0.3545	0.4710

Table F-1(b).

T^*	$\Omega^{(2,3)\star}$	$\Omega^{(2,4)\star}$	$\Omega^{(2,5)\star}$	$\Omega^{(2,6)\star}$	$\Omega^{(3,3)\star}$	$\Omega^{(4,4)\star}$
0.30	2.5812	2.3626	2.1704	2.0009	2.3989	2.5712
0.35	2.4096	2.1835	1.9898	1.8256	2.2249	2.3939
0.40	2.2577	2.0298	1.8407	1.6868	2.0799	2.2395
0.45	2.1241	1.8990	1.7185	1.5764	1.9572	2.1052
0.50	2.0070	1.7881	1.6181	1.4880	1.8523	1.9885
0.55	1.9046	1.6940	1.5350	1.4162	1.7618	1.8871
0.60	1.8150	1.6138	1.4658	1.3572	1.6833	1.7987
0.65	1.7364	1.5452	1.4075	1.3081	1.6147	1.7214
0.70	1.6674	1.4861	1.3581	1.2667	1.5544	1.6537
0.75	1.6065	1.4349	1.3157	1.2314	1.5011	1.5940
0.80	1.5526	1.3903	1.2791	1.2011	1.4538	1.5412
0.85	1.5048	1.3512	1.2471	1.1747	1.4115	1.4943
0.90	1.4620	1.3166	1.2191	1.1515	1.3737	1.4525
0.95	1.4238	1.2860	1.1944	1.1310	1.3396	1.4150
1.00	1.3893	1.2586	1.1723	1.1128	1.3089	1.3814
1.05	1.3582	1.2340	1.1526	1.0964	1.2809	1.3510
1.10	1.3301	1.2119	1.1348	1.0816	1.2555	1.3234
1.15	1.3044	1.1919	1.1186	1.0682	1.2323	1.2984
1.20	1.2810	1.1736	1.1040	1.0560	1.2110	1.2755
1.25	1.2596	1.1570	1.0906	1.0447	1.1914	1.2546
1.30	1.2399	1.1417	1.0783	1.0344	1.1733	1.2354
1.35	1.2218	1.1276	1.0669	1.0248	1.1567	1.2176
1.40	1.2050	1.1147	1.0564	1.0159	1.1412	1.2013
1.45	1.1895	1.1026	1.0466	1.0077	1.1268	1.1861
1.50	1.1750	1.0915	1.0376	0.9999	1.1134	1.1721
1.55	1.1616	1.0810	1.0291	0.9927	1.1008	1.1589
1.60	1.1490	1.0713	1.0211	0.9858	1.0891	1.1467
1.65	1.1373	1.0622	1.0136	0.9794	1.0781	1.1353
1.70	1.1262	1.0536	1.0066	0.9733	1.0678	1.1245
1.75	1.1159	1.0456	0.9999	0.9675	1.0580	1.1145
1.80	1.1061	1.0380	0.9936	0.9621	1.0488	1.1050
1.85	1.0969	1.0308	0.9876	0.9569	1.0401	1.0960
1.90	1.0882	1.0240	0.9820	0.9519	1.0319	1.0876
1.95	1.0800	1.0175	0.9765	0.9471	1.0241	1.0796
2.00	1.0722	1.0114	0.9714	0.9426	1.0167	1.0720
2.10	1.0577	0.9999	0.9618	0.9341	1.0029	1.0579
2.20	1.0446	0.9895	0.9529	0.9262	0.9904	1.0452
2.30	1.0326	0.9799	0.9447	0.9189	0.9790	1.0335
2.40	1.0216	0.9711	0.9372	0.9121	0.9684	1.0229
2.50	1.0115	0.9629	0.9301	0.9057	0.9587	1.0131
2.60	1.0022	0.9553	0.9235	0.8997	0.9497	1.0040
2.70	0.9935	0.9482	0.9173	0.8941	0.9414	0.9956
2.80	0.9854	0.9416	0.9114	0.8887	0.9336	0.9877
2.90	0.9778	0.9353	0.9059	0.8837	0.9263	0.9804
3.00	0.9707	0.9294	0.9007	0.8788	0.9194	0.9735
3.10	0.9640	0.9238	0.8957	0.8742	0.9130	0.9670
3.20	0.9577	0.9185	0.8910	0.8699	0.9069	0.9609

T^*	$\Omega^{(2,3)\star}$	$\Omega^{(2,4)\star}$	$\Omega^{(2,5)\star}$	$\Omega^{(2,6)\star}$	$\Omega^{(3,3)\star}$	$\Omega^{(4,4)\star}$
3.30	0.9518	0.9135	0.8865	0.8657	0.9011	0.9551
3.40	0.9461	0.9087	0.8821	0.8616	0.8956	0.9496
3.50	0.9407	0.9041	0.8780	0.8578	0.8904	0.9444
3.60	0.9356	0.8997	0.8740	0.8540	0.8855	0.9395
3.70	0.9308	0.8955	0.8702	0.8505	0.8808	0.9348
3.80	0.9261	0.8915	0.8666	0.8470	0.8763	0.9302
3.90	0.9216	0.8876	0.8630	0.8437	0.8720	0.9259
4.00	0.9174	0.8839	0.8596	0.8404	0.8678	0.9218
4.10	0.9133	0.8803	0.8563	0.8373	0.8639	0.9178
4.20	0.9093	0.8769	0.8531	0.8343	0.8600	0.9140
4.30	0.9056	0.8736	0.8500	0.8314	0.8564	0.9103
4.40	0.9019	0.8703	0.8470	0.8285	0.8528	0.9068
4.50	0.8984	0.8672	0.8441	0.8258	0.8494	0.9033
4.60	0.8950	0.8642	0.8413	0.8231	0.8461	0.9000
4.70	0.8917	0.8612	0.8386	0.8205	0.8429	0.8968
4.80	0.8885	0.8584	0.8359	0.8179	0.8398	0.8937
4.90	0.8854	0.8556	0.8333	0.8154	0.8368	0.8907
5.00	0.8824	0.8529	0.8308	0.8130	0.8339	0.8878
6.00	0.8566	0.8295	0.8086	0.7916	0.8090	0.8627
7.00	0.8363	0.8106	0.7906	0.7742	0.7893	0.8428
8.00	0.8195	0.7948	0.7754	0.7594	0.7731	0.8264
9.00	0.8052	0.7813	0.7623	0.7465	0.7593	0.8124
10.0	0.7927	0.7694	0.7508	0.7353	0.7474	0.8001
20.0	0.7165	0.6956	0.6786	0.6643	0.6743	0.7247
30.0	0.6751	0.6552	0.6389	0.6253	0.6349	0.6834
40.0	0.6469	0.6276	0.6118	0.5986	0.6081	0.6551
50.0	0.6256	0.6068	0.5914	0.5785	0.5879	0.6338
60.0	0.6086	0.5901	0.5751	0.5625	0.5718	0.6167
70.0	0.5945	0.5764	0.5616	0.5492	0.5585	0.6025
80.0	0.5825	0.5646	0.5501	0.5379	0.5471	0.5904
90.0	0.5720	0.5544	0.5401	0.5280	0.5372	0.5799
100	0.5628	0.5454	0.5312	0.5194	0.5285	0.5706
200	0.5050	0.4891	0.4761	0.4653	0.4740	0.5122
300	0.4735	0.4584	0.4461	0.4357	0.4444	0.4804
400	0.4521	0.4373	0.4250	0.4144	0.4245	0.4585

Table F-2. A *Mathematica®* program that may be used to compute values of the $\Omega^{\star}$-integrals for selected values of the reduced temperature, T^{*}. This program is currently configured to use the Lennard-Jones (6-12) potential model and the same set of reduced temperatures reported in Table F-1, but can be readily adapted to other potential models and alternative sets of reduced temperatures.

Evaluating Omega-Integrals:

INI, SKL, RVT 2005-2006

We have used the Lennard-Jones (6-12) potential model as an example. The results of this program compare well with those in:

1. Hirschfelder, J.O., Curtiss, C.F. and Bird, R.B., *Molecular Theory of Gases and Liquids* (John Wiley and Sons, New York, 1954). In the program we refer to this reference as HCB. Take special note of pages 557, 558, and 1132 in this text.

2. Maitland, G.C., Rigby, M., Smith, E.B. and Wakeham, W.A., *Intermolecular Forces* (Oxford University Press, Oxford, 1981). In the program we refer to this reference as MRSW. Take special note of pages 538-541 in this text.

The program consists of three parts:

1. Calculation of the scattering angle, chi (χ), and construction of an interpolation approximation to it.

2. Calculation of the Q-integrals using chi, and again interpolation approximations to these.

3. Calculation of the omega-integrals.

In the following, we have not focused on high accuracy. Our purpose has been to indicate how a simple program for calculating the omega-integrals can be constructed. We encourage the reader to explore the program. We have not included several comparative tests that were carried out during construction of the program.

We thank our students Ryan M. Meyer, Zebadiah Smith, and Earl Lynn Tipton for reviewing and helping us with improvements in parts of the program.

```
Needs["Graphics`Graphics`"];
Needs["Graphics`MultipleListPlot`"];
Needs["Graphics`Legend`"];
Off[NIntegrate::ncvb];
Off[NIntegrate::slwcon];
```

```
Off[General::spell];
Off[General::spell1];

(* Define the potential: In the following we have chosen
the Lennard-Jones (6-12) potential. *)

Clear[ustar, rstar];
ustar[rstar_]:=4(rstar^-12 - rstar^-6);

(* Construct the chi function, specialized form for the
Lennard-Jones (6-12) potential.  In the calculation of
chi we have used a standard transformation,
pstar=1/(rstar*2), which reduces the order of the
polynomial in the integrand, and also changes the
interval of integration to a finite interval.  For
calculating pstar0 we have use the Matrix Eigenvalue
method of finding zeros of a polynomial.  One can also
use the NSolve or FindRoot functions depending upon the
potential.  We have not accounted for the orbiting
condition in the scattering angle chi explicitly; we
have smoothed over it through the choice of a coarse
computational grid on bstar and estar.  A more careful
computation would be needed for precise work.  For using
the NSolve function, one can have a construct of the
type:

pstar0[bstar_,estar_]:=Min[Select[pstar/.NSolve[(1 -
  bstar^2 * pstar - 4(pstar^6 *
  pstar^3)/estar)== 0,pstar]//N, (Im[#]==0&&Re[#]>0) &]];

*)
Clear[pstar0, bstar, estar, chi, pstar00];
pstar0[bstar_,estar_]:=Min[Select[Eigenvalues[{{0, 0,
  0, 0, 0, estar/4}, {1, 0, 0, 0, 0, -estar/4 *
  bstar^2}, {0, 1, 0, 0, 0, 0}, {0, 0, 1, 0, 0, 1}, {0,
  0, 0, 1, 0, 0}, {0, 0, 0, 0, 1, 0}}]//N,
  (Im[#]==0&&Re[#]>0) &]];
chi[bstar_,estar_]:=Re[Pi - bstar *
  With[{pstar00=pstar0[bstar,estar]},
  NIntegrate[Evaluate[1/Sqrt[pstar(1 - bstar^2 * pstar
  - 4(pstar^6 - pstar^3)/estar)]], {pstar, 0, pstar00},
  MaxRecursion->30, AccuracyGoal->8,
  SingularityDepth->10]]];

(* We avoid the almost 0 imaginary part that drifts in
because of numerics. *)

(* Construct and test chiapprox which is based on
interpolation.  For higher precision or higher range on
tstar, the reduced temperature, the limits and the grids
below would need to be modified. *)
```

```
Clear[bstarlowlim, bstaruplim, estarlowlim, estaruplim,
  bstarlist, estarlist, reschi, chiapprox];
bstarlowlim=10^-5;
bstaruplim=5;
estarlowlim=10^-5;
estaruplim=1200;
bstarlist=bstarlowlim + Range[0.0,bstaruplim,0.025];
estarlist=estarlowlim + Range[0.0,estaruplim,0.025];
reschi=Outer[chi,bstarlist,estarlist];

(* The above evaluates chi on bstar, estar grid. *)

chiapprox=ListInterpolation[reschi,{{First[bstarlist],
  Last[bstarlist]}, {First[estarlist],
  Last[estarlist]}}];

(* The above constructs the interpolation function for
chi. *)

Print[chiapprox];
 InterpolatingFunction[{{0.00001,5.00001},{0.00001,1200.}},<>]

(* Plot (1-Cos[chi[bstar,estar]]), compare with HCB see
page 557. *)

Clear[chiapproxplotsHCB];
chiapproxplotsHCB=Map[Plot[1.0 -
  Cos[Re[chiapprox[Sqrt[bstarsq], #]]], {bstarsq,
  10^-10, 6}, PlotPoints→1000, PlotRange→{{10^-10, 6},
  {0, 2}}, Ticks→{{0, 1, 2, 3, 4, 5, 6}, {0, 0.5, 1.0,
  1.5, 2.0}}, Frame→True, RotateLabel→True,
  FrameLabel→{"b*2", "1-Cosχ"},
  DisplayFunction→Identity] &, {0.4, 0.8, 1.0, 10.0}];
Print["Plots of 1-Cosχapprox vs b*2 for E*={0.4,0.8,1.0
  and 10.0}", "\n "];
Show[GraphicsArray[Partition[chiapproxplotsHCB, 4]],
  DisplayFunction→$DisplayFunction];
```

Plots of 1-Cosχapprox vs
b^{*2} for E^{*}={0.4,0.8,1.0 and 10.0}

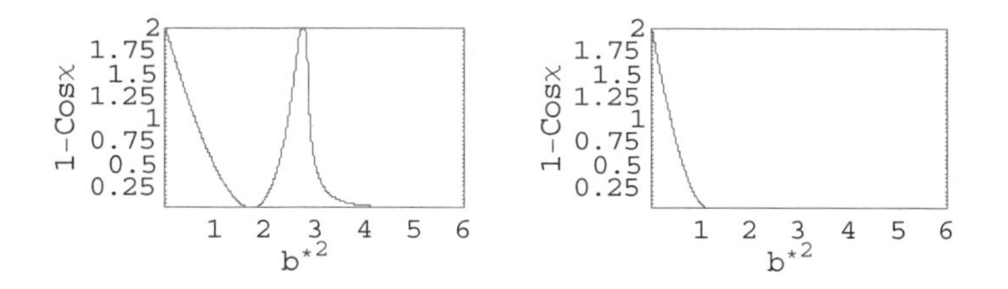


```
(* For the construction of qstar[[L]][estar], we first
evaluate the integrals on estar grid for given L (ℓ in
the textbook notation), and then construct an
interpolation function for each L.  We call the
interpolation functions qstarapprox2 which are elements
of the list qstarapproxlist.  Lhigh is the user assigned
integer, dictated by the orders of omega-integrals
needed. *)

Clear[qstarapprox1, qstarapprox2, qstarapproxlist,
  Lhigh];
qstarapprox1[L_,estar_]:=2(1-((1+(-1)^L)/(2(1+L))))^-1
  * NIntegrate[(1 - Cos[chiapprox[bstar,estar]]^L) *
  bstar, {bstar, First[bstarlist], Last[bstarlist]},
  MaxRecursion→5, AccuracyGoal→4];
qstarapprox2[L_]:=Interpolation[Map[{#,
  qstarapprox1[L,#]} &,estarlist]];
Lhigh=4;
qstarapproxlist=Map[qstarapprox2[#] &,Range[1,Lhigh]];
Print[qstarapproxlist];

{InterpolatingFunction[{{0.00001,1200.}},<>],InterpolatingFunc
tion[{{0.00001,1200.}},<>],InterpolatingFunction[{{0.00001,120
0.}},<>],InterpolatingFunction[{{0.00001,1200.}},<>]}

(* Tabulate and Plot qstar (L, estar).  Compare with HCB
page 558. *)

Print[TableForm[Table[Map[(Flatten[{N[#],
  Table[qstarapproxlist[[L]][#], {L, Lhigh}]}]) &,
  {10^-4, 10^-3, 10^-2, 10^-1, 1, 2, 3, 4, 5, 10, 100,
  1000}]], TableHeadings→{None, {"E* ", "Q*(1, E*)",
  "Q*(2, E*)", "Q*(3, E*)", "Q*(4, E*)"}}]];
Clear[qstarplot1];
qstarplot1=Plot[Evaluate[Chop[Table[
  qstarapproxlist[[L]][estar], {L, 1, Lhigh}], 10^-8]],
  {estar, 0.1, 4}, PlotPoints→1000, PlotRange→All,
  Ticks→{{0, 1, 2, 3, 4}, Automatic}, Frame→True,
  RotateLabel→True, FrameLabel→{"E* ", "Q*(ℓ, E*)"},
  AxesOrigin→{0, 0}, PlotStyle→Map[AbsoluteDashing[{5,
  #^2.5}] &,{1, 2, 3, 4}], PlotLegend→Map[ToString,
  N[{1, 2, 3, 4}]], LegendPosition→{1, -0.3},
  LegendSize→{0.8, 0.8}, LegendShadow→None,
  LegendLabel→StyleForm["ℓ", FontSize→14]];
```

E*	Q*(1,E*)	Q*(2,E*)	Q*(3,E*)	Q*(4,E*)
0.0001	34.8393	27.3578	29.8057	26.2607
0.001	33.1847	26.2679	28.5725	25.2987
0.01	19.8657	17.4243	18.5817	17.4597
0.1	5.38281	6.12706	6.23423	6.57795
1.	2.40659	2.80343	2.81556	3.02483
2.	1.46589	2.14097	1.94271	2.37249
3.	1.18881	1.61842	1.52617	1.84504
4.	1.06699	1.37593	1.30831	1.54584
5.	0.997701	1.24426	1.18465	1.37286
10.	0.855383	1.01136	0.95951	1.06798
100.	0.594234	0.7044	0.662703	0.73363
1000.	0.41265	0.495646	0.465145	0.519093

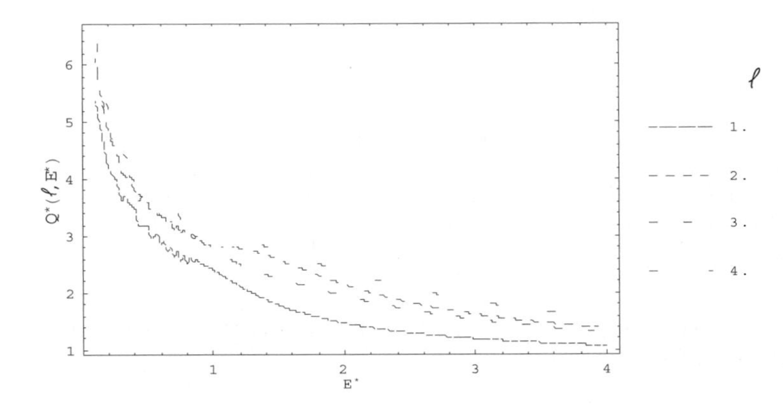


```
(* Compare again with HCB page 558, who have given a
LogLinear Plot of Q*(1,E*), Q*(2,E*), and Q*(4,E*). *)

Clear[qstarplot2];
qstarplot2=LogLinearPlot[Evaluate[Chop[Map[
  qstarapproxlist[[#]][estar] &, {1, 2, 4}], 10^-8]],
  {estar, 0.1, estaruplim}, PlotPoints→5000,
  PlotRange→All, Frame→True, RotateLabel→True,
  FrameLabel→{"E* ", "Q*(ℓ, E*)"}, AxesOrigin→{0.1, 0},
  AspectRatio→1.2, PlotStyle→Map[AbsoluteDashing[{5,
  #}] &,{1, 5, 15}], Epilog→{Text["Q*(ℓ, E*) for ℓ =
  1,2,4", {1.0, 5.0}], Text["ℓ = 1", {0.1, 2.0}],
  Text["ℓ = 2", {0.5, 1.5}], Text["ℓ = 4", {1.5,
  1.0}]}];
```

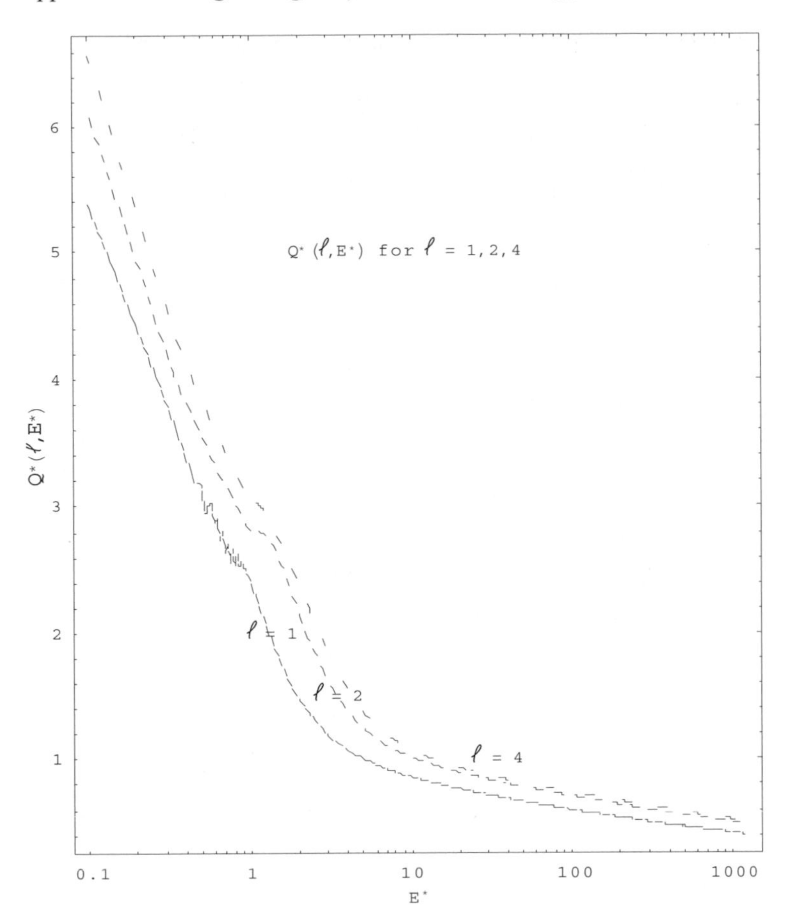


```
(* Evaluate the omega-integrals using
qstarapprox2[[L]][estar].  Compare with HCB pages 1126-
1127 and MRSW page 541. We have set AccuracyGoal high to
get reasonably accurate results for small Tstar.  We
have also evaluated the tail part of the integral
approximately as per MRSW. *)

Clear[omegastar];
Off[NIntegrate::ploss];
omegastar[L_,s_,tstar_]:=(NIntegrate[Chop[
  qstarapproxlist[[L]][estar]] * Exp[-estar/tstar] *
  estar^(s+1), {estar, First[estarlist],
  Last[estarlist]}, MinRecursion→10, MaxRecursion→30,
  AccuracyGoal→8]/(Factorial[(s+1)] * tstar^(s+2)) +
  Chop[qstarapproxlist[[L]][Last[estarlist]]] *
  NIntegrate[Exp[-estar/tstar] * estar^(s+1), {estar,
  Last[estarlist], Infinity}]/(Factorial[(s+1)] *
  tstar^(s+2)));
Print["Table of Omega Integrals", "\n"];
Do[Print["ℓ = ", L];
```

```
Print[TableForm[Chop[Map[{#, omegastar[L, 1, #],
omegastar[L, 2, #], omegastar[L, 3, #], omegastar[L,
4, #], omegastar[L, 5, #], omegastar[L, 6, #]} &,
{0.1, 0.2, 0.3, 0.5, 0.7, 1.0, 2.0, 3.0, 4.0, 5.0,
6.0, 7.0, 8.0, 9.0, 10.0, 50.0, 100.0}]],
TableHeadings→{None, {"T* ", "r=1 ", "r=2 ",
"r=3 ", "r=4 ", "r=5 ", "r=6 "}}]], {L, 1, Lhigh}];
```

Table of Omega Integrals

ℓ = 1

T*	r=1	r=2	r=3	r=4	r=5	r=6
0.1	4.03052	3.57351	3.25852	3.01939	2.82695	2.6648
0.2	3.14477	2.74038	2.44475	2.20647	2.00606	1.83615
0.3	2.6577	2.26061	1.96826	1.74315	1.5696	1.43693
0.5	2.06944	1.70836	1.47	1.31043	1.20216	1.12668
0.7	1.73095	1.42466	1.24203	1.12922	1.05596	1.0054
1.	1.44051	1.20433	1.07614	1.00046	0.951348	0.916617
2.	1.07552	0.951483	0.888318	0.849385	0.821901	0.800763
3.	0.950098	0.864833	0.819374	0.78935	0.766952	0.749051
4.	0.884639	0.817083	0.779017	0.752666	0.732417	0.715933
5.	0.842821	0.78494	0.75077	0.726388	0.70733	0.691661
6.	0.812858	0.760923	0.729126	0.705976	0.687689	0.672565
7.	0.789791	0.741824	0.711621	0.689324	0.67159	0.656868
8.	0.771157	0.726007	0.696951	0.675289	0.657978	0.643571
9.	0.755583	0.712527	0.684343	0.663177	0.646207	0.632057
10.	0.742235	0.700797	0.673301	0.65254	0.635852	0.621921
50.	0.575969	0.546111	0.524777	0.50829	0.494927	0.483736
100.	0.516794	0.48969	0.470324	0.455435	0.443509	0.433762

ℓ = 2

T*	r=1	r=2	r=3	r=4	r=5	r=6
0.1	4.56849	4.05719	3.719	3.47506	3.2892	3.14222
0.2	3.63283	3.24077	2.98294	2.78939	2.62689	2.47914
0.3	3.1838	2.83	2.57354	2.35902	2.16912	2.00066
0.5	2.64092	2.2811	2.00572	1.78801	1.61841	1.48839
0.7	2.27737	1.92119	1.66732	1.48643	1.35839	1.26698
1.	1.91273	1.59293	1.38953	1.2588	1.17244	1.11283
2.	1.37254	1.1759	1.07222	1.01133	0.971338	0.942531
3.	1.17317	1.0389	0.97065	0.929309	0.900594	0.878744
4.	1.07194	0.969964	0.917304	0.883842	0.859519	0.840357
5.	1.01019	0.926819	0.882326	0.852839	0.830714	0.812914
6.	0.967913	0.896217	0.85655	0.829375	0.808538	0.791551
7.	0.936631	0.872757	0.836199	0.810506	0.79051	0.774064
8.	0.912183	0.853822	0.819399	0.79473	0.775326	0.759271
9.	0.892297	0.837976	0.805097	0.781177	0.762215	0.746461
10.	0.875624	0.82437	0.79265	0.769302	0.750686	0.735172
50.	0.683549	0.649791	0.625553	0.606706	0.591349	0.578434
100.	0.616226	0.585166	0.562809	0.545531	0.531643	0.520264

ℓ = 3

T*	r=1	r=2	r=3	r=4	r=5	r=6
0.1	4.6529	4.12477	3.76508	3.49582	3.28207	3.10483
0.2	3.64653	3.19922	2.88425	2.63981	2.43999	2.27233
0.3	3.12014	2.70053	2.40109	2.17056	1.98643	1.83608
0.5	2.50274	2.11889	1.8529	1.65819	1.51165	1.39949
0.7	2.13222	1.78412	1.5546	1.39593	1.28309	1.20089
1.	1.7864	1.49032	1.30887	1.19164	1.1125	1.05667
2.	1.29264	1.11282	1.01663	0.95883	0.920299	0.892353
3.	1.10901	0.984202	0.919351	0.879455	0.851593	0.830375
4.	1.01433	0.918331	0.867753	0.835342	0.811756	0.793201
5.	0.955952	0.876791	0.833868	0.805309	0.783903	0.766717
6.	0.915696	0.847243	0.808919	0.782632	0.762519	0.746158
7.	0.885775	0.824573	0.789249	0.764437	0.745173	0.729366
8.	0.862323	0.806281	0.773037	0.749251	0.73059	0.715185
9.	0.843213	0.790986	0.759258	0.736227	0.718018	0.702922
10.	0.827177	0.777862	0.747282	0.724831	0.706977	0.692127
50.	0.643025	0.610888	0.587878	0.570016	0.555479	0.543265
100.	0.579091	0.54966	0.528516	0.512193	0.499083	0.488348

ℓ = 4

T*	r=1	r=2	r=3	r=4	r=5	r=6
0.1	4.9391	4.38923	4.02342	3.75748	3.55326	3.39054
0.2	3.92571	3.49768	3.215	3.00503	2.83367	2.68347
0.3	3.43691	3.05538	2.78624	2.56878	2.38056	2.21396
0.5	2.86481	2.49314	2.21284	1.98837	1.80744	1.6625
0.7	2.48993	2.12111	1.85296	1.65362	1.5054	1.3948
1.	2.10675	1.76422	1.53592	1.38128	1.27473	1.19933
2.	1.50388	1.27437	1.14761	1.07189	1.02261	0.987894
3.	1.26756	1.10635	1.02299	0.973386	0.940009	0.915372
4.	1.14572	1.023	0.960207	0.921683	0.894636	0.873884
5.	1.07167	0.972207	0.920578	0.887715	0.863818	0.844987
6.	1.02153	0.937115	0.892193	0.862615	0.840527	0.822804
7.	0.984951	0.910811	0.870234	0.842741	0.821803	0.804792
8.	0.956765	0.889963	0.852368	0.826292	0.806141	0.789623
9.	0.934145	0.872772	0.83732	0.812258	0.792677	0.776523
10.	0.915416	0.858177	0.824323	0.800019	0.780869	0.764997
50.	0.71244	0.677832	0.65302	0.633697	0.617927	0.604644
100.	0.643407	0.611512	0.588504	0.570684	0.556337	0.544568

REFERENCES

1. Ferziger, J.H. and Kaper, H.G., *Mathematical Theory of Transport Processes in Gases* (North-Holland, Amsterdam-London, 1972).
2. Maitland, G.C., Rigby, M., Smith, E.B., and Wakeham, W.A., *Intermolecular Forces: Their Origin and Determination* (Oxford University Press, New York, 1981, reprinted and corrected edition, 1987).
3. Hirschfelder, J.O., Curtiss, C.F., and Bird, R.B., *Molecular Theory of Gases and Liquids* (John Wiley and Sons, New York, 1954).

Author Index

In this author index, bold page entries signify a citation in the references. Unbolded page entries signify a citation by name in the text.

Subject Index

–L–

–M–

Mechanics

FLUID MECHANICS AND ITS APPLICATIONS

Series Editor: R. Moreau

Aims and Scope of the Series

The purpose of this series is to focus on subjects in which fluid mechanics plays a fundamental role. As well as the more traditional applications of aeronautics, hydraulics, heat and mass transfer etc., books will be published dealing with topics which are currently in a state of rapid development, such as turbulence, suspensions and multiphase fluids, super and hypersonic flows and numerical modeling techniques. It is a widely held view that it is the interdisciplinary subjects that will receive intense scientific attention, bringing them to the forefront of technological advancement. Fluids have the ability to transport matter and its properties as well as to transmit force, therefore fluid mechanics is a subject that is particularly open to cross fertilization with other sciences and disciplines of engineering. The subject of fluid mechanics will be highly relevant in domains such as chemical, metallurgical, biological and ecological engineering. This series is particularly open to such new multidisciplinary domains.

1. M. Lesieur: *Turbulence in Fluids.* 2nd rev. ed., 1990 ISBN 0-7923-0645-7
2. O. Métais and M. Lesieur (eds.): *Turbulence and Coherent Structures.* 1991 ISBN 0-7923-0646-5
3. R. Moreau: *Magnetohydrodynamics.* 1990 ISBN 0-7923-0937-5
4. E. Coustols (ed.): *Turbulence Control by Passive Means.* 1990 ISBN 0-7923-1020-9
5. A.A. Borissov (ed.): *Dynamic Structure of Detonation in Gaseous and Dispersed Media.* 1991 ISBN 0-7923-1340-2
6. K.-S. Choi (ed.): *Recent Developments in Turbulence Management.* 1991 ISBN 0-7923-1477-8
7. E.P. Evans and B. Coulbeck (eds.): *Pipeline Systems.* 1992 ISBN 0-7923-1668-1
8. B. Nau (ed.): *Fluid Sealing.* 1992 ISBN 0-7923-1669-X
9. T.K.S. Murthy (ed.): *Computational Methods in Hypersonic Aerodynamics.* 1992 ISBN 0-7923-1673-8
10. R. King (ed.): *Fluid Mechanics of Mixing.* Modelling, Operations and Experimental Techniques. 1992 ISBN 0-7923-1720-3
11. Z. Han and X. Yin: *Shock Dynamics.* 1993 ISBN 0-7923-1746-7
12. L. Svarovsky and M.T. Thew (eds.): *Hydroclones.* Analysis and Applications. 1992 ISBN 0-7923-1876-5
13. A. Lichtarowicz (ed.): *Jet Cutting Technology.* 1992 ISBN 0-7923-1979-6
14. F.T.M. Nieuwstadt (ed.): *Flow Visualization and Image Analysis.* 1993 ISBN 0-7923-1994-X
15. A.J. Saul (ed.): *Floods and Flood Management.* 1992 ISBN 0-7923-2078-6
16. D.E. Ashpis, T.B. Gatski and R. Hirsh (eds.): *Instabilities and Turbulence in Engineering Flows.* 1993 ISBN 0-7923-2161-8
17. R.S. Azad: *The Atmospheric Boundary Layer for Engineers.* 1993 ISBN 0-7923-2187-1
18. F.T.M. Nieuwstadt (ed.): *Advances in Turbulence IV.* 1993 ISBN 0-7923-2282-7
19. K.K. Prasad (ed.): *Further Developments in Turbulence Management.* 1993 ISBN 0-7923-2291-6
20. Y.A. Tatarchenko: *Shaped Crystal Growth.* 1993 ISBN 0-7923-2419-6
21. J.P. Bonnet and M.N. Glauser (eds.): *Eddy Structure Identification in Free Turbulent Shear Flows.* 1993 ISBN 0-7923-2449-8
22. R.S. Srivastava: *Interaction of Shock Waves.* 1994 ISBN 0-7923-2920-1
23. J.R. Blake, J.M. Boulton-Stone and N.H. Thomas (eds.): *Bubble Dynamics and Interface Phenomena.* 1994 ISBN 0-7923-3008-0

Mechanics

FLUID MECHANICS AND ITS APPLICATIONS

Series Editor: R. Moreau

24. R. Benzi (ed.): *Advances in Turbulence V.* 1995 ISBN 0-7923-3032-3
25. B.I. Rabinovich, V.G. Lebedev and A.I. Mytarev: *Vortex Processes and Solid Body Dynamics.* The Dynamic Problems of Spacecrafts and Magnetic Levitation Systems. 1994 ISBN 0-7923-3092-7
26. P.R. Voke, L. Kleiser and J.-P. Chollet (eds.): *Direct and Large-Eddy Simulation I.* Selected papers from the First ERCOFTAC Workshop on Direct and Large-Eddy Simulation. 1994 ISBN 0-7923-3106-0
27. J.A. Sparenberg: *Hydrodynamic Propulsion and its Optimization.* Analytic Theory. 1995 ISBN 0-7923-3201-6
28. J.F. Dijksman and G.D.C. Kuiken (eds.): *IUTAM Symposium on Numerical Simulation of Non-Isothermal Flow of Viscoelastic Liquids.* Proceedings of an IUTAM Symposium held in Kerkrade, The Netherlands. 1995 ISBN 0-7923-3262-8
29. B.M. Boubnov and G.S. Golitsyn: *Convection in Rotating Fluids.* 1995 ISBN 0-7923-3371-3
30. S.I. Green (ed.): *Fluid Vortices.* 1995 ISBN 0-7923-3376-4
31. S. Morioka and L. van Wijngaarden (eds.): *IUTAM Symposium on Waves in Liquid/Gas and Liquid/Vapour Two-Phase Systems.* 1995 ISBN 0-7923-3424-8
32. A. Gyr and H.-W. Bewersdorff: *Drag Reduction of Turbulent Flows by Additives.* 1995 ISBN 0-7923-3485-X
33. Y.P. Golovachov: *Numerical Simulation of Viscous Shock Layer Flows.* 1995 ISBN 0-7923-3626-7
34. J. Grue, B. Gjevik and J.E. Weber (eds.): *Waves and Nonlinear Processes in Hydrodynamics.* 1996 ISBN 0-7923-4031-0
35. P.W. Duck and P. Hall (eds.): *IUTAM Symposium on Nonlinear Instability and Transition in Three-Dimensional Boundary Layers.* 1996 ISBN 0-7923-4079-5
36. S. Gavrilakis, L. Machiels and P.A. Monkewitz (eds.): *Advances in Turbulence VI.* Proceedings of the 6th European Turbulence Conference. 1996 ISBN 0-7923-4132-5
37. K. Gersten (ed.): *IUTAM Symposium on Asymptotic Methods for Turbulent Shear Flows at High Reynolds Numbers.* Proceedings of the IUTAM Symposium held in Bochum, Germany. 1996 ISBN 0-7923-4138-4
38. J. Verhás: *Thermodynamics and Rheology.* 1997 ISBN 0-7923-4251-8
39. M. Champion and B. Deshaies (eds.): *IUTAM Symposium on Combustion in Supersonic Flows.* Proceedings of the IUTAM Symposium held in Poitiers, France. 1997 ISBN 0-7923-4313-1
40. M. Lesieur: *Turbulence in Fluids.* Third Revised and Enlarged Edition. 1997 ISBN 0-7923-4415-4; Pb: 0-7923-4416-2
41. L. Fulachier, J.L. Lumley and F. Anselmet (eds.): *IUTAM Symposium on Variable Density Low-Speed Turbulent Flows.* Proceedings of the IUTAM Symposium held in Marseille, France. 1997 ISBN 0-7923-4602-5
42. B.K. Shivamoggi: *Nonlinear Dynamics and Chaotic Phenomena.* An Introduction. 1997 ISBN 0-7923-4772-2
43. H. Ramkissoon, *IUTAM Symposium on Lubricated Transport of Viscous Materials.* Proceedings of the IUTAM Symposium held in Tobago, West Indies. 1998 ISBN 0-7923-4897-4
44. E. Krause and K. Gersten, *IUTAM Symposium on Dynamics of Slender Vortices.* Proceedings of the IUTAM Symposium held in Aachen, Germany. 1998 ISBN 0-7923-5041-3
45. A. Biesheuvel and G.J.F. van Heyst (eds.): *In Fascination of Fluid Dynamics.* A Symposium in honour of Leen van Wijngaarden. 1998 ISBN 0-7923-5078-2

Mechanics

FLUID MECHANICS AND ITS APPLICATIONS

Series Editor: R. Moreau

46. U. Frisch (ed.): *Advances in Turbulence VII.* Proceedings of the Seventh European Turbulence Conference, held in Saint-Jean Cap Ferrat, 30 June–3 July 1998. 1998 ISBN 0-7923-5115-0
47. E.F. Toro and J.F. Clarke: *Numerical Methods for Wave Propagation.* Selected Contributions from the Workshop held in Manchester, UK. 1998 ISBN 0-7923-5125-8
48. A. Yoshizawa: *Hydrodynamic and Magnetohydrodynamic Turbulent Flows.* Modelling and Statistical Theory. 1998 ISBN 0-7923-5225-4
49. T.L. Geers (ed.): *IUTAM Symposium on Computational Methods for Unbounded Domains.* 1998 ISBN 0-7923-5266-1
50. Z. Zapryanov and S. Tabakova: *Dynamics of Bubbles, Drops and Rigid Particles.* 1999 ISBN 0-7923-5347-1
51. A. Alemany, Ph. Marty and J.P. Thibault (eds.): *Transfer Phenomena in Magnetohydrodynamic and Electroconducting Flows.* 1999 ISBN 0-7923-5532-6
52. J.N. Sørensen, E.J. Hopfinger and N. Aubry (eds.): *IUTAM Symposium on Simulation and Identification of Organized Structures in Flows.* 1999 ISBN 0-7923-5603-9
53. G.E.A. Meier and P.R. Viswanath (eds.): *IUTAM Symposium on Mechanics of Passive and Active Flow Control.* 1999 ISBN 0-7923-5928-3
54. D. Knight and L. Sakell (eds.): *Recent Advances in DNS and LES.* 1999 ISBN 0-7923-6004-4
55. P. Orlandi: *Fluid Flow Phenomena.* A Numerical Toolkit. 2000 ISBN 0-7923-6095-8
56. M. Stanislas, J. Kompenhans and J. Westerveel (eds.): *Particle Image Velocimetry.* Progress towards Industrial Application. 2000 ISBN 0-7923-6160-1
57. H.-C. Chang (ed.): *IUTAM Symposium on Nonlinear Waves in Multi-Phase Flow.* 2000 ISBN 0-7923-6454-6
58. R.M. Kerr and Y. Kimura (eds.): *IUTAM Symposium on Developments in Geophysical Turbulence* held at the National Center for Atmospheric Research, (Boulder, CO, June 16–19, 1998) 2000 ISBN 0-7923-6673-5
59. T. Kambe, T. Nakano and T. Miyauchi (eds.): *IUTAM Symposium on Geometry and Statistics of Turbulence.* Proceedings of the IUTAM Symposium held at the Shonan International Village Center, Hayama (Kanagawa-ken, Japan November 2–5, 1999). 2001 ISBN 0-7923-6711-1
60. V.V. Aristov: *Direct Methods for Solving the Boltzmann Equation and Study of Nonequilibrium Flows.* 2001 ISBN 0-7923-6831-2
61. P.F. Hodnett (ed.): *IUTAM Symposium on Advances in Mathematical Modelling of Atmosphere and Ocean Dynamics.* Proceedings of the IUTAM Symposium held in Limerick, Ireland, 2–7 July 2000. 2001 ISBN 0-7923-7075-9
62. A.C. King and Y.D. Shikhmurzaev (eds.): *IUTAM Symposium on Free Surface Flows.* Proceedings of the IUTAM Symposium held in Birmingham, United Kingdom, 10–14 July 2000. 2001 ISBN 0-7923-7085-6
63. A. Tsinober: *An Informal Introduction to Turbulence.* 2001 ISBN 1-4020-0110-X; Pb: 1-4020-0166-5
64. R.Kh. Zeytounian: *Asymptotic Modelling of Fluid Flow Phenomena.* 2002 ISBN 1-4020-0432-X
65. R. Friedrich and W. Rodi (eds.): *Advances in LES of Complex Flows.* Prodeedings of the EUROMECH Colloquium 412, held in Munich, Germany, 4-6 October 2000. 2002 ISBN 1-4020-0486-9
66. D. Drikakis and B.J. Geurts (eds.): *Turbulent Flow Computation.* 2002 ISBN 1-4020-0523-7
67. B.O. Enflo and C.M. Hedberg: *Theory of Nonlinear Acoustics in Fluids.* 2002 ISBN 1-4020-0572-5

Mechanics

FLUID MECHANICS AND ITS APPLICATIONS

Series Editor: R. Moreau

68. I.D. Abrahams, P.A. Martin and M.J. Simon (eds.): *IUTAM Symposium on Diffraction and Scattering in Fluid Mechanics and Elasticity*. Proceedings of the IUTAM Symposium held in Manchester, (UK, 16-20 July 2000). 2002 ISBN 1-4020-0590-3
69. P. Chassaing, R.A. Antonia, F. Anselmet, L. Joly and S. Sarkar: *Variable Density Fluid Turbulence*. 2002 ISBN 1-4020-0671-3
70. A. Pollard and S. Candel (eds.): *IUTAM Symposium on Turbulent Mixing and Combustion*. Proceedings of the IUTAM Symposium held in Kingston, Ontario, Canada, June 3-6, 2001. 2002 ISBN 1-4020-0747-7
71. K. Bajer and H.K. Moffatt (eds.): *Tubes, Sheets and Singularities in Fluid Dynamics*. 2002 ISBN 1-4020-0980-1
72. P.W. Carpenter and T.J. Pedley (eds.): *Flow Past Highly Compliant Boundaries and in Collapsible Tubes*. IUTAM Symposium held at the Univerity of Warwick, Coventry, United Kingdom, 26-30 March 2001. 2003 ISBN 1-4020-1161-X
73. H. Sobieczky (ed.): *IUTAM Symposium Transsonicum IV*. Proceedings of the IUTAM Symposium held in Göttingen, Germany, 2-6 September 2002. 2003 ISBN 1-4020-1608-5
74. A.J. Smits (ed.): *IUTAM Symposium on Reynolds Number Scaling in Turbulent Flow*. Proceedings of the IUTAM Symposium held in Princeton, NJ, U.S.A., September 11-13, 2002. 2003 ISBN 1-4020-1775-8
75. H. Benaroya and T. Wei (eds.): *IUTAM Symposium on Integrated Modeling of Fully Coupled Fluid Structure Interactions Using Analysis, Computations and Experiments*. Proceedings of the IUTAM Symposium held in New Jersey, U.S.A., 2-6 June 2003. 2003 ISBN 1-4020-1806-1
76. J.-P. Franc and J.-M. Michel: *Fundamentals of Cavitation*. 2004 ISBN 1-4020-2232-8
77. T. Mullin and R.R. Kerswell (eds.): *IUTAM Symposium on Laminar Turbulent Transition and Finite Amplitude Solutions*. 2005 ISBN 1-4020-4048-2
78. R. Govindarajan (ed.): *Sixth IUTAM Symposium on Laminar-Turbulent Transition*. Proceedings of the Sixth IUTAM Symposium on Laminar-Turbulent Transition, Bangalore, India, 2004. 2006 ISBN 1-4020-3459-8
79. S. Kida (ed.): *IUTAM Symposium on Elementary Vortices and Coherent Structures: Significance in Turbulence Dynamics*. Proceedings of the IUTAM Symposium held at Kyoto International Community House, Kyoto, Japan, 26-28 October, 2004. 2006 ISBN 1-4020-4180-2
80. S.S. Molokov, R. Moreau and H.K. Moffatt (eds.): *Magnetohydrodynamics*. Historical Evolution and Trends. 2006 ISBN 1-4020-4832-7
81. S. Balachandar and A. Prosperetti (eds.): *IUTAM Symposium on Computational Approaches to Multiphase Flow*. Proceedings of an IUTAM Symposium held at Argonne National Laboratory, October 4-7, 2004. 2006 ISBN 1-4020-4976-5
82. A. Gyr and K. Hoyer: *Sediment Transport*. A Geophysical Phenomenon. 2006 ISBN 1-4020-5015-1
83. I.N. Ivchenko, S.K. Loyalka and R.V. Tompson, Jr.: *Analytical Methods for Problems of Molecular Transport*. 2007 ISBN 1-4020-5864-0